Javed Ahsan, John N. Mordeson, and Muhammad Shabir

Fuzzy Semirings with Applications to Automata Theory

Studies in Fuzziness and Soft Computing, Volume 278

Editor-in-Chief

Prof. Janusz Kacprzyk
Systems Research Institute
Polish Academy of Sciences
ul. Newelska 6
01-447 Warsaw
Poland
E-mail: kacprzyk@ibspan.waw.pl

Further volumes of this series can be found on our homepage: springer.com

Vol. 264. Hak-Keung Lam and
Frank Hung-Fat Leung
Stability Analysis of Fuzzy-Model-Based Control Systems, 2011
ISBN 978-3-642-17843-6

Vol. 265. Ronald R. Yager, Janusz Kacprzyk, and Prof. Gleb Beliakov (eds.)
Recent Developments in the Ordered Weighted Averaging Operators: Theory and Practice, 2011
ISBN 978-3-642-17909-9

Vol. 266. Edwin Lughofer
Evolving Fuzzy Systems – Methodologies, Advanced Concepts and Applications, 2011
ISBN 978-3-642-18086-6

Vol. 267. Enrique Herrera-Viedma, José Luis García-Lapresta, Janusz Kacprzyk, Mario Fedrizzi, Hannu Nurmi, and Sławomir Zadrożny
Consensual Processes, 2011
ISBN 978-3-642-20532-3

Vol. 268. Olga Poleshchuk and
Evgeniy Komarov
Expert Fuzzy Information Processing, 2011
ISBN 978-3-642-20124-0

Vol. 269. Kasthurirangan Gopalakrishnan, Siddhartha Kumar Khaitan, and Soteris Kalogirou (Eds.)
Soft Computing in Green and Renewable Energy Systems, 2011
ISBN 978-3-642-22175-0

Vol. 270. Christer Carlsson and Robert Fullér
Possibility for Decision, 2011
ISBN 978-3-642-22641-0

Vol. 271. Enric Trillas, P. Bonissone, Luis Magdalena and Janusz Kacprzyk
Combining Experimentation and Theory, 2011
ISBN 978-3-642-24665-4

Vol. 272. Oscar Castillo
Type-2 Fuzzy Logic in Intelligent Control Applications, 2012
ISBN 978-3-642-24662-3

Vol. 273. Rudolf Seising and
Veronica Sanz (Eds.)
Soft Computing in Humanities and Social Sciences, 2012
ISBN 978-3-642-24671-5

Vol. 274. Irina Georgescu
Possibility Theory and the Risk, 2012
ISBN 978-3-642-24739-2

Vol. 275. Alexander P. Rotshtein
and Hanna B. Rakytyanska
Fuzzy Evidence in Identification, Forecasting and Diagnosis, 2012
ISBN 978-3-642-25785-8

Vol. 276. Jaime Gil Aluja and
Ana Maria Gil Lafuente
Towards an Advanced Modelling of Complex Economic Phenomena, 2012
ISBN 978-3-642-24811-5

Vol. 277. Lotfi A. Zadeh
Computing with Words, 2012
ISBN 978-3-642-27472-5

Vol. 278. Javed Ahsan, John N. Mordeson, and Muhammad Shabir
Fuzzy Semirings with Applications to Automata Theory, 2012
ISBN 978-3-642-27640-8

Javed Ahsan, John N. Mordeson,
and Muhammad Shabir

Fuzzy Semirings with Applications to Automata Theory

Springer

Authors
Prof. Dr. Javed Ahsan
Department of Mathematics
Wah Engineering College
University of Wah
Wah Cantt
Pakistan

and

Ex. Professor of Mathematics
King Fahd University of Petroleum and Minerals
Dhahran
Kingdom of Saudi Arabia

Prof. Dr. John N. Mordeson
Department of Mathematics
Creighton University
Omaha
Nebraska
USA

Prof. Dr. Muhammad Shabir
Department of Mathematics
Quad-i-Azam University
Islamabad
Pakistan

ISSN 1434-9922
e-ISSN 1860-0808
ISBN 978-3-642-44003-8
ISBN 978-3-642-27641-5 (eBook)
DOI 10.1007/978-3-642-27641-5
Springer Heidelberg New York Dordrecht London

Printed on acid-free paper

Springer is part of Springer Science+Business Media (www.springer.com)

To our wives Hajra, Patricia, Nafisa Sarwat,
with our appreciation for their
patience, support, and love.

Foreword

My real joy, once again, is to see another significant book written in the area of mathematics of uncertainty by Javed Ahsan, John Mordeson, and Muhammad Shabir entitled "Fuzzy Semirings with Applications to Fuzzy Automata Theory." This book is another testimony for their contribution in Springer's series which consists of various topics in fuzzy mathematics. The publication of this volume is a serious endeavor in the game changing, paradigm shifting effort to target college education regarding the need of mathematics of uncertainty as a vehicle for scientific advancement.

The study of fuzzy semirings was initiated by Professor Ahsan in 1993 and since has had major contributions by Professors Ahsan and Shabir. The subject attracted the attention and interest of many authors and today it has taken shape as a viable theory with immediate application to automata theory and theoretical computer science.

One should not forget that it took no less than five centuries for probability theory to enter the main stream in college research and education. This process can become a reality if and only if economic incentive and jobs creation becomes a reality. In that vein, Professor John Mordeson's tireless efforts have been needed and they are to be admired. More importantly, fuzzy mathematics, under his and others' leadership has really flourished.

For years, Professor Mordeson has cared so much about the research and the education of mathematics of uncertainty to such an extent that he is willing for an all out effort in promoting mathematics of uncertainty and, literally, taking the mission under his wings. It gives me even more gratification as well as real joy to be his colleague in the founding of the Society for Mathematics of Uncertainty, highlighting the importance of the mathematics of uncertainty. One may rest assured it is not our intention at all to muddy or to confuse the landscape of the mathematics, because many believe rightfully and correctly that there is one and only one mathematics. We did this only with the good intention of shortening the evolution period which has been historically too lengthy.

We salute Professors Ahsan, Mordeson, and Shabir for their adding more coals to the stove and their determination to work closely with colleagues-at-large who are working in the application of mathematics.

Paul P. Wang
Professor Emeritus
Pratt School of Engineering
Duke University
Durham, North Carolina

Preface

In 1965, Lotfi A. Zadeh introduced the notion of a fuzzy subset of an ordinary set as a way of representing uncertainty. Fuzzy set theory was then formulated in view of the fact that the classical sets are not appropriate in describing real-life problems. Fuzzy set theory has greater richness in applications than the ordinary set theory. In fact, the field grew enormously, and applications were found in areas as diverse as washing machines to handwriting recognition.

Following the discovery of fuzzy sets, much attention has been paid to generalize the basic concepts of classical algebra in a fuzzy framework, and thus developing a theory of fuzzy algebra. In recent years, much interest is shown to generalize algebraic structures of groups, rings, modules, vector spaces, etc. However, this process of fuzzification in the case of *semirings* has been some what slow. On the other hand, semirings have proven to be useful in some areas of applied mathematics and Computer Science. Semirings have also proved useful in studying automata and formal languages. Recently, the notions of automata and formal languages, themselves, have been generalized and extensively studied in a fuzzy framework. It is interesting to note that the unit interval $[0, 1]$ itself admits the structure of a semiring under the binary operations of $+ = \max$ and $\cdot =$ usual multiplication of numbers or under max and min with appropriate meanings and thus it is very natural to study fuzzy automata and fuzzy formal languages in the context of fuzzy semirings.

Motivated by these observations, the authors conceived the idea of writing a comprehensive book on fuzzy semirings covering various aspects of theory and applications. Our first aim in this direction is to study some important classes of semirings using techniques already developed in the current literature and by developing new methods to investigate the structure of semirings in a fuzzy context. We then make a thorough study of the ideal theory of fuzzy semirings. Subsequently, we extend these investigations to the more general setting of fuzzy semimodules over semirings. The scheme of the book is described more explicitly in the following paragraphs.

Chapter 1 begins with a succint introduction of semirings, includes a very brief description of fuzzy set theory, and the developments made towards the study of various algebraic structures such as groups, semigroups, rings, modules etc. in a fuzzy context. In this chapter, we assemble all basic definitions and preliminary results that are needed in the subsequent chapters. Chapter 2 is to set the stage for a thorough study of fuzzy ideals of semirings. We also make a detailed study of semirings which are regular (in the sense of von Neumann), weakly regular, and fully idempotent. Characterizations of these important classes of semirings in terms of their fuzzy ideals is given in this chapter. Chapter 3 is devoted to a study of fuzzy subsemimodules of a semimodule over a semiring. We also define the notions of prime and semiprime subsemimodules and characterize semirings all of whose fuzzy ideals are prime (semiprime). In chapter 4, we initiate the idea of "k-sum" and "k-product" of fuzzy k-ideals of a semiring and characterize "k-regular" semirings in terms of these ideals. We then very briefly study "k-semirings" and also include a section on fuzzy congruences of semirings. The goal of chapter 5 is to study fuzzy quasi-ideals and bi-ideals in semirings and conclude this chapter by characterizing regular and intra-regular semirings in terms of these ideals. The object of chapter 6 is to investigate "$(\in, \in \vee q)$-fuzzy ideals" in semirings. The concept of $(\in, \in \vee q)$-fuzzy ideals in semiring was introduced by Dudek et. al and Ma and Zhan. A detailed study of $(\in, \in \vee q)$-fuzzy k-ideals, k-quasi-ideals and k-bi-ideals and k-regular, and k-intra-regular semirings is also made in this chapter. Chapter 7 is devoted to a study of $(\overline{\in}, \overline{\in} \vee \overline{q})$-fuzzy ideals, fuzzy quasi-ideals and fuzzy bi-ideals of semirings on the pattern of chapter 6. Chapter 8 is modelled on the pattern of the previous two chapters and a similar study of fuzzy ideals, fuzzy quasi-ideals and fuzzy bi-ideals with thresholds is made in this chapter.

A special feature of this book is its second part which consists of three chapters written by the invited authors. These authors are prominent experts in the theory of fuzzy semirings and their applications. Chapter 9 deals with fuzzy "*LD*-bigroupoids" by Hee Sik Kim and J. Neggers.

In this chapter the authors introduce a generalization of semirings called the class of "*LD*-bigroupoids" and "*D*-bigroupoids". The authors develop several basic properties of these structures, and then they consider the fuzzified versions of these algebraic systems. This study yields a generalization of the theory of fuzzy semirings and demonstrates the possibility of further investigations in the area of fuzzy semirings and related topics. Chapter 10 is the next invited chapter by Yudong Liu. This chapter deals with syntactic parsing which is an important task in natural language processing (NLP). In this chapter, the author introduces an application of semiring theory in parsing called a. k. a. semiring parsing.

We conclude the book by its final invited chapter which is written by Manfred Droste, Ingmar Meinecke, Branimir Seselja, and Andreja Tepavcevic. In this chapter the authors consider coverings and decompositions of semiring weighted finite transition system (WTS) with weights from naturally ordered semirings. Such semirings comprise the natural numbers with ordinary addition and multiplication as well as distributive lattices and the max-plus semiring. For these systems the concepts of "covering" and "cascade product" are explored. A cascade decomposition result for such WTS using special partitions of the states of the system is shown. This study extends a classical result of automata theory to the weighted setting.

Finally, the authors would like to express their sincere thanks to the authors of each invited chapter in the second part of this book for their valuable contributions.

Acknowledgements

Javed Ahsan is most grateful to his wife Hajra for providing the right environment at home and her constant support which enabled him to devote all his time in working on this book. He is also thankful to his two beloved children Nasir and Samar for their moral support.

Muhammad Shabir acknowledges the moral support provided by his beloved children Umair, Farwa and Ali to establish his scholarly work presented here by. He is also grateful to his students and colleagues for working and collaborating with him in the field of Fuzzy Algebra at the department of Mathematics, Quaid-i-Azam University, Islamabad, Pakistan. His special thanks are to Dr. Imran Rashid, Department of Mathematics, COMSATS Institute of Information Technology ABBOTTABAD, Pakistan for his assistance in developing the presented work.

John Mordeson dedicates the book to his wonderful granddaughter Isabelle. He also is grateful for the generous invitation of Professors Ahsan and Shabir to work with them on the book. At Creighton University, Professor Mordeson is indebted to Dr. Robert Lueger, Dean of the College of Arts and Sciences, for his support to undergraduate research and to Dr. George and Mrs. Sally Haddix for their generous endowments.

Contents

Part I

Acronyms

K	A set of cardinal numbers
Id(R)	Set of all ideals of a ring R
$\mathbb{N}$	Set of all natural numbers
$\mathbb{Q}^+$	Set of all positive rational numbers
$\mathbb{R}^+$	Set of all positive real numbers
$\mathbb{N}_0$	Set of all non-negative integers
$\mathscr{P}(X)$	Power set of a set X
$\mathscr{M}_n(R)$	Set of all $n \times n$ matrices over a semiring R
$\mathbb{R}$	Set of all real numbers
${}_RM$	M is a left semimodule over R
${}_RR$	R is a left semimodule over R
R_R	R is a right semimodule over R
$< V >$	Subsemimodule generated by V
χ_A	The characteristic function of A
Hom_R	Set of all right R-homomorphisms
End_R	Set of all right R-endomorphisms
R-Cong	Set of all right R-congruences
$\equiv_r$	Trivial congruence
$\equiv_\varpi$	Universal congruence
$\equiv_N$	Bourne relation
CKY	Cocke-Kasami-Younger
GHR	Graham-Harrison-Ruzzo
PCFGs	Probabilistic context free grammars
WTS	Weighted finite transition system
TS	Finite transition system
NLP	Natural language processing

Part I

Chapter 1
Fundamental Concepts

This introductory chapter comprises 6 sections. In section 1, we give a general introduction to the algebraic structure of semirings. Section 2 contains basic definitions, examples, and important applications of semirings. In section 3, we assemble preliminary definitions and results, and section 4 provides a summary of algebraic preliminaries related to the structure of semirings. In section 5, we present the concept of fuzzy sets which was introduced by Lotfi A. Zadeh in 1965, collect basic definitions, examples, and applications. In the final section 6 of this chapter, we give a brief review of various Fuzzy algebraic structures.

1.1 Genesis and a Brief Introduction of Semirings

In 1894, Dedekind introduced the modern definition of the ideal of a ring. He considered the family *Id(R)* of all the ideals of a ring *R*, defined the 'sum' $(+)$ and the 'product' $(\cdot)$ on this family and observed that the system $(Id(R), +, \cdot)$ obeyed most of the rules that the system $(R, +, \cdot)$ did, but the algebraic system $(Id(R), +, \cdot)$ was *not* a ring, since $(Id(R), +)$ was not a group, it was only a commutative monoid. The system $(Id(R), +, \cdot)$ had, however, the additional structure of a 'complete lattice' [25], and the interaction of these two structures promised interesting results. In fact, the system $(Id(R), +, \cdot)$ had, however all the properties of an important algebraic structure, which was later termed a "semiring". We briefly outline a historical introduction of this algebraic structure.

In 1934, H. S. Vandiver published a short paper [153] in which he constructed an algebraic system, which consisted of a nonempty set S with two binary operations addition $(+)$ and multiplication $(\cdot)$ such that S was a semigroup under both operations. The system $(S, +, \cdot)$ satisfied both distributive laws but did not satisfy the cancelation law of addition. The system he constructed was ring-like but not exactly a ring. Vandiver called this system a 'semiring'. Vandiver was later informed by R. Brauer that this concept, but not the same name, appeared in Dedekind [40].

It was natural to ask whether semirings can be embedded into rings. Vandiver in his subsequent papers observed that there are semirings which cannot be embedded

J. Ahsan et al.: Fuzzy Semirings with Applications, STUDFUZZ 278, pp. 3–13.
springerlink.com

in any ring, showing that the ring is not the fundamental system for associative algebras with two binary operations. The structure of semirings was later investigated by many authors in the 1950's and in the subsequent decades. The theory produced several hundred research papers. In recent years, semirings have proved to be an important tool in many areas of applied mathematics and Computer Science. For an up-to-date and comprehensive bibliography on semirings, covering all aspects of its development, we refer to the excellent bibliographic source due to Glazek [59]. For the algebraic theory of semirings and applications to Computer Science, we refer to Hebisch and Weinert [68], Golan [62], and Droste et.al [43].

1.2 Basic Definitions, Some Examples, and Important Applications of Semirings

A *semiring* as considered throughout this book unless stated otherwise, is a set R together with two binary operations addition $(+)$ and multiplication $(\cdot)$ such that $(R,+)$ is a commutative semigroup, and $(R,\cdot)$ is a (not necessarily commutative) semigroup, where both algebraic structures are connected by the two distributive laws:

$$a(b+c)=ab+ac, \text{ and } (a+b)c=ac+bc,$$

for all $a, b, c \in R$. If the semigroup $(R,\cdot)$ has a (uniquely determined) identity element e, it is called the *identity* of the semiring $(R,+,\cdot)$. If the semigroup $(R,+)$ has (again a uniquely determined) identity element this element is called the *zero* of the semiring $(R,+,\cdot)$ and is usually denoted by 0. We shall say that a semiring $(R,+,\cdot)$ has an *absorbing zero* 0 if $a+0=0+a=a$ and $a\cdot 0=0\cdot a=0$ for all $a\in R$. A semiring R is *commutative* if the multiplication in R is commutative, that is, $ab=ba$ for all $a, b \in R$. A commutative semiring with identity in which each nonzero element has a multiplicative inverse is called a *semifield*. Thus all rings are semirings. Moreover, if $(L,\vee,\wedge)$ is a distributive lattice, then L is a semiring with $+=\vee$ and $\cdot=\wedge$. In particular the unit interval $[0, 1]$ of real numbers is a semiring with $+=\max$ and $\cdot=\min$, or with $+=\min$ and $\cdot=\max$, or even with $+=\max$ and $\cdot=$ usual product of real numbers. These semirings are of course very different from rings. In the sequel we will review some more examples of semirings.

Example 1.1. The set of natural numbers $\mathbb{N}$, the non negative integers $\mathbb{N}_0$, the positive rational numbers $\mathbb{Q}^+$, the positive real numbers $\mathbb{R}^+$, under the usual addition and multiplication are familiar examples of semirings. Note that none of them is a ring.

Example 1.2. Let m be a transfinite cardinal number and K be the set of all cardinal numbers less than or equal to m. Then, under the usual cardinal sum $(+)$ and cardinal product $(\cdot)$, $(K,+,\cdot)$ is a semiring containing $(\mathbb{N}_0,+,\cdot)$ as a 'subsemiring'.

Example 1.3. For any set X, the system $(\mathscr{P}(X),\bigcup,\bigcap)$ consisting of the power set $\mathscr{P}(X)$ of X under the binary operations of $\bigcup$ and $\bigcap$ is a semiring.

Example 1.4. For each positive integer n, the set $\mathscr{M}_n(R)$ of all $n \times n$ matrices over a semiring R is a semiring under the usual operations of matrix addition and multiplication.

Note that here we are dealing with a concept of semiring which includes commutativity of addition, as in rings. However, in the current literature on semirings, one finds many examples of semirings with non commutative addition which are *regular* in the sense of von Neumann. Recall that a semiring R is called *von Neumann regular* or simply *regular* if for each $a \in R$, there exists $x \in R$ such that $a = axa$. Regular semirings have been investigated by many authors (see for example [59, 83, 84, 162, 163]) and they play an important role in the study of semirings from an algebraic point of view [12, 162, 163].

Example 1.5. Let R be a nonempty set. Define on R, $+$ by $a+b=b$ and $\cdot$ by $a \cdot b = a$ for all $a, b \in R$. Then $(R, +, \cdot)$ is a semiring with non commutative addition. Clearly, R is regular in the sense of von Neumann, in which each element is both 'additively and multiplicatively idempotent' but not 'central'.

Next, we include some examples of semirings which admit the familiar order relation on their elements.

Example 1.6. Consider the totally ordered set $(\mathbb{R}, \leq)$, where $\leq$ denotes the usual order on the set $\mathbb{R}$ of all real numbers. Define the two binary operations addition $\oplus$ and multiplication $\odot$ on $\mathbb{R}$ as follows:

$$a \oplus b = \max\{a, b\} \text{ and } a \odot b = \min\{a, b\}.$$

Then $(\mathbb{R}, \oplus, \odot)$ is a commutative semiring which has neither a zero nor an identity element. The corresponding statements also hold for $a \oplus b = \min\{a, b\}$ and $a \odot b = \max\{a, b\}$.

Example 1.7. Consider the set $\mathbb{R}$ of all real numbers and define addition $\oplus$ and multiplication $\odot$ on R as $a \oplus b = \min\{a, b\}$ with respect to the usual total order on $\mathbb{R}$, and consider the usual addition as the multiplication on $\mathbb{R}$, that is, $a \odot b = a + b$. Then the system $(\mathbb{R}, \oplus, \odot) = (\mathbb{R}, \min, +)$ is a commutative semiring and it has no zero element. The number 0 is the identity element of $(\mathbb{R}, \oplus, \odot)$. Note that this semiring is, in fact, a semifield. However, it has no absorbing zero. Let us adjoin an absorbing zero element, denoted by ∞, to $(\mathbb{R}, \oplus, \odot)$ ([68], Lemma 2.16, p.14) with the property that $\infty \notin \mathbb{R}$, and for all $x \in R$, $x < \infty$. The semifield $(\mathbb{R} \cup \{\infty\}, \min, +)$ is called a *path algebra* and has many important applications such as optimization problems and has become a standard tool in hundreds of papers on optimization. Later a group of Russian mathematicians led by Victor P. Maslov created a new probability theory which was based on this structure and produced interesting applications in Quantum Physics.

If $S = \mathbb{R}^+ \cup \{\infty\}$, then $(S, \min, +)$ is a 'simple' subsemiring of $(\mathbb{R} \cup \{\infty\}, \min, +)$. For applications of this semiring in optimization theory, we refer to Gondran and Minoux [64]. This semiring S has a subsemiring $(\mathbb{N} \cup \{\infty\}, \min, +)$ which has important applications in the theory of formal languages and automata theory [112, 147]. Likewise, we obtain the semifields $(\mathbb{R}, \max, +)$ and $(\mathbb{R} \cup \{-\infty\}, \max, +)$. This semiring is called the *Schedule algebra.* Cuninghame-Green [37] illustrates how it can be used in the analysis of the behavior of industrial processes. For the use of this semiring in finding critical paths, we refer to [34].

1.3 Preliminary Definitions and Related Concepts

As stated in 1.1 the concept of a semiring can essentially be found in the studies of Dedekind [40], and was formally introduced by H. S. Vandiver [153]. The foundations of its algebraic theory were laid by Samuel Bourne [26, 27, 28, 31, 29, 30] and others in the 1950's. In the subsequent decades, the theory was further developed by a number of authors including P. J. Allen, E. Barbut, W. H. Cornish, L. Dale, R. P. Dilworth, K. Glazek, M. P. Grillet, V. Hebisch, M. Henrikson, P. H. Karvellas, D. R. LaTorre, J. R. Mosher, H. E. Stone, M. V. Subramanian, H. J. Weinert, J. Zelzenikov and others. These authors have investigated various aspects of the algebraic theory of semirings including embedding of semirings into richer semirings, general ideal theory and congruences in semirings, some variants of homomorphism theorems, semiring analogue of the classical Wederburn-Artin theorem on semisimple artinian rings, semifields of quotients of semirings, semirings with chain conditions, partially ordered semirings, semigroup semirings, semimodules over semirings and the homological classification of semirings. For details of these investigations and related references, we refer to Glazek [59], Hebisch and Weinert [68], and Golan [62, 63]. Here we give a succint review of the basic concepts of ideals, congruences and homomorphisms in the context of semirings.

A subset I of a semiring R is a *right* (resp. *left*) *ideal* of R if for $a, b \in I$ and $r \in R$, $a + b \in I$ and ar (resp. ra) $\in I$; I is a *two-sided ideal* or simply *ideal* if it is both a right and a left ideal. Although this definition of "ideals" in semirings is useful for many purposes it does not, in general, coincide with the usual ring ideals if the semiring R is a ring. For this reason, its use is somewhat limited in trying to obtain analogues of ring theorems for semirings. Indeed, using this notion of ideals many results in rings have apparently no analogues in semirings. M. Henrikson [69, 70] defined a more restricted class of ideals in semirings which he called *k-ideals* with the property that if the semiring R is a ring then a subset in R is a *k-ideal* if and only if it is a ring ideal. Equivalently, a right (left or two-sided) ideal I of a semiring R is a right (left, two-sided) *k-ideal* provided that $a \in I$ and $a + x \in I$ implies $x \in I$. For the semiring $\mathbb{N}_0$ of non negative integers under the usual addition and multiplication and each $n \in \mathbb{N}_0$, the set $n\mathbb{N}_0$ of multiples of n is a k-ideal of $\mathbb{N}_0$. Unfortunately, the k-ideals of a semiring do not have nice properties for example neither the sum nor the product of k-ideals is necessarily a k-ideal. Observe that, for example, the sum $2\mathbb{N}_0 + 3\mathbb{N}_0$ is not a k-ideal of $\mathbb{N}_0$. Nevertheless, one can talk about principal, finitely generated, semiprime, prime and maximal ideals and also k-ideals in an obvious way and can obtain a fair number of familiar results from ring theory. Now we recall the definition of semimodules over semirings. In what follows, R will denote a semiring which contains an identity 1 and an absorbing zero 0.

An additively written commutative semigroup M with a neutral element 0 is a *right R-semimodule* written M_R, if there is a function $f : M \times R \to M$ where $f(m,r)$ is denoted by mr, such that the following conditions hold:

(a) $(m+m')\,r = mr + m'r$,
(b) $m(r+r') = mr + mr'$,
(c) $m(rr') = (mr)\,r'$,
(d) $m1 = m$,
(e) $0r = m0 = 0$, for all $m, m' \in M$, and $r, r' \in R$.

One can similarly define a left R-semimodule $_RM$. A semiring R is a right (resp. left) semimodule over itself which is denoted by R_R (resp. $_RR$). An *R-subsemimodule N* of a right R-semimodule M is a subsemigroup of $(M,+)$ such that $nr \in N$ for all $n \in N$ and $r \in R$. The R-subsemimodules of R_R (resp. $_RR$) are the right (resp. left) ideals of the semiring R.

Let V be a subset of a right R-semimodule M. By M_0 we denote the set of all elements of the form $\Sigma u r_u$ ($r_u \in R$) such that all but a finite number of the terms in the sum are zero, that is, $r_u = 0$ except for a finite number of $u \in V$. Then M_0 is an R-subsemimodule of M, containing V. Symbolically, we write $M_0 = \langle V \rangle$. If $\langle V \rangle = M$, then M is said to be *generated* by V. If there is a finite subset V of M such that $M = \langle V \rangle$, then M is called *finitely generated*. In particular, if $|V| = 1$ and $M = \langle V \rangle$ then M is *cyclic*. If $M = R_R$ (resp. $_RR$), cyclic (or finitely generated) R-subsemimodules of M are called *principal* (or *finitely generated*) *right* (resp. *left*) ideals of the semiring R.

Let M and N be the right R-semimodules. A function $f : M \to N$ is called a *right R-homomprphism* if

(a) $f(m+m') = f(m) + f(m')$,
(b) $f(mr) = f(m)r$, for all $m, m' \in M$ and $r \in R$.

The set of all right R-homomorphisms from M to N is denoted by $Hom_R(M,N)$. By $End_R(M)$ we shall mean the set of all right R-endomorphisms of M. Using standard arguments it can be shown that for each right R-semimodule M, $End_R(M)$ is a semiring.

An equivalence relation $\equiv$ defined on a right R-semimodule M is a *right R-congruence* if and only if $m \equiv m'$ and $n \equiv n'$ in M imply that $m+n \equiv m'+n'$ and $mr \equiv m'r$ for all $r \in R$. The set of all right R-congruences on M is denoted by R-Cong(M). This set contains the *trivial congruence* $\equiv_r$ defined by $m \equiv_r m' \Leftrightarrow m = m'$, and the *universal congruence* $\equiv_\varpi$ defined by $m \equiv_\varpi m'$ for all $m, m' \in M$. If $M \neq \{0_M\}$ and these are the only two elements of R-Cong(M), then M is called *simple*. Each R-subsemimodule N of M determines a right R-congruence $\equiv_N$ on M, called the *Bourne relation*, defined by setting $m \equiv_N m' \Leftrightarrow$ there exist elements n and n' of N such that $m+n = m'+n'$. Moreover, for every right R-congruence $\equiv$ on M, the congruence classes $[a]_\equiv$ form a right R-semimodule $M/\equiv$ such that the natural mapping g defined by $g(a) = [a]_\equiv$ is a surjective R-homomorphism.

Now let $f : M \to N$ be an R-homomorphism between right R -semimodules M and N. Then f defines a right R-congruence $\equiv_f$ on M by $m \equiv_f m' \Leftrightarrow f(m) = f(m')$, and the homomorphic image $f(M)$, an R-subsemimodule of N, is R-isomorphic to $M/\equiv_f$. However, a crucial point in the context of R-semimodules over semirings is that there is in general no way of describing $M/\equiv_f$ by the help of R-subsemimodules of M. Illustrative examples are the $\mathbb{N}_0$-congruences and the $\mathbb{N}_0$-subsemimodules of "free" $\mathbb{N}_0$-semimodules $\mathbb{N}_{0\mathbb{N}_0}$. In particular, by defining the kernel of $f: M \to N$ by $\ker f = \{m \in M : f(m) = 0\}$ one obtains that its Bourne congruence $\equiv_{\ker f}$ is contained in $\equiv_f$, and, in general, properly.

Dealing with right R-semimodules, N is said to be a *retract* of M if there exist right R-homomorphisms $g: N \to M$ and $p: M \to N$ such that $pg = i_N$. This, clearly, is the general concept in categories, applied to the category of right R-semimodules for a fixed semiring R with right R-homomorphisms as morphisms.

1.4 Algebraic Preliminaries of Semirings

A semiring as defined in section 1.2 is a set R together with two binary operations addition $(+)$ and multiplication $(\cdot)$ such that $(R,+)$ is a commutative semigroup, and $(R,\cdot)$ is a (not necessarily commutative) semigroup, where both algebraic structures are connected by the two distributive laws:

$$a(b+c) = ab+ac, \text{ and } (a+b)c = ac+bc,$$

for all $a,b,c \in R$. If the semigroup $(R,\cdot)$ has a (uniquely determined) identity element e, it is called the *identity* of the semiring $(R,+,\cdot)$. If the semigroup $(R,+)$ has (again a uniquely determined) identity element this element is called the *zero* of the semiring $(R,+,\cdot)$ and is usually denoted by 0. We shall say that a semiring $(R,+,\cdot)$ has an *absorbing zero* 0 if $a+0 = 0+a = a$ and $a\cdot 0 = 0\cdot a = 0$ for all $a \in R$. A semiring R is *commutative* if the multiplication in R is commutative, that is, $ab = ba$ for all $a,b \in R$. A commutative semiring in which each nonzero element has a multiplicative inverse is called a *semifield*. Thus all rings are semirings. Moreover, if $(L,\vee,\wedge)$ is a distributive lattice, then L is a semiring with $+ = \vee$ and $\cdot = \wedge$. In particular the unit interval $[0,1]$ of real numbers is a semiring with $+ = \max$ and $\cdot = \min$, or with $+ = \min$ and $\cdot = \max$, or even with $+ = \max$ and $\cdot =$ usual product of real numbers. These semirings are of course very different from rings. We will review some more examples of semirings.

Example 1.8. Let $(S,\cdot)$ be a semigroup and $\mathscr{P}(S)$ the power set of S. Then for $A,B \in \mathscr{P}(S)$, $A \cup B$ may be considered as an addition on $\mathscr{P}(S)$ and $A\cdot B = \{a\cdot b : a \in A \text{ and } b \in B\}$, as a multiplication on $\mathscr{P}(S) \setminus \{\emptyset\}$, where $\emptyset$ denotes the empty set. Then it is easy to check that $(\mathscr{P}(S),\cup,\cdot)$ is a semiring. This semiring has an identity E if and only if $(S,\cdot)$ has an identity e, namely, $E = \{e\}$. If one applies the rule $A\cdot B = \{a\cdot b : a \in A, b \in B\}$ also to empty sets, one obtains $A\cdot B = \emptyset$ for $A = \emptyset$ or $B = \emptyset$. Then the system $(\mathscr{P}(S),\cup,\cdot)$ is a semiring with $\emptyset$ as an absorbing zero. The finite subsets of S form a subsemiring of $(\mathscr{P}(S),\cup,\cdot)$.

Semirings provide an important algebraic tool in the study of various areas of theoretical Computer Science, in particular, in the study of automata theory and formal languages. For earlier studies in this subject we refer to [34] and [54] and for more recent treatment of automata theory and formal languages based on semirings, we refer to [68]. In order to present the next example which is related to the semiring of formal languages, we include some preliminary definitions and notations borrowed from [68].

Let X be a nonempty set. Define a set F_X by

$$F_X = \underset{n\in\mathbb{N}}{\cup} X^n = \underset{n\in\mathbb{N}}{\cup} \{(x_1,\cdots,x_n) : x_i \in X\}.$$

Then $(F_X,\cdot)$ is a semigroup under the binary operation given by

$$(x_1,\cdots,x_n)\cdot(y_1,\cdots,y_m) = (x_1,\cdots,x_n,y_1,\cdots,y_m).$$

Simplifying the notation by $(x_i) = x_i$ for all $x_i \in X$, we obtain $(x_1,x_2\cdots,x_n) = (x_1)(x_2)\cdots(x_n) = x_1.x_2\cdots x_n = x_1x_2\cdots x_n$. Thus $x_1x_2\cdots x_n = x'_1x'_2\cdots x'_m$ is equivalent with $n = m$ and $x_i = x'_i$ for $i = 1,2,\cdots n$. We read the multiplication $(x_1\cdots x_n)\cdot(y_1\cdots y_m) = x_1\cdots x_ny_1\cdots y_m$. The semigroup $(F_X,\cdot)$ described above is called the *free semigroup* on the set $X \neq \emptyset$ (or on the *alphabet* X), each element $w = x_1\cdots x_n \in F_X$ is a *word* on X and $l(w) = n$, the *length* of w. We also introduce $l_x(w) \in \mathbb{N}_0$ for each $x \in X$, which counts how often the letter x occurs in w. The multiplication in F_X is also called *juxtaposition* or *concatenation.* If one adjoins to $(F_X,\cdot)$ an element $\lambda \notin F_X$ as identity, the resulting semigroup $(F_X \cup \{\lambda\},\cdot) = (F_X^\lambda,\cdot)$ is called the *free semigroup with identity* (or the *free monoid*) on X. Because of $l(w_1w_2) = l(w_1) + l(w_2)$ for all $w_i \in F_X$, it follows that $l(\lambda) = 0$ for the identity λ of $(F_X,\cdot)$ and interpret λ as the *empty word* of F_X. If $X = \{x\}$ then $F_{\{x\}} = \{x^n : n \in \mathbb{N}\}$ and $F^\lambda_{\{x\}} = \{x^n : n \in \mathbb{N}_0\}$ with $x^0 = \lambda$ and $x^n \neq x^m$ for $n \neq m$. Thus it follows that the free semigroup (with identity) on each singleton set $X = \{x\}$ is just the infinite cyclic semigroup (with identity) generated by x. Clearly the free semigroups $(F_X,\cdot)$ and $(F_X^\lambda,\cdot)$ are commutative for $|X| = 1$. However, for $|X| \geq 2$, these semigroups are not commutative, since $x_1x_2 \neq x_2x_1$ holds for all $x_1 \neq x_2$ belonging to X. We now add an important definition. Let X be a nonempty set (usually assumed to be finite in this context) and , $(F_X^\lambda,\cdot)$ be the free monoid on X. Then each subset L of F_X, that is, each set of words formed by letters of X, is called a (*formal language*) L on the alphabet X. One writes $\mathscr{L}_X = \mathscr{P}(F_X^\lambda)$ for the *set of all languages* on the alphabet X. (including the empty language $\emptyset$ and the language $\{\lambda\}$ consisting merely of the empty word). Further a language $L \in \mathscr{L}_X$ satisfying $\lambda \notin L$ is called λ *free* or *proper* and we denote by $L_X = \mathscr{P}(F_X)$ the set of all proper languages and by $\mathscr{F}_X$ the set of all finite languages on X. Now using the binary operations $\cup$ and $\cdot$ (as described in Example 1.8) in the free monoid $(F_X^\lambda,\cdot)$, we obtain the following important example of a semiring ([68], Theorem 5.14, p.243).

Example 1.9. The set $\mathscr{L}_X = \mathscr{P}(F_X^\lambda)$ of all languages on X with operations $L_1 \cup L_2$ and $L_1 \cdot L_2 = \{w_1 \cdot w_2 : w_1 \in L_1 \text{and } w_2 \in L_2\}$ is a semiring $(\mathscr{L}_X,\cup,\cdot) = (\mathscr{P}(F_X^\lambda),\cup,\cdot)$. The empty language $\emptyset$ is the absorbing zero and the language $\{\lambda\}$ is the identity of this semiring. As one will expect, $\mathscr{L} = \mathscr{L}_X$ contains languages with different properties, whose investigation for the case when the alphabet X are finite, is the topic of the theory of formal languages [68].

1.5 Fuzzy Sets, Basic Definitions, Examples and Their Applications

In 1965, Lotfi A. Zadeh initiated the concept of a fuzzy set. In his definitive paper [160], Zadeh formulated the definition of *fuzzy subset* of a nonempty set as a collection of objects with a 'grade' or 'degree of membership', each object being assigned a value between 0 and 1. In other words, a fuzzy set is a generalization of "characteristic function", where in the degree of member of an element is not only 0 or 1, but it may take any value between 0 and 1 or in the unit interval $[0,1]$ of real numbers. Thus each *fuzzy subset* is precisely determined by a particular membership function. Fuzzy set theory was developed on the basis that the crisp (or ordinary) sets were not appropriate or natural in describing the real world problems. Fuzzy set theory has, in fact, greater scope and richness in applications than the ordinary set theory.

The theory of fuzzy sets has attracted the attention of researchers in a wide variety of fields. The theory is growing enormously and finding applications in such diverse areas as mathematics, computer science, artificial intelligence, pattern recognition, robotics, medical sciences, social sciences, engineering, and many other areas. In the present section we mainly give some basic definitions and properties of fuzzy sets which concerns the algebraic operations.

Definition 1.1. Let X be a nonempty (ordinary) set. A *fuzzy set* (or more precisely *fuzzy subset*) μ of the set X is a function $\mu : X \to [0, 1]$.

Note that Goguen [60] has generalized the fuzzy subsets of X to L -fuzzy subsets, as a function from X to a "complete distributive lattice" L. If L is the unit interval $[0,1]$ of real numbers, L -fuzzy subsets are the usual fuzzy subsets in the sense of Zadeh as defined above.

A fuzzy subset μ of X is *empty* $\Leftrightarrow$ μ is identically zero on X. Thus μ is *nonempty* if it is not the constant function always taking the value 0.

Two fuzzy subsets μ and λ of a set X are said to be *disjoint* if there is no $x \in X$ such that $\mu(x) = \lambda(x)$. If $\lambda(x) = \mu(x)$ for each $x \in X$, then we say that λ and μ are *equal* and write $\lambda = \mu$. If λ and μ are fuzzy subsets of X, then λ is said to be *contained*in μ, written as $\lambda \subseteq \mu \Leftrightarrow \lambda(x) \leq \mu(x)$ for all $x \in X$, and $\lambda \subset \mu \Leftrightarrow \lambda \subseteq \mu$ and $\lambda \neq \mu$, that is, λ is *properly contained* in μ. The *union* of two fuzzy subsets λ and μ, denoted by $\lambda \cup \mu$, is a fuzzy subset of the set X defined as

$$(\lambda \cup \mu)(x) = \max\{\lambda(x), \mu(x)\}$$

for every $x \in X$. The *union* of any family $\{\mu_i : i \in I\}$ of fuzzy subsets μ_i of X is defined by $\left(\bigcup_{i\in I} \mu_i\right)(x) = \sup_{i\in I}\{\mu_i(x)\}$, for all $x \in X$. It may be noted that the union of λ and μ is the "smallest" fuzzy subset containing both λ and μ. More precisely, if δ is any fuzzy subset of X which contains both λ and μ, then δ also contains their union. Similarly, the *intersection* of two fuzzy subsets λ and μ of a set X, denoted by $\lambda \cap \mu$, is a fuzzy subset of X defined as

$$(\lambda \cap \mu)(x) = \min\{\lambda(x), \mu(x)\}$$

for all $x \in X$. The intersection of any family $\{\lambda_i : i \in I\}$ of fuzzy subsets of X is defined by

$$\left(\bigcap_{i \in I} \lambda_i \right)(x) = \inf_{i \in I} \{\lambda_i(x)\}, \text{ for all } x \in X.$$

It can be easily seen that the intersection of λ and μ is the "largest" fuzzy subset which is contained in both λ and μ. Finally, the *complement* of a fuzzy subset μ of a set X, denoted by μ^c, is defined as $\mu^c(x) = 1 - \mu(x)$, for all $x \in X$.

The properties of fuzzy subsets λ, μ and ν of a set X, relative to commutativity, associativity, idempotence, distributivity, absorption, DeMorgan's laws and involution can be summarized as follows:

(a) Commutativity: $\lambda \cup \mu = \mu \cup \lambda$ and $\lambda \cap \mu = \mu \cap \lambda$.

(b) Associativity: $\lambda \cup (\mu \cup \nu) = (\lambda \cup \mu) \cup \nu$ and $\lambda \cap (\mu \cap \nu) = (\lambda \cap \mu) \cap \nu$.

(c) Idempotence: $\lambda \cup \lambda = \lambda$ and $\lambda \cap \lambda = \lambda$.

(d) Distributivity: $\lambda \cup (\mu \cap \nu) = (\lambda \cup \mu) \cap (\lambda \cup \nu)$ and $\lambda \cap (\mu \cup \nu) = (\lambda \cap \mu) \cup (\lambda \cap \nu)$.

(e) Absorption: $\mu \cap (\mu \cup \lambda) = \mu$ and $\mu \cup (\mu \cap \lambda) = \mu$.

(f) DeMorgan's laws: $(\mu \cap \lambda)^c = \mu^c \cup \lambda^c$ and $(\mu \cup \lambda)^c = \mu^c \cap \lambda^c$.

(g) Involution: $(\mu^c)^c = \mu$.

Let χ_A denote the characteristic function of a subset A of a set X.

Remark 1.1. The following properties which are true in ordinary set theory are, generally speaking, no longer true in fuzzy subsets or more conveniently, in fuzzy set theory:

(a$'$) $A \cap A^c = \emptyset$ but $\mu \cap \mu^c \neq \chi_\emptyset$, the empty fuzzy subset of a set X.

(b$'$) $A \cup A^c = X$ but $\mu \cup \mu^c \neq \chi_X$, where A is any subset of a set X and μ is any fuzzy subset of X; of course, $\chi_X(x) = 1$ for all $x \in X$.

We now define *level subsets* of a fuzzy subset. Let μ be any fuzzy subset of a set X and let $t \in [0,1]$. The set

$$\mu_t = \{x \in X : \mu(x) \geq t\}$$

is called *level subset* of μ. Clearly, $\mu_t \subseteq \mu_s$, whenever $t > s$. A fuzzy subset μ of X is said to be a *normalized fuzzy subset* if there exists $x \in X$ such that $\mu(x) = 1$. Finally, let '$\cdot$' be a binary operation on a set X and λ, μ any two fuzzy subsets of X. Then the *product* $\lambda \circ \mu$, is defined by

$$(\lambda \circ \mu)(x) = \begin{cases} \sup\limits_{x = y \cdot z} \{\min(\lambda(y), \mu(z))\} & \text{for } x = y \cdot z \text{ where } y, z \in X. \\ 0 & \text{if } x \text{ is not expressible as } x = y \cdot z \text{ for all } y, z \in X. \end{cases}$$

Clearly, $\lambda \circ \mu$ is a fuzzy subset of X.

Example 1.10. (a) Let X be the set $\mathbb{R}$ of real numbers and μ be a fuzzy subset of real numbers which are "much greater" than 1. It is possible to give a subjective characterization of μ by defining a function μ on X. Representative values of such a function might be $\mu(0) = 0$, $\mu(1) = 0$, $\mu(5) = 0.1$, $\mu(10) = 0.2$, $\mu(100) = 0.95$, $\mu(500) = 1$.

(b) Let $\mathbb{N}$ be the set of natural numbers and consider the fuzzy subset of "small" natural numbers:

$$\lambda(1) = 1, \lambda(2) = 0.8, \lambda(3) = 0.4, \lambda(4) = 0.2, \lambda(5) = 0, \lambda(6) = 0, \cdots .$$

Lastly, we include a Proposition in this section:

Proposition 1.1. *Let $\mathscr{F}(X)$ be the set of all fuzzy subsets of X. Then $(\mathscr{F}(X), \leq, \cap, \cup)$ is a complete distributive lattice with a least element $\chi_\emptyset$ and a greatest element χ_X which easily follows from the above definitions and property (d) above.*

1.6 Fuzzy Algebraic Structures: A Brief Review

The concept of fuzzy sets was applied to generalize different algebraic structures, like other branches of mathematics. In this connection the first attempt was made in 1971 by A. Rosenfeld [127], where he defined a fuzzy subgroupoid of a groupoid and a fuzzy subgroup of a group. Many authors followed this line of research (cf. Das [38], Bhattacharya and Mukharjee [24], J. M. Anthony and H. Sherwood [14] and others).

We now define a *fuzzy subgroupoid* of a groupoid. Let S be a groupoid. A fuzzy set $\mu : S \to [0,1]$ is called a *fuzzy subgroupoid* of S if, for $x, y \in S$, $\mu(xy) \geq \min(\mu(x), \mu(y))$. If S is a group, a fuzzy subgroupoid μ of S is called a *fuzzy subgroup* of S if $\mu(x^{-1}) \geq \mu(x)$, for all $x \in S$.

In [72, 73] Wang-Jin Liu introduced and developed basic properties concerning the notions of fuzzy subrings as well as fuzzy ideals of a ring, defined as follows:

A nonempty fuzzy subset μ of a ring R is called a *fuzzy subring* of R, if for all $x, y \in R$, the following conditions hold:

(i) $\mu(x - y) \geq \min(\mu(x), \mu(y))$, and

(ii) $\mu(xy) \geq \min(\mu(x), \mu(y))$

μ is called a *fuzzy right ideal* if $\mu(x - y) \geq \min(\mu(x), \mu(y))$, and $\mu(xy) \geq \mu(x)$ and a *fuzzy left ideal* is defined similarly. Moreover, μ is called a *fuzzy ideal* if μ is both a fuzzy right and a fuzzy left ideal, i.e. if $\mu(xy) \geq \max(\mu(x), \mu(y))$.

The properties of fuzzy ideals and fuzzy prime ideals of a ring have also been studied by many authors including Mukherjee and Sen [115], Zhang Yue [166], Swamy and Swamy [152], Malik and Mordeson [111], Dixit et al. [42] and others.

Fuzzy vector spaces has been studied by Katosaras and Liu [86], and others.

The concept of fuzzy modules was introduced by Negoita and Rafescu [119], which was subsequently studied by Pan [122, 123], Golan [61], Lopez-Permouth [104], S. Nanda [119], S. Majumdar [109], J. Ahsan [6] and others.

In 1979, N. Kuroki [97] laid the foundation of a theory of fuzzy semigroups. Subsequently, many authors including Kuroki himself, Ahsan et al. [10], M. Shabir [131] and others studied this structure. Later, Ahsan, Khan and Saifullah [9] initiated the study of *fuzzy semirings* and *fuzzy semimodules* which will be discussed more thoroughly in chapter 2.

Chapter 2
Fuzzy Ideals of Semirings

Section 1 of this chapter begins with the definitions of sums and products of fuzzy sets and fuzzy ideals of a semiring together with some basic results related to these definitions. It is shown that the sum of fuzzy left (right) ideals, product of left (right) ideals is a fuzzy left (right) ideal. It is further shown that a fuzzy subset of a semiring R is a fuzzy left (right) ideal of R if and only if each nonempty level subset (as defined in 2.1) is a left (right) ideal of R. In section 2, we characterize Regular semirings in terms of their fuzzy left and fuzzy right ideals. Characterizations of weakly regular semirings in terms of fuzzy left (right) ideals are given in section 3. Finally in section 4, we collect various characterizations of fully idempotent semirings in terms of their fuzzy ideals. Here we also give characterizations of fully idempotent semirings in terms of their fuzzy prime ideals (cf. Theorem 2.13).

2.1 Sum and Product of Fuzzy Sets

We begin with some basic definitions and various preliminary results. A fuzzy subset λ of a semiring R is called a *fuzzy subsemiring* of R if

1. $\lambda(x+y) \geq \lambda(x) \wedge \lambda(y)$
2. $\lambda(xy) \geq \lambda(x) \wedge \lambda(y)$

for all $x, y \in R$.

A fuzzy subset λ of a semiring R is called a *fuzzy left (right) ideal* of R if

1. $\lambda(x+y) \geq \lambda(x) \wedge \lambda(y)$
2. $\lambda(xy) \geq \lambda(y) (\lambda(xy) \geq \lambda(x))$

for all $x, y \in R$.

A fuzzy subset λ of a semiring R is called a *fuzzy two-sided ideal* or simply a *fuzzy ideal* of R if it is both a fuzzy left ideal and a fuzzy right ideal of R. If λ is a fuzzy left (right) ideal of a semiring R and R contains an absorbing zero element 0, then $\lambda(0) \geq \lambda(x)$ for all $x \in R$. It is obvious from above definitions that a fuzzy left (right) ideal of a semiring R is a fuzzy subsemiring of R but the converse is not

J. Ahsan et al.: Fuzzy Semirings with Applications, STUDFUZZ 278, pp. 15–29.
springerlink.com

true, in general. A fuzzy left (right, two-sided) ideal λ of a semiring R is called a fuzzy left (right) *k-ideal* (right, two-sided) *k-ideal* of R if $x+a=b$ implies $\lambda(x) \geq \lambda(a) \wedge \lambda(b)$ for all x, a, $b \in R$. Recall that, if λ is a fuzzy subset of R and $t \in [0,1]$, then the set $\{x \in R : \lambda(x) \geq t\}$ is called the t-level subset of λ.

Theorem 2.1. *A fuzzy subset λ of a semiring R is a fuzzy subsemiring (left, right, two-sided ideal) of R if and only if each nonempty level subset λ_t of λ is a subsemiring (left, right, two-sided ideal) of R, for all $t \in [0,1]$.*

Proof. Suppose λ be a fuzzy subsemiring of R and $\lambda_t \neq \emptyset$ for $t \in [0,1]$. Let $x, y \in \lambda_t$. Then $\lambda(x) \geq t$ and $\lambda(y) \geq t$. Since $\lambda(x+y) \geq \lambda(x) \wedge \lambda(y) \geq t$ and $\lambda(xy) \geq \lambda(x) \wedge \lambda(y) \geq t$, we have $x+y \in \lambda_t$ and $xy \in \lambda_t$. Hence λ_t is a subsemiring of R.

Conversely, assume that each nonempty λ_t is a subsemiring of R. Let $x, y \in R$ be such that $\lambda(x+y) < \lambda(x) \wedge \lambda(y)$. Select $t \in [0,1]$ such that $\lambda(x+y) < t \leq \lambda(x) \wedge \lambda(y)$. Then $x, y \in \lambda_t$ but $x+y \notin \lambda_t$, a contradiction. Hence $\lambda(x+y) \geq \lambda(x) \wedge \lambda(y)$ for all x, $y \in R$. Similarly if there exist $x, y \in R$ such that $\lambda(xy) < \lambda(x) \wedge \lambda(y)$, then we can select $t \in [0,1]$ such that $\lambda(xy) < t \leq \lambda(x) \wedge \lambda(y)$. This implies x, $y \in \lambda_t$ but $xy \notin \lambda_t$. This contradicts our hypothesis. Hence $\lambda(xy) \geq \lambda(x) \wedge \lambda(y)$ for all x, $y \in R$. This shows that λ is a fuzzy subsemiring of R. The argument for ideals follows in a similar manner.

Corollary 2.1. *A fuzzy subset λ of a semiring R is a fuzzy left (right)* k-ideal *of R if and only if each nonempty level subset λ_t of λ is a left (right)* k-ideal *of R, for all $t \in [0,1]$.*

Proof. Suppose λ be a fuzzy left k-ideal of R, then each nonempty λ_t is a left ideal of R. Let $x \in R$ and a, $b \in \lambda_t$ such that $x+a=b$. Then $\lambda(a) \geq t$ and $\lambda(b) \geq t$. Since $\lambda(x) \geq \lambda(a) \wedge \lambda(b) \geq t$, we have $x \in \lambda_t$. Hence λ_t is a left k-ideal of R.

Conversely, assume that each nonempty level subset λ_t is a left k-ideal of R. Then λ is a fuzzy left ideal of R. Let x, a, $b \in R$ be such that $x+a=b$, but $\lambda(x) < \lambda(a) \wedge \lambda(b)$. Select $t \in [0,1]$ such that $\lambda(x) < t \leq \lambda(a) \wedge \lambda(b)$. Then $a, b \in \lambda_t$ but $x \notin \lambda_t$, a contradiction. Hence $\lambda(x) \geq \lambda(a) \wedge \lambda(b)$. This implies λ is a fuzzy left k-ideal of R.

Corollary 2.2. *Let A be a nonempty subset of a semiring R and $t_1, t_2 \in [0,1]$ be such that $t_1 < t_2$. Define a fuzzy subset λ_A of R as following:*

$$\lambda_A(x) = \begin{cases} t_2 & \text{if } x \in A \\ t_1 & \text{if } x \notin A \end{cases}$$

for all $x \in R$. Then λ_A is a fuzzy left (right) ideal [k-ideal] of R if and only if A is a left (right) ideal [k-ideal] of R.

Corollary 2.3. *Let A be a nonempty subset of a semiring R. Then A is a left (right) ideal [k-ideal] of R if and only if the characteristic function of A is a fuzzy left (right) ideal [k-ideal] of R.*

Proof. Follows from above Corollary.

Example 2.1. Consider the semiring $R = \{0,1,a,b,c\}$ defined by the following tables:

$+$	0	1	a	b	c
0	0	1	a	b	c
1	1	b	1	a	1
a	a	1	a	b	a
b	b	a	b	1	b
c	c	1	a	b	c

$\cdot$	0	1	a	b	c
0	0	0	0	0	0
1	0	1	a	b	c
a	0	a	a	a	c
b	0	b	a	1	c
c	0	c	c	c	0

It can be verified that the subsemirings of R are, $\{0\},\{a\},\{0,a\},\{0,c\}$, $\{0,a,c\},\{1,a,b\},R$, and the ideals of R are,$\{0\},\{0,c\},\{0,a,c\},R$.

Let us define a fuzzy subset λ of R as follows:

$\lambda(0) = 0.1, \lambda(c) = 0.1, \lambda(b) = 0.5, \lambda(1) = 0.5, \lambda(a) = 0.6.$

Then

$$\lambda_t = \begin{cases} R & \text{if } t \leq 0.1 \\ \{1,a,b\} & \text{if } 0.1 < t \leq 0.5 \\ \{a\} & \text{if } 0.5 < t \leq 0.6 \\ \emptyset & \text{if } t > 0.6 \end{cases}$$

Thus by Theorem 2.1, λ is a fuzzy subsemiring of R but not a fuzzy ideal of R, because $\lambda_{0.5} = \{1,a,b\}$ and $\lambda_{0.6} = \{a\}$ are not ideals of R.

Suppose $t_1,t_2,t_3,t_4 \in [0,1]$ are such that $t_1 > t_2 > t_3 > t_4$. Define a fuzzy subset μ of R as follows:

$$\mu(0) = t_1, \mu(1) = t_4, \mu(a) = t_3, \mu(b) = t_4, \mu(c) = t_2.$$

Then

$$\mu_t = \begin{cases} R & \text{if } t \leq t_4 \\ \{0,a,c\} & \text{if } t_4 < t \leq t_3 \\ \{0,c\} & \text{if } t_3 < t \leq t_2 \\ \{0\} & \text{if } t_2 < t \leq t_1 \\ \emptyset & \text{if } t > t_1 \end{cases}$$

Thus by Theorem 2.1, μ is a fuzzy ideal of R.

Remark 2.1. If R is a semiring containing the absorbing zero 0 and if we define the subsemiring of R as a nonempty subset of R containing 0 which is closed with respect to both operations, then Theorem 2.1 is not true.

Example 2.2. Consider the semiring $R = \{0,x,1\}$ defined by the following tables:

$+$	0	x	1
0	0	x	1
x	x	x	x
1	1	x	1

$\cdot$	0	x	1
0	0	0	0
x	0	x	x
1	0	x	1

The ideals of R are $\{0\},\{0,x\}$ and $\{0,x,1\}$ but $\{0,x\}$ is not a k-ideal because $1+x = x$, but $1 \notin \{0,x\}$. Let λ be the fuzzy subset of R defined by

$$\lambda(0) = t_1, \lambda(x) = t_2 \text{ and } \lambda(1) = t_3,$$

where $t_1, t_2, t_3 \in [0,1]$ and $t_1 > t_2 > t_3$. Then

$$\lambda_t = \begin{cases} \{0,x,1\} & \text{if } t \leq t_3 \\ \{0,x\} & \text{if } t_3 < t \leq t_2 \\ \{0\} & \text{if } t_2 < t \leq t_1 \\ \emptyset & \text{if } t > t_1 \end{cases}$$

Thus by Theorem 2.1, λ is a fuzzy ideal of R but not a fuzzy k-ideal of R, since $\lambda_{t_2} = \{0,x\}$, which is not a k-ideal of R. Define a fuzzy subset μ of R as follows:

$$\mu(0) = 0, \mu(x) = 0.5 \text{ and } \mu(1) = 0.6$$

Then simple calculations show that μ is a fuzzy subsemiring of R, but $\mu_{0.5} = \{x,1\}$ does not contain an absorbing zero.

Below we define "sum" and "product" of fuzzy subsets of a semiring R.

Definition 2.1. Let λ and μ be fuzzy subsets of a semiring R. Define the fuzzy subset $\lambda + \mu$ of R by, for all $x \in R$

$$(\lambda + \mu)(x) = \begin{cases} \bigvee\limits_{x=a+b} \{\lambda(a) \wedge \mu(b)\} \\ 0 \quad \text{if } x \text{ is not expressible as } x = a+b. \end{cases}$$

We now define the "product" of two fuzzy subsets of a semiring.

Definition 2.2. Let λ and μ be fuzzy subsets of a semiring R. Define the fuzzy subset $\lambda \circ \mu$ of R by, for all $x \in R$

$$(\lambda \circ \mu)(x) = \begin{cases} \bigvee\limits_{x=\sum_{i=1}^{n} a_i b_i} \left\{ \bigwedge\limits_{i=1}^{n} (\lambda(a_i) \wedge \mu(b_i)) \right\} \\ 0 \quad \text{if } x \text{ is not expressible as } x = \sum\limits_{i=1}^{n} a_i b_i. \end{cases}$$

The following Proposition can be proved using standard arguments.

Proposition 2.1. *If A and B are nonempty subsets of a semiring R, then*
(i) $\chi_A + \chi_B = \chi_{A+B}$.
(ii) $\chi_A \circ \chi_B = \chi_{AB}$.
(iii) $\chi_A \wedge \chi_B = \chi_{A\cap B}$.
where χ_A is the characteristic function of A.

The following Lemma can be proved using standard arguments.

Lemma 2.1. *Let λ, μ, ν be fuzzy subsets of a semiring R. Then*
(i) $\lambda \circ (\mu \circ \nu) = (\lambda \circ \mu) \circ \nu$.
(ii) If $\lambda \leq \mu$, then $\lambda \circ \nu \leq \mu \circ \nu$ and $\nu \circ \lambda \leq \nu \circ \mu$.
(iii) $\lambda + \mu = \mu + \lambda$.
(iv) $\lambda + (\mu + \nu) = (\lambda + \mu) + \nu$.

Theorem 2.2. *Let λ,μ be fuzzy left (right) ideals of a semiring R. Then $\lambda+\mu$ is a fuzzy left (right) ideal of R.*

Proof. Suppose λ,μ are fuzzy left ideals of a semiring R and $x,y\in R$. If $(\lambda+\mu)(x)=0$ or $(\lambda+\mu)(y)=0$,
then

$$(\lambda+\mu)(x)\wedge(\lambda+\mu)(y)=0\leq(\lambda+\mu)(x+y).$$

If $(\lambda+\mu)(x)\neq 0$ and $(\lambda+\mu)(y)\neq 0$ then,

$$(\lambda+\mu)(x)=\bigvee_{x=a+b}\{\lambda(a)\wedge\mu(b)\} \text{ and } (\lambda+\mu)(y)=\bigvee_{y=c+d}\{\lambda(c)\wedge\mu(d)\}.$$

Thus

$$\begin{aligned}(\lambda+\mu)(x)\wedge(\lambda+\mu)(y)&=\left(\bigvee_{x=a+b}\{\lambda(a)\wedge\mu(b)\}\right)\wedge\left(\bigvee_{y=c+d}\{\lambda(c)\wedge\mu(d)\}\right)\\&=\bigvee_{x=a+b}\ \bigvee_{y=c+d}\{(\lambda(a)\wedge\mu(b))\wedge(\lambda(c)\wedge\mu(d))\}\\&=\bigvee_{x=a+b}\ \bigvee_{y=c+d}\{(\lambda(a)\wedge\lambda(c))\wedge(\mu(b)\wedge\mu(d))\}\\&\leq\bigvee_{x=a+b}\ \bigvee_{y=c+d}\{\lambda(a+c)\wedge\mu(b+d)\}\\&\leq(\lambda+\mu)(x+y).\end{aligned}$$

Again, if $(\lambda+\mu)(x)=0$ then $(\lambda+\mu)(x)\leq(\lambda+\mu)(yx)$. If $(\lambda+\mu)(x)\neq 0$, then

$$\begin{aligned}(\lambda+\mu)(x)&=\bigvee_{x=a+b}\{\lambda(a)\wedge\mu(b)\}\\&\leq\bigvee_{x=a+b}\{\lambda(ya)\wedge\mu(yb)\}\\&\leq\bigvee_{yx=c+d}\{\lambda(c)\wedge\mu(d)\}\\&=(\lambda+\mu)(yx).\end{aligned}$$

Hence $\lambda+\mu$ is a fuzzy left ideal of R.

Theorem 2.3. *If λ,μ are fuzzy left (right) ideals of a semiring R, then $\lambda\circ\mu$ is a fuzzy left (right) ideal of R.*

Proof. Suppose λ,μ are fuzzy left ideals of a semiring R and $x,y\in R$.
If $(\lambda\circ\mu)(x)=0$ or $(\lambda\circ\mu)(y)=0$, then $(\lambda\circ\mu)(x)\wedge(\lambda\circ\mu)(y)=0\leq(\lambda\circ\mu)(x+y)$.
If $(\lambda\circ\mu)(x)\neq 0$ and $(\lambda\circ\mu)(y)\neq 0$, then

$$(\lambda \circ \mu)(x) = \bigvee_{x=\sum_{i=1}^{n} a_i b_i} \left\{ \bigwedge_{i=1}^{n} (\lambda(a_i) \wedge \mu(b_i)) \right\}$$

$$(\lambda \circ \mu)(y) = \bigvee_{y=\sum_{j=1}^{m} c_j d_j} \left\{ \bigwedge_{j=1}^{m} (\lambda(c_j) \wedge \mu(d_j)) \right\}$$

$$\begin{aligned}
(\lambda \circ \mu)(x) \wedge (\lambda \circ \mu)(y) &= \left[\bigvee_{x=\sum_{i=1}^{n} a_i b_i} \left\{ \bigwedge_{i=1}^{n} (\lambda(a_i) \wedge \mu(b_i)) \right\} \right] \wedge \left[\bigvee_{y=\sum_{j=1}^{m} c_j d_j} \left\{ \bigwedge_{j=1}^{m} (\lambda(c_j) \wedge \mu(d_j)) \right\} \right] \\
&= \bigvee_{x=\sum_{i=1}^{n} a_i b_i} \ \bigvee_{y=\sum_{j=1}^{m} c_j d_j} \left[\bigwedge_{i=1}^{n} (\lambda(a_i) \wedge \mu(b_i)) \right] \wedge \left[\bigwedge_{j=1}^{m} (\lambda(c_j) \wedge \mu(d_j)) \right] \\
&\leq \bigvee_{x+y=\sum_{k=1}^{p} e_k f_k} \left[\bigwedge_{k=1}^{p} (\lambda(e_k) \wedge \mu(f_k)) \right] \\
&= (\lambda \circ \mu)(x+y).
\end{aligned}$$

Again, if $(\lambda \circ \mu)(x) = 0$ then $(\lambda \circ \mu)(x) \leq (\lambda \circ \mu)(yx)$.
If $(\lambda \circ \mu)(x) \neq 0$, then

$$\begin{aligned}
(\lambda \circ \mu)(x) &= \bigvee_{x=\sum_{i=1}^{n} a_i b_i} \left\{ \bigwedge_{i=1}^{n} (\lambda(a_i) \wedge \mu(b_i)) \right\} \\
&\leq \bigvee_{x=\sum_{i=1}^{n} a_i b_i} \left\{ \bigwedge_{i=1}^{n} (\lambda(y a_i) \wedge \mu(b_i)) \right\} \\
&\leq \bigvee_{yx=\sum_{j=1}^{m} c_j d_j} \left\{ \bigwedge_{j=1}^{m} (\lambda(c_j) \wedge \mu(d_j)) \right\} = (\lambda \circ \mu)(yx).
\end{aligned}$$

Hence $\lambda \circ \mu$ is a fuzzy left ideal of R.

Theorem 2.4. *A fuzzy subset λ of a semiring R is a fuzzy left (right) ideal of R if and only if $\lambda + \lambda \leq \lambda$ and $\chi_R \circ \lambda \leq \lambda$ $(\lambda \circ \chi_R \leq \lambda)$, where χ_R is the characteristic function of R.*

Proof. Suppose λ is a fuzzy left ideal of R and $x \in R$. If $(\lambda + \lambda)(x) = 0$, then $(\lambda + \lambda)(x) \leq \lambda(x)$.

Otherwise $(\lambda + \lambda)(x) = \bigvee_{x=a+b} \{\lambda(a) \wedge \lambda(b)\} \leq \bigvee_{x=a+b} \lambda(a+b) = \lambda(x)$.

Thus $\lambda+\lambda \leq \lambda$.

Again, if $(\chi_R \circ \lambda)(x)=0$ then $(\chi_R \circ \lambda)(x) \leq \lambda(x)$.

Otherwise

$$
\begin{aligned}
(\chi_R \circ \lambda)(x) &= \bigvee_{x=\sum_{i=1}^{n} a_i b_i} \left\{\bigwedge_{i=1}^{n} (\chi_R(a_i) \wedge \lambda(b_i))\right\} \\
&= \bigvee_{x=\sum_{i=1}^{n} a_i b_i} \left\{\bigwedge_{i=1}^{n} (1 \wedge \lambda(b_i))\right\} \\
&= \bigvee_{x=\sum_{i=1}^{n} a_i b_i} \left\{\bigwedge_{i=1}^{n} \lambda(b_i)\right\} \\
&\leq \bigvee_{x=\sum_{i=1}^{n} a_i b_i} \left\{\bigwedge_{i=1}^{n} \lambda(a_i b_i)\right\} \\
&\leq \bigvee_{x=\sum_{i=1}^{n} a_i b_i} \left\{\lambda\left(\sum_{i=1}^{n} a_i b_i\right)\right\} = \lambda(x).
\end{aligned}
$$

Thus $\chi_R \circ \lambda \leq \lambda$.

Conversely, assume that $\lambda+\lambda \leq \lambda$ and $\chi_R \circ \lambda \leq \lambda$. Let $x, y \in R$. Then

$$
\begin{aligned}
\lambda(x+y) &\geq (\lambda+\lambda)(x+y) \\
&= \bigvee_{x=a+b} \{\lambda(a) \wedge \lambda(b)\} \\
&\geq \lambda(x) \wedge \lambda(y).
\end{aligned}
$$

Also

$$
\begin{aligned}
\lambda(xy) \geq (\chi_R \circ \lambda)(xy) &= \bigvee_{xy=\sum_{i=1}^{n} a_i b_i} \left\{\bigwedge_{i=1}^{n} (\chi_R(a_i) \wedge \lambda(b_i))\right\} \\
&\geq \chi_R(x) \wedge \lambda(y) \\
&= 1 \wedge \lambda(y) \\
&= \lambda(y).
\end{aligned}
$$

Thus λ is a fuzzy left ideal of R.

The next theorem shows that product of a fuzzy right ideal and a fuzzy left ideal of a semiring is contained into their intersection.

Theorem 2.5. *Let λ be a fuzzy right ideal and μ a fuzzy left ideal of a semiring R. Then $\lambda \circ \mu \leq \lambda \wedge \mu$.*

Proof. Since λ is a fuzzy right ideal of R, we have $\lambda \circ \chi_R \leq \lambda$. Since $\mu \leq \chi_R$, we have $\lambda \circ \mu \leq \lambda \circ \chi_R \leq \lambda$. Similarly $\lambda \circ \mu \leq \chi_R \circ \mu \leq \mu$. Hence $\lambda \circ \mu \leq \lambda \wedge \mu$.

Example 2.3. Consider the semiring $\mathbb{N}$ of all positive integers. Let λ and μ be the fuzzy subsets of $\mathbb{N}$ defined by for all $x \in \mathbb{N}$

$$\lambda(x) = \begin{cases} 0.6 & \text{if } x \text{ is even} \\ 0 & \text{otherwise} \end{cases}$$

$$\mu(x) = \begin{cases} 0.7 & \text{if } x \text{ is a multiple of 3} \\ 0 & \text{otherwise} \end{cases}$$

Then λ, μ are fuzzy ideals of $\mathbb{N}$.

$(\lambda+\mu)(2) = \bigvee_{2=a+b} \{\lambda(a) \wedge \mu(b)\} = \lambda(1) \wedge \mu(1) = 0$

$(\lambda+\mu)(3) = \bigvee_{3=a+b} \{\lambda(a) \wedge \mu(b)\} = \bigvee\{\lambda(1) \wedge \mu(2), \lambda(2) \wedge \mu(1)\} = 0 \vee 0 = 0.$

Thus $\lambda \nleq \lambda+\mu$ and $\mu \nleq \lambda+\mu$.

Remark 2.2. If R is a semiring with absorbing zero and λ, μ are fuzzy ideals of R, then $\lambda \leq \lambda+\mu$ and $\mu \leq \lambda+\mu$.
In general sum of fuzzy k-ideals is not necessarily a fuzzy k-ideal, as shown in the following example.

Example 2.4. Consider the semiring $\mathbb{N}_0$ of all non negative integers. Let λ and μ be the fuzzy subsets of $\mathbb{N}_0$ defined by for all $x \in \mathbb{N}_0$

$$\lambda(x) = \begin{cases} 0.6 & \text{if } x = 2k \text{ for } k \in \mathbb{N}_0 \\ 0 & \text{otherwise} \end{cases}$$

$$\mu(x) = \begin{cases} 0.7 & \text{if } x = 3k \text{ for } k \in \mathbb{N}_0 \\ 0 & \text{otherwise} \end{cases}$$

Then λ, μ are fuzzy k-ideals of $\mathbb{N}_0$.

$(\lambda+\mu)(2) = 0.6$, $(\lambda+\mu)(9) = 0.6$, $(\lambda+\mu)(7) = 0$.

Now $7+2=9$, but $(\lambda+\mu)(7) \ngeq (\lambda+\mu)(2) \wedge (\lambda+\mu)(9)$. Thus $\lambda+\mu$ is not a fuzzy k-ideal of $\mathbb{N}_0$.

Similarly we can show that $\lambda \circ \mu$ is not a fuzzy k-ideal of $\mathbb{N}_0$. However if we define *k-sum* and *k-product* of fuzzy k-ideals λ and μ of a semiring R by, for all $x \in R$

$$(\lambda +_k \mu)(x) = \begin{cases} \bigvee\limits_{x+(a+b)=a'+b'} \{\lambda(a) \wedge \mu(b) \wedge \lambda(a') \wedge \mu(b')\} \\ 0 \quad \text{if } x \text{ is not expressible as } x+(a+b)=a'+b' \end{cases}$$

$$(\lambda \circ_k \mu)(x) = \begin{cases} \bigvee\limits_{x+\sum_{i=1}^{n}(a_i+b_i)=\sum_{j=1}^{m}a'+b'} \left\{\left(\bigwedge\limits_{i=1}^{n}\lambda(a_i)\right) \wedge \left(\bigwedge\limits_{i=1}^{n}\mu(b_i)\right) \wedge \left(\bigwedge\limits_{i=1}^{m}\lambda(a'_i)\right) \wedge \left(\bigwedge\limits_{i=1}^{m}\mu(b'_i)\right)\right\} \\ 0 \quad \text{if } x \text{ is not expressible as } x+\sum\limits_{i=1}^{n}(a_i+b_i)=\sum\limits_{j=1}^{m}a'+b' \end{cases}$$

then we have the following result.

Theorem 2.6. *If λ, μ are fuzzy k-ideals of a semiring R, then*
(i) $\lambda +_k \mu$ is a fuzzy k-ideal of R.
(ii) $\lambda \circ_k \mu$ is a fuzzy k-ideal of R.

2.2 Regular Semirings

Recall that a semiring R is a *regular semiring* if for each $x \in R$ there exists $a \in R$ such that $x = xax$. It is well known that:

Theorem 2.7. *A semiring R is regular if and only if $A \cap B = AB$ for all left ideals B and right ideals A of R.*

The fuzzy analogue of this result is also true and is proved below.

Theorem 2.8. *A semiring R is regular if and only if $\lambda \wedge \mu = \lambda \circ \mu$ for all fuzzy left ideals μ and fuzzy right ideals λ of R.*

Proof. Suppose R is a regular semiring and λ, μ are fuzzy right and fuzzy left ideals of R, respectively. Let $x \in R$. Then there exists $a \in R$ such that $x = xax$. Now

$$\begin{aligned}(\lambda \circ \mu)(x) &= \bigvee_{x=\sum_{i=1}^{n}a_ib_i} \left\{\bigwedge_{i=1}^{n}(\lambda(a_i) \wedge \mu(b_i))\right\} \\ &\geq (\lambda(xa) \wedge \mu(x)) \geq (\lambda(x) \wedge \mu(x)) = (\lambda \wedge \mu)(x).\end{aligned}$$

Thus $\lambda \wedge \mu \leq \lambda \circ \mu$. By Theorem 2.5, $\lambda \wedge \mu \geq \lambda \circ \mu$. Hence $\lambda \wedge \mu = \lambda \circ \mu$.

Conversely, assume that $\lambda \wedge \mu = \lambda \circ \mu$ for every fuzzy left ideal μ and fuzzy right ideal λ of R. Let A be a right ideal and B be a left ideal of R. Then by Corollary 2.3 , χ_A is a fuzzy right ideal and χ_B is a fuzzy left ideal of R. Hence by hypothesis $\chi_A \wedge \chi_B = \chi_A \circ \chi_B$. By Proposition 2.1, then $\chi_{A\cap B} = \chi_{AB}$. This implies $AB = A \cap B$. Hence by Theorem 2.7, R is a regular semiring.

2.3 Weakly Regular Semirings

A ring R is called *right weakly regular* if $x \in (xR)^2$ for each $x \in R$. Brown and Mc Coy [32], introduced these rings. Later Ramamurthy [126], Camillo and Xiao [33] investigated these rings. Adopting this definition, Ahsan et al. [8] introduced right weakly regular semirings. A semiring R is right weakly regular if for each $x \in R$, $x \in (xR)^2$.

Every regular semiring is right weakly regular, but the converse need not be true. If R is commutative, then the notions of regular and weakly regular semirings coincide.

In this section, R denotes a semiring with an absorbing zero 0 and a multiplicative identity 1.

Theorem 2.9. *The following conditions are equivalent for a semiring R.*
(i) R is right weakly regular.
(ii) All right ideals of R are idempotent.
(iii) $BA = B \cap A$ *for all right ideals B and two-sided ideals A of R.*
(iv) All fuzzy right ideals of R are idempotent (A fuzzy right ideal λ *of R is idempotent if* $\lambda \circ \lambda = \lambda$*).*
(v) $\lambda \circ \mu = \lambda \wedge \mu$ *for all fuzzy right ideals* λ *and all fuzzy two-sided ideals* μ *of R.*
If R is commutative, then the above conditions are equivalent to:
(vi) R is regular.

Proof. $(i) \Leftrightarrow (ii) \Leftrightarrow (iii)$ are due to (cf. [8]).
$(i) \Rightarrow (iv)$ Let λ be a fuzzy right ideal of R and $x \in R$. Then

$$\begin{aligned}(\lambda \circ \lambda)(x) &= \bigvee_{x=\sum_{i=1}^{n} a_i b_i} \left\{ \bigwedge_{i=1}^{n} (\lambda(a_i) \wedge \lambda(b_i)) \right\} \\ &\leq \bigvee_{x=\sum_{i=1}^{n} a_i b_i} \left\{ \bigwedge_{i=1}^{n} \lambda(a_i) \right\} \leq \bigvee_{x=\sum_{i=1}^{n} a_i b_i} \left\{ \bigwedge_{i=1}^{n} \lambda(a_i b_i) \right\} \\ &\leq \bigvee_{x=\sum_{i=1}^{n} a_i b_i} \left\{ \lambda\left(\sum_{i=1}^{n} a_i b_i \right) \right\} = \lambda(x).\end{aligned}$$

Now, since R is right weakly regular, we have $x \in (xR)^2$. This implies that there exist $r_i, s_i \in R$ such that $x = \sum_{i=1}^{n} x r_i x s_i$. Thus
$\lambda(x) = \lambda(x) \wedge \lambda(x) \leq \lambda(x r_i) \wedge \lambda(x s_i)$ for $1 \leq i \leq n$.

This implies

$$\lambda(x) \leq \bigwedge_{i=1}^{n} (\lambda(x r_i) \wedge \lambda(x s_i)) \leq \bigvee_{x=\sum_{j=1}^{m} a_j b_j} \left\{ \bigwedge_{j=1}^{m} (\lambda(a_j) \wedge \lambda(b_j)) \right\} = (\lambda \circ \lambda)(x).$$

Hence $\lambda(x) = (\lambda \circ \lambda)(x)$, that is $\lambda \circ \lambda = \lambda$.

$(iv) \Rightarrow (ii)$ Let A be a fuzzy right ideal of R. Then by Corollary 2.3, χ_A is a fuzzy right ideal of R. Thus by the hypothesis $\chi_A \circ \chi_A = \chi_A$. By Proposition 2.1, $\chi_{AA} = \chi_A$. This implies $A^2 = A$.

$(i) \Rightarrow (v)$ Let λ be a fuzzy right ideal and μ a fuzzy ideal of R. Then by Theorem 2.5, $\lambda \circ \mu \leq \lambda \wedge \mu$. Let $x \in R$. Since R is right weakly regular, there exist $r_i, s_i \in R$ such that $x = \sum_{i=1}^{n} xr_ixs_i$. Now

$(\lambda \wedge \mu)(x) = \lambda(x) \wedge \mu(x) \leq \lambda(xr_i) \wedge \mu(xs_i)$ for $1 \leq i \leq n$. This implies

$$(\lambda \wedge \mu)(x) \leq \bigwedge_{i=1}^{n} (\lambda(xr_i) \wedge \mu(xs_i)) \leq \bigvee_{x=\sum_{j=1}^{m} a_jb_j} \left\{ \bigwedge_{j=1}^{m} (\lambda(a_j) \wedge \mu(b_j)) \right\}$$

$= (\lambda \circ \mu)(x)$. Thus $\lambda \wedge \mu = \lambda \circ \mu$.

$(v) \Rightarrow (iii)$ Let B be a right ideal and A be a two-sided ideal of R. Then χ_B is a fuzzy right ideal and χ_A is a fuzzy two-sided ideal of R, and by hypothesis $\chi_B \circ \chi_A = \chi_B \wedge \chi_A$. By Proposition 2.1, $\chi_{BA} = \chi_{B \cap A}$. This implies $BA = B \cap A$.

Finally if R is commutative, then (vi) is equivalent to (iii).

2.4 Fully Idempotent Semirings

A semiring R is said to be *fully idempotent* if each ideal of R is idempotent. These semirings are studied in [3]. Regular and weakly regular semirings are fully idempotent, but the converse is not true in general. If R is commutative, then the notions of regular, weakly regular and fully idempotent semirings are identical.

Theorem 2.10. *The following conditions are equivalent for a semiring R.*
(i) R is fully idempotent.
(ii) $BA = B \cap A$ for each pair of ideals A, B of R.
(iii) All fuzzy ideals of R are idempotent (A fuzzy ideal λ of R is idempotent if $\lambda \circ \lambda = \lambda$).
(iv) $\lambda \circ \mu = \lambda \wedge \mu$ for each pair of fuzzy ideals λ, μ of R.
If R is commutative, then the above conditions are equivalent to:
(v) R is regular.

Proof. $(i) \Leftrightarrow (ii)$ are due to (cf. [3]).
$(i) \Rightarrow (iii)$ Let λ be a fuzzy ideal of R. Then $\lambda \circ \lambda \leq \lambda$. Now, because each ideal of R is idempotent, we have for each $x \in R$, $<x> = <x>^2$, $<x>$ denotes the principal ideal of R generated by x. Hence, there exist $r_i, r'_i, s_i, s'_i \in R$ such that $x = \sum_{i=1}^{n} r_ixr'_is_ixs'_i$.
Thus

$$\lambda(x) = \lambda(x) \wedge \lambda(x) \leq \lambda(r_ixr'_i) \wedge \lambda(s_ixs'_i) \text{ for } 1 \leq i \leq n.$$

This implies

$$\lambda(x) \leq \bigwedge_{i=1}^{n} \lambda(r_ixr'_i) \wedge \lambda(s_ixs'_i) \leq \bigvee_{x=\sum_{j=1}^{m} a_jb_j} \left\{ \bigwedge_{j=1}^{m} (\lambda(a_j) \wedge \lambda(b_j)) \right\} = (\lambda \circ \lambda)(x).$$

Hence $\lambda(x) = (\lambda \circ \lambda)(x)$, that is $\lambda \circ \lambda = \lambda$.

$(iii) \Rightarrow (i)$ Let A be an ideal of R. Then χ_A is the fuzzy ideal of R. Then by hypothesis, $\chi_A \circ \chi_A = \chi_A$. By Proposition 2.1, $\chi_{AA} = \chi_A$. This implies $A^2 = A$.
$(i) \Rightarrow (iv)$ Let λ and μ be any pair of fuzzy ideals of R. Then by Theorem 2.5,

$\lambda \circ \mu \leq \lambda \wedge \mu$. Let $x \in R$. Since each ideal of R is idempotent, for each $x \in R$, $<x>=<x>^2$. Hence, there exist $r_i, r_i', s_i, s_i' \in R$ such that $x = \sum_{i=1}^{n} r_i x r_i' s_i x s_i'$. Thus

$$(\lambda \wedge \mu)(x) = \lambda(x) \wedge \mu(x) \leq \lambda\left(r_i x r_i'\right) \wedge \mu\left(s_i x s_i'\right) \text{ for } 1 \leq i \leq n.$$

This implies

$$(\lambda \wedge \mu)(x) \leq \bigwedge_{i=1}^{n} (\lambda(r_i x r_i') \wedge \mu(s_i x s_i')) \leq \bigvee_{x=\sum_{j=1}^{m} a_j b_j} \left\{ \bigwedge_{j=1}^{m} (\lambda(a_j) \wedge \mu(b_j)) \right\}$$

$= (\lambda \circ \mu)(x)$.
Thus $\lambda \wedge \mu = \lambda \circ \mu$.

$(iv) \Rightarrow (iii)$ Let λ and μ be any pair of fuzzy ideals of R. We have $\lambda \wedge \mu = \lambda \circ \mu$. Take $\mu = \lambda$. Thus $\lambda \wedge \lambda = \lambda \circ \lambda$, that is $\lambda = \lambda^2$.

If R is commutative, then (i) is equivalent to (v).

Theorem 2.11. *The following assertions are equivalent for a semiring R.*
(i) R is fully idempotent.
(ii) The set of all fuzzy ideals of R (ordered by $\leq$) form a distributive lattice under the sum and intersection of fuzzy ideals with $\lambda \wedge \mu = \lambda \circ \mu$, for each pair of fuzzy ideals λ, μ of R.

Proof. $(i) \Rightarrow (ii)$ Let $\mathscr{L}_R$ be the set of all fuzzy ideals of R (ordered by $\leq$). Then clearly $\mathscr{L}_R$ is a lattice under the sum and intersection of fuzzy ideals. Since R is fully idempotent so $\lambda \wedge \mu = \lambda \circ \mu$, for each pair of fuzzy ideals λ, μ of R. We now show that $\mathscr{L}_R$ is distributive. Let λ, μ, ν be fuzzy ideals of R and $x \in R$. Then

$$\begin{aligned}
[(\lambda \wedge \mu) + \nu](x) &= \bigvee_{x=a+b} \{(\lambda \wedge \mu)(a) \wedge \nu(b)\} \\
&= \bigvee_{x=a+b} \{\lambda(a) \wedge \mu(a) \wedge \nu(b)\} \\
&= \bigvee_{x=a+b} \{(\lambda(a) \wedge \nu(b)) \wedge (\mu(a) \wedge \nu(b))\} \\
&\leq \bigvee_{x=a+b} \{(\lambda + \nu)(x) \wedge (\mu + \nu)(x)\} \\
&= (\lambda + \nu)(x) \wedge (\mu + \nu)(x) \\
&= [(\lambda + \nu) \wedge (\mu + \nu)](x).
\end{aligned}$$

Again

$$\begin{aligned}
[(\lambda + \nu) \wedge (\mu + \nu)](x) &= [(\lambda + \nu) \circ (\mu + \nu)](x) \\
&= \bigvee_{x=\sum_{j=1}^{m} a_j b_j} \left\{ \bigwedge_{j=1}^{m} ((\lambda + \nu)(a_j) \wedge (\mu + \nu)(b_j)) \right\}
\end{aligned}$$

$$= \bigvee_{x=\sum_{j=1}^{m} a_j b_j} \left\{ \bigwedge_{j=1}^{m} \left[\left[\bigvee_{a_j=a'_j+b'_j} \{\lambda(a'_j) \wedge \nu(b'_j)\} \right] \wedge \left[\bigvee_{b_j=c_j+d_j} (\mu(c_j) \wedge \nu(d_j)) \right] \right] \right\}$$

$$= \bigvee_{x=\sum_{j=1}^{m} a_j b_j} \left\{ \bigwedge_{j=1}^{m} \left[\bigvee_{\substack{a_j=a'_j+b'_j \\ b_j=c_j+d_j}} \{\lambda(a'_j) \wedge \nu(b'_j) \wedge (\mu(c_j) \wedge \nu(d_j))\} \right] \right\}$$

$$= \bigvee_{x=\sum_{j=1}^{m} a_j b_j} \left\{ \bigwedge_{j=1}^{m} \left[\bigvee_{\substack{a_j=a'_j+b'_j \\ b_j=c_j+d_j}} \{\lambda(a'_j) \wedge \nu(b'_j) \wedge \nu(b'_j) \wedge (\mu(c_j) \wedge \nu(d_j))\} \right] \right\}$$

$$\leq \bigvee_{x=\sum_{j=1}^{m} a_j b_j} \left\{ \bigwedge_{j=1}^{m} \left[\bigvee_{\substack{a_j=a'_j+b'_j \\ b_j=c_j+d_j}} \{\lambda(a'_j c_j) \wedge \nu(b'_j c_j) \wedge \nu(b'_j d_j) \wedge (\mu(a'_j c_j) \wedge \nu(a'_j d_j))\} \right] \right\}$$

$$\leq \bigvee_{x=\sum_{j=1}^{m} a_j b_j} \left\{ \bigwedge_{j=1}^{m} \left[\bigvee_{\substack{a_j=a'_j+b'_j \\ b_j=c_j+d_j}} \{\lambda(a'_j c_j) \wedge \mu(a'_j c_j) \wedge \nu(b'_j c_j) \wedge \nu(b'_j d_j) \wedge (\wedge \nu(a'_j d_j))\} \right] \right\}$$

$$\leq \bigvee_{x=\sum_{j=1}^{m} a_j b_j} \left\{ \bigwedge_{j=1}^{m} \left[\bigvee_{\substack{a_j=a'_j+b'_j \\ b_j=c_j+d_j}} \{(\lambda \wedge \mu)(a'_j c_j) \wedge \nu(b'_j c_j + b'_j d_j + a'_j d_j)\} \right] \right\}$$

$$\leq \bigvee_{x=\sum_{j=1}^{m} a_j b_j} \left\{ \bigwedge_{j=1}^{m} \left[\bigvee_{\substack{a_j=a'_j+b'_j \\ b_j=c_j+d_j}} [(\lambda \wedge \mu) + \nu](a_j b_j) \right] \right\}$$

$$\leq \bigvee_{x=\sum_{j=1}^{m} a_j b_j} [(\lambda \wedge \mu) + \nu](x)$$

$$= [(\lambda \wedge \mu) + \nu](x).$$

Thus $(\lambda + \nu) \wedge (\mu + \nu) = (\lambda \wedge \mu) + \nu$.

$(ii) \Rightarrow (i)$ Let λ, μ be any pair of fuzzy ideals of R. Then by hypothesis $\lambda \wedge \mu = \lambda \circ \mu$. Take $\mu = \lambda$. Then $\lambda = \lambda \circ \lambda$. Hence R is fully idempotent.

Definition 2.3. A fuzzy ideal λ of a semiring R is called a *fuzzy prime ideal* of R if for any fuzzy ideals μ, ν of R, $\mu \circ \nu \leq \lambda$ implies $\mu \leq \lambda$ or $\nu \leq \lambda$; λ is called *fuzzy irreducible* if for any fuzzy ideals μ, ν of R, $\mu \wedge \nu = \lambda$ implies $\mu = \lambda$ or $\nu = \lambda$.

Theorem 2.12. *For a fuzzy ideal λ of a fully idempotent semiring R, the following conditions are equivalent:*

(i) λ *is a fuzzy prime ideal.*
(ii) λ *is a fuzzy irreducible ideal.*

Proof. $(i) \Rightarrow (ii)$ Assume λ is a fuzzy prime ideal of R and let μ, ν be any fuzzy ideals of R such that $\lambda = \mu \wedge \nu$. Then $\lambda \leq \mu$ and $\lambda \leq \nu$. Since R is fully idempotent, so $\mu \wedge \nu = \mu \circ \nu$. Hence $\lambda = \mu \circ \nu$. Since λ is fuzzy prime, we have either $\mu \leq \lambda$ or $\nu \leq \lambda$. Thus either $\lambda = \mu$ or $\lambda = \nu$.

$(ii) \Rightarrow (i)$ Suppose λ is a fuzzy irreducible ideal of R and let μ, ν be any fuzzy ideals of R such that $\lambda \geq \mu \circ \nu$. Since R is a fully idempotent semiring, we have $\mu \circ \nu = \mu \wedge \nu$. Thus $\mu \wedge \nu \leq \lambda$. So by Theorem 2.11 $\lambda = (\mu \wedge \nu) + \lambda = (\mu + \lambda) \wedge (\nu + \lambda)$. Since λ is fuzzy irreducible, we have $(\mu + \lambda) = \lambda$ or $(\nu + \lambda) = \lambda$. This implies that $\mu \leq \lambda$ or $\nu \leq \lambda$.

Lemma 2.2. *Let R be a fully idempotent semiring. If λ is a fuzzy ideal of R and $\lambda(a) = t \in (0,1]$, where $a \in R$, then there exists a fuzzy prime ideal μ of R such that $\lambda \leq \mu$ and $\mu(a) = t$.*

Proof. Let $\mathscr{A} = \{\nu : \nu \text{ is a fuzzy ideal of } R,\ \nu(a) = t, \text{ and } \lambda \leq \nu\}$. Then $\mathscr{A} \neq \emptyset$, since $\lambda \in \mathscr{A}$. Let $\mathscr{F}$ be a totally ordered subset of $\mathscr{A}$, say $\mathscr{F} = \{\delta_i : i \in I\}$. Let $x, y \in R$, then

$$\begin{aligned}
\left(\bigvee_{i\in I}\delta_i\right)(x) \wedge \left(\bigvee_{i\in I}\delta_i\right)(y) &= \left(\bigvee_{i\in I}\delta_i(x)\right) \wedge \left(\bigvee_{i\in I}\delta_i(y)\right) \\
&= \bigvee_{j\in I}\left(\bigvee_{i\in I}(\delta_i(x) \wedge \delta_j(y))\right) \\
&\leq \bigvee_{j\in I}\left(\bigvee_{i\in I}\left(\delta_i^j(x) \wedge \delta_i^j(y)\right)\right) \quad \textit{where } \delta_i^j = \max\{\delta_i, \delta_j\} \\
&\leq \bigvee_{j\in I}\left(\bigvee_{i\in I}\left(\delta_i^j(x+y)\right)\right) \\
&\leq \bigvee_{i,j\in I}\left(\delta_i^j(x+y)\right) \\
&\leq \bigvee_{i\in I}(\delta_i(x+y)).
\end{aligned}$$

Also $\left(\bigvee_{i\in I}\delta_i\right)(x) = \left(\bigvee_{i\in I}\delta_i(x)\right) \leq \left(\bigvee_{i\in I}\delta_i(xr)\right) = \left(\bigvee_{i\in I}\delta_i\right)(xr)$.

Similarly $\left(\bigvee_{i\in I}\delta_i\right)(x) \leq \left(\bigvee_{i\in I}\delta_i\right)(rx)$.

Thus $\left(\bigvee_{i\in I}\delta_i\right)$ is a fuzzy ideal of R. Clearly $\lambda \leq \left(\bigvee_{i\in I}\delta_i\right)$ and $\left(\bigvee_{i\in I}\delta_i\right)(a) = t$. Thus $\bigvee_{i\in I}\delta_i$ is the l.u.b. of $\mathscr{F}$. Hence by Zorn's Lemma there exists a fuzzy ideal ν of R which is maximal with respect to the property that $\lambda \leq \nu$ and $\nu(a) = t$. We now show that ν is a fuzzy irreducible ideal of R. Suppose γ, η are fuzzy ideals of R such that $\nu = \gamma \wedge \eta$. This implies $\nu \leq \gamma$ and $\nu \leq \eta$. We claim that either $\nu = \gamma$ or $\nu = \eta$. Suppose on the contrary, $\nu \neq \gamma$ and $\nu \neq \eta$. Then $\gamma(a) \neq t$ and $\eta(a) \neq t$. Hence $t = \nu(a) = (\gamma \wedge \eta)(a) = \gamma(a) \wedge \eta(a) \neq t$, which is impossible. Hence either $\nu = \gamma$ or $\nu = \eta$. Thus ν is fuzzy irreducible and hence by Theorem 2.12 ,ν is fuzzy prime ideal.

Theorem 2.13. *The following conditions are equivalent for a semiring R:*

(i) R is fully idempotent.
(ii) The set of all fuzzy ideals of R (ordered by $\leq$) form a distributive lattice under the sum and intersection of fuzzy ideals with $\lambda \wedge \mu = \lambda \circ \mu$, for each pair of fuzzy ideals λ, μ of R.
(iii) Each fuzzy ideal is the intersection of those fuzzy prime ideals of R which contain it.
If R is commutative, then the above conditions are equivalent to:
(iv) R is regular.

Proof. $(i) \Leftrightarrow (ii)$ This is Theorem 2.11.
$(ii) \Rightarrow (iii)$ Let λ be a fuzzy ideal of R and $\{\delta_i : i \in I\}$ be a family of all fuzzy prime ideals of R which contain λ. Then $\lambda \leq \bigwedge_{i \subset I} \delta_i$. Let $a \in R$. Then by Lemma 2.2 there exists a fuzzy prime ideal δ such that $\lambda \leq \delta$ and $\delta(a) = \lambda(a)$. Thus $\delta \in \{\delta_i : i \in I\}$. Hence $\left(\bigwedge_{i \in I} \delta_i\right)(a) \leq \delta(a) = \lambda(a)$. This implies that $\bigwedge_{i \in I} \delta_i \leq \lambda$. Hence $\bigwedge_{i \in I} \delta_i = \lambda$.
$(iii) \Rightarrow (i)$ Let λ be a fuzzy ideal of R. Then $\lambda \circ \lambda$ is also a fuzzy ideal of R. Hence by hypothesis $\lambda \circ \lambda = \bigwedge_{i \in I} \delta_i$ where $\{\delta_i : i \in I\}$ is a family of all fuzzy prime ideals of R which contain $\lambda \circ \lambda$. Now $\lambda \circ \lambda \leq \delta_i$ for each $i \in I$, since δ_i is fuzzy prime, so $\lambda \leq \delta_i$ for all $i \in I$. Thus $\lambda \leq \bigwedge_{i \in I} \delta_i = \lambda \circ \lambda$. But $\lambda \circ \lambda \leq \lambda$ always holds. Hence $\lambda \circ \lambda = \lambda$.
$(i) \Leftrightarrow (iv)$ Immediate.

Chapter 3
Fuzzy Subsemimodules over Semirings

It is well-known that modules are a generalization of vector spaces of linear algebra in which the "scalars" are allowed to be from an arbitrary ring, rather than a field. This rather modest weakening of the axioms is quite far reaching, including, for example, the theory of rings and ideals and the theory of abelian groups as special cases.

It is then natural to define "semimodules over semirings" as a generalization to modules over rings. The aim of this chapter is to give a brief summary of basic results concerning semimodules over semirings in a fuzzy context. Note that semirings have proven to be useful in studying automata and formal languages (cf. [11, 62, 68, 102]). The notions of automata and formal languages have been generalized and extensively studied in a fuzzy framework (cf. [102, 113, 114, 154]). Thus it is very natural to expect that semirings and semimodules over them will prove as important tools in studying fuzzy automata and fuzzy formal languages. Nevertheless an extensive study of semimodules over semirings in a fuzzy context still awaits investigation.

The aim of Chapter 3 is to define fuzzy subsemimodule of a semimodule and discuss some basic properties of these subsemimodules. We also define "pure fuzzy subsemimodule" and "normal fuzzy semimodule" and show that if M is a cyclic semimodule over a right weakly regular semiring R then M is fuzzy normal semimodule over R. Later in section 2 of this chapter, definitions of fuzzy prime (semiprime) subsemimodules of a semimodule and a study of basic properties of these subsemimodules is given in this section. In section 3, we study and characterize semirings all of whose fuzzy ideals are prime (semiprime). Section 4 provides an example of a semiring which is fully idempotent with a non prime fuzzy ideal.

3.1 Fuzzy Subsemimodules

Throughout, R, as usual will denote a semiring with an identity element 1 and an absorbing zero 0. An additively written commutative semigroup M with a neutral element 0 is called a *right R-semimodule*, M_R, if R is a semiring and there is a

J. Ahsan et al.: Fuzzy Semirings with Applications, STUDFUZZ 278, pp. 31–52.
springerlink.com

function $\phi : M \times R \to M$ such that if $\phi(m,a)$ is denoted by ma, then the following conditions hold:

(i) $(m+m')a = ma + m'a$,
(ii) $m(a+a') = ma + ma'$,
(iii) $m(aa') = (ma)a'$,
(iv) $m \cdot 1 = m$,
(v) $0 \cdot a = m \cdot 0 = 0$, for all $a, a' \in R$ and $m, m' \in M$ (cf. [9, 62]).

Similarly, one can define a *left* R-semimodule ${}_RM$. A semiring R is a right (left) R-semimodule over itself which is denoted by R_R (${}_RR$). A *subsemimodule* N of a right R-semimodule M is a subsemigroup of M such that $na \in N$ for all $n \in N$ and $a \in R$. By a right (left) ideal of R, we shall mean a subsemimodule of R_R (${}_RR$). The word 'ideal' will always mean a two-sided ideal of R, that is, an ideal which is both a left and a right ideal of R. An ideal generated by an element x will be denoted by $< x >$. The sum and product of ideals of semirings are defined as in rings.

Definition 3.1. Let M be a right R-semimodule over a semiring R. A function $\lambda : M \to [0,1]$ is called a *fuzzy subsemimodule* of M_R, if the following conditions hold:

(i) $\lambda(m+m') \geq \lambda(m) \wedge \lambda(m')$, for all $m, m' \in M$,
(ii) $\lambda(ma) \geq \lambda(m)$, for all $m \in M$ and $a \in R$.

In the sequel, fuzzy subsemimodules of R_R are called *fuzzy right ideals* of the semiring R. Fuzzy left ideals of R are defined analogously. By a *fuzzy ideal* of R we mean a fuzzy subset of R which is both a fuzzy right and a fuzzy left ideal of R.

We now prove a result which provides a necessary and sufficient condition on an arbitrary fuzzy subset of a right R-semimodule to be a fuzzy subsemimodule.

Theorem 3.1. *A fuzzy subset λ of a right R-semimodule M is a fuzzy subsemimodule of M if and only if each nonempty level subset λ_t of λ is a subsemimodule of M, for all $t \in [0,1]$.*

Proof. Suppose λ is a fuzzy subsemimodule of M and $\lambda_t \neq \emptyset$ for $t \in [0,1]$. Let $m_1, m_2 \in \lambda_t$. Then $\lambda(m_1) \geq t$ and $\lambda(m_2) \geq t$. Since $\lambda(m_1+m_2) \geq \lambda(m_1) \wedge \lambda(m_2) \geq t$ and $\lambda(m_1 r) \geq \lambda(m_1) \geq t$ for all $r \in R$, we have $m_1 + m_2 \in \lambda_t$ and $m_1 r \in \lambda_t$. Hence λ_t is a subsemimodule of M.

Conversely, assume that each nonempty λ_t is a subsemimodule of M. Let $m_1, m_2 \in M$ be such that $\lambda(m_1+m_2) < \lambda(m_1) \wedge \lambda(m_2)$. Select $t \in [0,1]$ such that $\lambda(m_1+m_2) < t \leq \lambda(m_1) \wedge \lambda(m_2)$. Then $m_1, m_2 \in \lambda_t$ but $m_1 + m_2 \notin \lambda_t$, which is a contradiction. Hence $\lambda(m_1+m_2) \geq \lambda(m_1) \wedge \lambda(m_2)$. Similarly, if there exist $m \in M$ and $r \in R$ such that $\lambda(mr) < \lambda(m)$, then we can select $t \in [0,1]$ such that $\lambda(mr) < t \leq \lambda(m)$. This implies that $m \in \lambda_t$ but $mr \notin \lambda_t$. This contradicts our hypothesis. Hence $\lambda(mr) \geq \lambda(m)$ for all $m \in M$ and $r \in R$. Thus we conclude that λ is a fuzzy subsemimodule of M.

Corollary 3.1. *Let N be a nonempty subset of a right R-semimodule M and $t_1, t_2 \in [0,1]$ be such that $t_1 < t_2$. Define a fuzzy subset λ_N of M as follows:*

$$\lambda_N(m) = \begin{cases} t_2 & \text{if } m \in N \\ t_1 & \text{if } m \notin N \end{cases}$$

for all $m \in M$. Then λ_N is a fuzzy subsemimodule of M if and only if N is a subsemimodule of M.

Proof. Follows from the above theorem.

Corollary 3.2. *Let N be a nonempty subset of a right R-semimodule M. Then N is a subsemimodule of M if and only if the characteristic function of N is a fuzzy subsemimodule of M.*

Proof. Follows from the above corollary.

Below we define the "sum" of fuzzy subsets of a right R-semimodule M and the "product" of a fuzzy subset of M and a fuzzy subset of R.

Definition 3.2. Let λ and μ be fuzzy subsets of a right R-semimodule M. Define the fuzzy subset $\lambda + \mu$ of M as follows:

$$(\lambda + \mu)(m) = \bigvee_{m = m_1 + m_2} \{\lambda(m_1) \wedge \mu(m_2)\}.$$

Let λ be a fuzzy subset of a right R-semimodule M and μ be a fuzzy subset of R. Define the fuzzy subset $\lambda \circ \mu$ of M as follows:

$$(\lambda \circ \mu)(m) = \bigvee_{m = \sum_{i=1}^{n} m_i r_i} \left[\bigwedge_{i=1}^{n} [\lambda(m_i) \wedge \mu(r_i)] \right],$$

where $m \in M, m_i \in M, r_i \in R$ and $n \in \mathbb{N}$.

Theorem 3.2. *Let λ, μ be fuzzy subsemimodules of a right R-semimodule M. Then $\lambda + \mu$ is a fuzzy subsemimodule of M.*

Proof. Suppose λ, μ are fuzzy subsemimodules of a right R-semimodule M and $m_1, m_2 \in M$. Then

$$(\lambda + \mu)(m_1) = \bigvee_{m_1 = a + b} \{\lambda(a) \wedge \mu(b)\} \text{ and } (\lambda + \mu)(m_2) = \bigvee_{m_2 = c + d} \{\lambda(c) \wedge \mu(d)\}.$$

Thus

$$
\begin{aligned}
(\lambda+\mu)(m_1)\wedge(\lambda+\mu)(m_2) &= \left(\bigvee_{m_1=a+b}\{\lambda(a)\wedge\mu(b)\}\right)\bigwedge\left(\bigvee_{m_2=c+d}\{\lambda(c)\wedge\mu(d)\}\right)\\
&= \bigvee_{m_1=a+b}\bigvee_{m_2=c+d}\{\lambda(a)\wedge\mu(b)\wedge\lambda(c)\wedge\mu(d)\}\\
&= \bigvee_{m_1=a+b}\bigvee_{m_2=c+d}\{\lambda(a)\wedge\lambda(c)\wedge\mu(b)\wedge\mu(d)\}\\
&\le \bigvee_{m_1=a+b}\bigvee_{m_2=c+d}\{\lambda(a+c)\wedge\mu(b+d)\}\\
&\le \bigvee_{m_1+m_2=x+y}\{\lambda(x)\wedge\mu(y)\}\\
&= (\lambda+\mu)(m_1+m_2)\,.
\end{aligned}
$$

Again,

$$
\begin{aligned}
(\lambda+\mu)(m_1) &= \bigvee_{m_1=a+b}\{\lambda(a)\wedge\mu(b)\}\\
&\le \bigvee_{m_1=a+b}\{\lambda(ar)\wedge\mu(br)\}\\
&\le \bigvee_{m_1r=x+y}\{\lambda(x)\wedge\mu(y)\}\\
&= (\lambda+\mu)(m_1r)\,.
\end{aligned}
$$

Hence $\lambda+\mu$ is a fuzzy subsemimodule of M.

By employing similar arguments, we can also prove:

Theorem 3.3. *If λ is a fuzzy subset of a right R-semimodule M and μ is a fuzzy right ideal of R, then $\lambda\circ\mu$ is a fuzzy subsemimodule of M.*

We will now prove the following lemma;

Lemma 3.1. *If λ is a fuzzy left ideal and μ a fuzzy right ideal of a semiring R then $\lambda\circ\mu$ is a fuzzy ideal of R.*

Proof. Let $a,b\in R$. Then

$$
(\lambda\circ\mu)(a) = \bigvee_{a=\sum_{i=1}^{n}x_iy_i}\left[\bigwedge_{i=1}^{n}[\lambda(x_i)\wedge\mu(y_i)]\right],
$$

$$
(\lambda\circ\mu)(b) = \bigvee_{b=\sum_{j=1}^{m}x'_jy'_j}\left[\bigwedge_{j=1}^{m}[\lambda(x'_j)\wedge\mu(y'_j)]\right].
$$

Therefore

$$(\lambda \circ \mu)(a) \wedge (\lambda \circ \mu)(b) = \left[\bigvee_{a=\sum_{i=1}^{n} x_i y_i} \left[\bigwedge_{i=1}^{n} [\lambda(x_i) \wedge \mu(y_i)] \right] \right] \wedge \left[\bigvee_{b=\sum_{j=1}^{m} x'_j y'_j} \left[\bigwedge_{j=1}^{m} [\lambda(x'_j) \wedge \mu(y'_j)] \right] \right]$$

$$= \bigvee_{a=\sum_{i=1}^{n} x_i y_i} \bigvee_{b=\sum_{j=1}^{m} x'_j y'_j} \left[\left[\bigwedge_{i=1}^{n} [\lambda(x_i) \wedge \mu(y_i)] \right] \wedge \left[\bigwedge_{j=1}^{m} [\lambda(x'_j) \wedge \mu(y'_j)] \right] \right]$$

$$\leq \bigvee_{a+b=\sum_{k=1}^{q} x''_k y''_k} \left[\bigwedge_{k=1}^{q} [\lambda(x''_k) \wedge \mu(y''_k)] \right] = (\lambda \circ \mu)(a+b).$$

On the other hand, for $a, r \in R$, we have

$$(\lambda \circ \mu)(a) = \bigvee_{a=\sum_{i=1}^{n} x_i y_i} \left[\bigwedge_{i=1}^{n} [\lambda(x_i) \wedge \mu(y_i)] \right]$$

$$\leq \bigvee_{a=\sum_{i=1}^{n} x_i y_i} \left[\bigwedge_{i=1}^{n} [\lambda(rx_i) \wedge \mu(y_i)] \right]$$

$$\leq \bigvee_{ra=\sum_{j=1}^{m} x'_j y'_j} \left[\bigwedge_{j=1}^{m} [\lambda(x'_j) \wedge \mu(y'_j)] \right] = (\lambda \circ \mu)(ra)$$

and similarly $(\lambda \circ \mu)(a) \leq (\lambda \circ \mu)(ar)$. Thus $\lambda \circ \mu$ is a fuzzy ideal of R.

Definition 3.3. A two-sided ideal I of a semiring R is called *right t-pure* if, for each $x \in I$, there exists $y \in I$ such that $x = xy$.

Using standard arguments we can prove:

Proposition 3.1. *A two-sided ideal I of a semiring R is right t-pure if and only if $J \cap I = JI$ for any right ideal J of R.*

Extending the above notion to arbitrary semimodules, we obtain the following definition.

Definition 3.4. A subsemimodule N of a right R-semimodule M is called *pure* in M if and only if $N \cap MI = NI$ for each ideal I of R. M is called *normal* if each subsemimodule of M is pure in M.

Definition 3.5. A fuzzy ideal λ of a semiring R is called a *right t-pure fuzzy ideal* of R if $\mu \wedge \lambda = \mu \circ \lambda$ for each fuzzy right ideal μ of R.

Definition 3.6. A fuzzy subsemimodule λ of a right R-semimodule M is called a *pure fuzzy subsemimodule* of M if for each fuzzy ideal μ of R, $\lambda \wedge \mathscr{M} \circ \mu = \lambda \circ \mu$, where $\mathscr{M}$ is the fuzzy subsemimodule of M defined by $\mathscr{M}(m) = 1$ for each $m \in M$. M is called *fuzzy normal* if each fuzzy subsemimodule of M is a pure fuzzy subsemimodule of M. In particular, R is called *fuzzy normal* if R_R is fuzzy normal.

Proposition 3.2. *The following conditions for an ideal I of a semiring R are equivalent:*
(*i*) *I is right t-pure in R.*
(*ii*) *The characteristic function of I is a right t-pure fuzzy ideal of R.*

Proof. (*i*) $\Rightarrow$ (*ii*) Assume that I is right t-pure in R. Since I is a two-sided ideal of R, χ_I is a fuzzy ideal of R. Let ν be a fuzzy right ideal of R, we show that $\nu \wedge \chi_I = \nu \circ \chi_I$. Let $x \in R$. Then

$$\begin{aligned}
\nu \circ \chi_I(x) &= \bigvee_{x=\sum_{i=1}^n a_i b_i} \left[\bigwedge_{i=1}^n [\nu(a_i) \wedge \chi_I(b_i)]\right] \\
&\le \bigvee_{x=\sum_{i=1}^n a_i b_i} \left[\bigwedge_{i=1}^n [\nu(a_i b_i) \wedge \chi_I(a_i b_i)]\right] \\
&= \bigvee_{x=\sum_{i=1}^n a_i b_i} \left[\left(\bigwedge_{i=1}^n \nu(a_i b_i)\right) \wedge \left(\bigwedge_{i=1}^n \chi_I(a_i b_i)\right)\right] \\
&\le \bigvee_{x=\sum_{i=1}^n a_i b_i} [\nu(x) \wedge \chi_I(x)] \\
&= \nu(x) \wedge \chi_I(x) = (\nu \wedge \chi_I)(x).
\end{aligned}$$

Thus $\nu \circ \chi_I \le \nu \wedge \chi_I$. For the reverse inclusion, if $x \notin I$, then

$$(\nu \wedge \chi_I)(x) = \nu(x) \wedge \chi_I(x) = 0 \le (\nu \circ \chi_I)(x).$$

If $x \in I$, then $(\nu \wedge \chi_I)(x) = \nu(x) \wedge \chi_I(x) = \nu(x) \wedge \chi_I(t)$ for every $t \in I$ such that

$$\begin{aligned}
x &= xt \\
&\le \bigvee_{x=\sum_{i=1}^n a_i b_i} \left[\bigwedge_{i=1}^n [\nu(a_i) \wedge \chi_I(b_i)]\right] = \nu \circ \chi_I(x).
\end{aligned}$$

Thus for any fuzzy right ideal ν of R, $\nu \wedge \chi_I = \nu \circ \chi_I$. Hence χ_I is a right t-pure fuzzy ideal of R.

$(ii) \Rightarrow (i)$ Suppose χ_I is a right t-pure fuzzy ideal of R, then I is an ideal of R. Let J be a right ideal of R. Then χ_J is a fuzzy right ideal of R. Then by the hypothesis, $\chi_J \wedge \chi_I = \chi_J \circ \chi_I$ and this implies that $\chi_{J\cap I} = \chi_{JI}$. From this it follows that $J \cap I = JI$. Hence, I is a right t-pure ideal of R.

Proposition 3.3. *The following statements are true.*

(i) *If λ and μ are t-pure fuzzy ideals of a semiring R, then so is $\lambda \wedge \mu$.*

(ii) *If $\{\lambda_i : i \in I\}$ is a family of right t-pure fuzzy ideals of R, then $\mu \wedge \left(\bigvee_{i\in I} \lambda_i \right) = \mu \circ \left(\bigvee_{i\in I} \lambda_i \right)$ for each fuzzy right ideal μ of R.*

Proof. (i) Suppose λ and μ are t-pure fuzzy ideals of R. Let ν be a fuzzy right ideal of R. In order to show that $\lambda \wedge \mu$ is a t-pure fuzzy ideal of R, we have to show that $\nu \wedge (\lambda \wedge \mu) = \nu \circ (\lambda \wedge \mu)$. Now

$$\nu \wedge (\lambda \wedge \mu) = (\nu \wedge \lambda) \wedge (\nu \wedge \mu) = \nu \circ \lambda \wedge \nu \circ \mu. \tag{3.1}$$

(because λ and μ are t-pure fuzzy ideals)

Also, since $\lambda \wedge \mu \leq \lambda$ and $\lambda \wedge \mu \leq \mu$, we have $\nu \circ (\lambda \wedge \mu) \leq \nu \circ \lambda$ and $\nu \circ (\lambda \wedge \mu) \leq \nu \circ \mu$. Thus

$$\nu \circ (\lambda \wedge \mu) \leq \nu \circ \lambda \wedge \nu \circ \mu. \tag{3.2}$$

Then from 3.1 and 3.2 we get

$$\nu \circ (\lambda \wedge \mu) \leq \nu \wedge (\lambda \wedge \mu). \tag{3.3}$$

On the other hand, we have

$$\nu \wedge (\lambda \wedge \mu) = (\nu \wedge \lambda) \wedge \mu = \nu \circ \lambda \wedge \mu = (\nu \circ \lambda) \circ \mu,$$

(because λ is a t-pure fuzzy ideal, and $\nu \circ \lambda$ is a fuzzy right ideal and μ is a t-pure fuzzy ideal)

$$\nu \wedge (\lambda \wedge \mu) = \nu \circ (\lambda \circ \mu). \tag{3.4}$$

(by the associativity of the operation involved)

Since $\lambda \circ \mu \leq \lambda \wedge \mu$, we have

$$\nu \circ (\lambda \circ \mu) \leq \nu \circ (\lambda \wedge \mu). \tag{3.5}$$

Thus from 3.4 and 3.5, we get

$$\nu \wedge (\lambda \wedge \mu) \leq \nu \circ (\lambda \wedge \mu). \tag{3.6}$$

Hence, a combination of 3.3 and 3.6 yields $\nu \wedge (\lambda \wedge \mu) = \nu \circ (\lambda \wedge \mu)$.

(ii) Suppose $\{\lambda_i : i \in I\}$ is a family of right t-pure fuzzy ideals of R. Let μ be any fuzzy right ideal of R. Then $\mu \wedge \left(\bigvee_{i\in I}\lambda_i\right) = \bigvee_{i\in I}(\mu \wedge \lambda_i) = \bigvee_{i\in I}(\mu \circ \lambda_i)$, since λ_i's are t-pure.

Also, since $\mu \circ \lambda_i \leq \mu \circ \left(\bigvee_{i\in I}\lambda_i\right)$, we have $\bigvee_{i\in I}(\mu \circ \lambda_i) \leq \mu \circ \left(\bigvee_{i\in I}\lambda_i\right)$. Thus

$$\mu \wedge \left(\bigvee_{i\in I}\lambda_i\right) \leq \mu \circ \left(\bigvee_{i\in I}\lambda_i\right). \tag{3.7}$$

On the other hand, for each $x \in R$, we have

$$\begin{aligned}
\left[\mu \circ \left(\bigvee_{i\in I}\lambda_i\right)\right](x) &= \bigvee_{x=\sum_{i=1}^{m} a_i b_i} \left[\bigwedge_{i=1}^{m}\left[\mu(a_i) \wedge \left(\bigvee_{i\in I}\lambda_i\right)(b_i)\right]\right] \\
&\leq \bigvee_{x=\sum_{i=1}^{m} a_i b_i} \left[\bigwedge_{i=1}^{m}\left[\mu(a_ib_i) \wedge \left(\bigvee_{i\in I}\lambda_i\right)(a_ib_i)\right]\right] \\
&= \bigvee_{x=\sum_{i=1}^{m} a_i b_i} \left[\left(\bigwedge_{i=1}^{m}\mu(a_ib_i)\right) \wedge \left(\bigwedge_{i=1}^{m}\left(\bigvee_{i\in I}\lambda_i\right)(a_ib_i)\right)\right] \\
&= \bigvee_{x=\sum_{i=1}^{m} a_i b_i} \left[\mu(x) \wedge \left(\bigvee_{i\in I}\lambda_i\right)(x)\right] \\
&= \mu(x) \wedge \left(\bigvee_{i\in I}\lambda_i\right)(x) = \left[\mu \wedge \left(\bigvee_{i\in I}\lambda_i\right)\right](x).
\end{aligned}$$

Thus

$$\mu \circ \left(\bigvee_{i\in I}\lambda_i\right) \leq \mu \wedge \left(\bigvee_{i\in I}\lambda_i\right). \tag{3.8}$$

Hence 3.7 and 3.8 yield

$$\mu \circ \left(\bigvee_{i\in I}\lambda_i\right) = \mu \wedge \left(\bigvee_{i\in I}\lambda_i\right).$$

Using Theorem 2.9, and Definitions 3.4 and 3.5, we obtain the following theorem.

Theorem 3.4. *The following statements for a semiring R are equivalent:*

(i) R is right weakly regular.
(ii) All right ideals of R are idempotent.
(iii) Each two-sided ideal of R is right t-pure.
(iv) R_R is normal.
(v) All fuzzy right ideal of R are idempotent.
(vi) All fuzzy ideals of R are right t-pure fuzzy ideals.
If R is commutative, then the above statements are equivalent to:
(vii) R is von Neumann regular.

Lemma 3.2. *If M is a semimodule over a right weakly regular semiring R, then for any fuzzy subsemimodule λ of M and any fuzzy ideal μ of R,*

$$(\lambda \circ \mu)(m) = \bigvee_{m=\sum_{i=1}^{n} m_i r_i} \left[\bigwedge_{i=1}^{n} [\lambda(m_i r_i) \wedge \mu(r_i)] \right] \text{ for all } m \in M.$$

Proof. For any $m \in M$, we have, by definition $(\lambda \circ \mu)(m) = \bigvee_{m=\sum_{i=1}^{n} m_i r_i} \left[\bigwedge_{i=1}^{n} [\lambda(m_i) \wedge \mu(r_i)] \right]$. Since R is right weakly regular, for each $r_i \in R$, there exist $a_i, b_i \in R$ such that $r_i = r_i a_i r_i b_i$. Note thar $\lambda(m_i r_i) \leq \lambda(m_i r_i a_i) \leq \lambda(m_i r_i a_i r_i b_i) = \lambda(m_i r_i)$. Thus $\lambda(m_i r_i) = \lambda(m_i r_i a_i)$. Moreover, $\mu(r_i) \leq \mu(r_i b_i) \leq \mu(r_i a_i r_i b_i) = \mu(r_i)$. Thus $\mu(r_i) = \mu(r_i b_i)$.

Hence, we have

$$\begin{aligned}(\lambda \circ \mu)(m) &\leq \bigvee_{m=\sum_{i=1}^{n} m_i r_i} \left[\bigwedge_{i=1}^{n} [\lambda(m_i r_i) \wedge \mu(r_i)] \right] \\ &= \bigvee_{m=\sum_{i=1}^{n} m_i r_i} \left[\bigwedge_{i=1}^{n} [\lambda(m_i r_i a_i) \wedge \mu(r_i b_i)] \right] \\ &\leq \bigvee_{m=\sum_{j=1}^{p} n_j s_j} \left[\bigwedge_{j=1}^{p} [\lambda(n_j) \wedge \mu(s_j)] \right] \\ &= (\lambda \circ \mu)(m).\end{aligned}$$

Thus it follows that $(\lambda \circ \mu)(m) = \bigvee_{m=\sum_{i=1}^{n} m_i r_i} \left[\bigwedge_{i=1}^{n} [\lambda(m_i r_i) \wedge \mu(r_i)] \right]$.

Lemma 3.3. *Let M be a cyclic semimodule over a right weakly regular semiring R. If $\sum_{i=1}^{n} m_i r_i$, where $m_i \in M$ and $r_i \in R$, is any expression form of an arbitrary element* *of M, then it can be written as mr, where m is a generator of the cyclic semimodule M and r is any element of R, which satisfies the inequality*

$$\mu(r) \geq \bigwedge_{i=1}^{n} \mu(r_i)$$

for all fuzzy ideals μ of R.

Proof. As m is a generator of M, we have $m_i = mr'_i$ for some $r'_i \in R$. Thus $\sum_{i=1}^{n} m_i r_i = \sum_{i=1}^{n} m(r'_i r_i) = m \sum_{i=1}^{n} r'_i r_i = mr$, where $r = \sum_{i=1}^{n} r'_i r_i \in R$.

Also, $\mu(r) = \mu\left(\sum_{i=1}^{n} r'_i r_i\right) \geq \bigwedge_{i=1}^{n} \mu(r'_i r_i) \geq \bigwedge_{i=1}^{n} \mu(r_i)$.

Theorem 3.5. *If M is a cyclic semimodule over a right weakly regular semiring R, then M is fuzzy normal over R.*

Proof. Let λ be any fuzzy subsemimodule of M and μ any fuzzy ideal of R. Then from Lemma 3.2, we have

$$\begin{aligned}
(\lambda \circ \mu)(m) &= \bigvee_{m=\sum_{i=1}^{n} m_i r_i} \left[\bigwedge_{i=1}^{n} [\lambda(m_i r_i) \wedge \mu(r_i)]\right] \quad \text{for all } m \in M \\
&= \bigvee_{m=\sum_{i=1}^{n} m_i r_i} \left[\left(\bigwedge_{i=1}^{n} \lambda(m_i r_i)\right) \wedge \left(\bigwedge_{i=1}^{n} \mu(r_i)\right)\right] \\
&\leq \bigvee_{m=\sum_{i=1}^{n} m_i r_i} \left[\lambda(m) \wedge \left(\bigwedge_{i=1}^{n} \mu(r_i)\right)\right].
\end{aligned}$$

Now using Lemma 3.3, we have

$$(\lambda \circ \mu)(m) \leq \bigvee_{m=m'r} [\lambda(m'r) \wedge \mu(r)] \text{ where } m = m'r \text{ and } \mu(r) \geq \bigwedge_{i=1}^{n} \mu(r_i).$$

Finally,

$$(\lambda \circ \mu)(m) \leq \bigvee_{m=m'r} [\lambda(m'r) \wedge \mu(r)] \leq \bigvee_{m=\sum_{i=1}^{n} m_i r_i} \left[\bigwedge_{i=1}^{n} [\lambda(m_i r_i) \wedge \mu(r_i)]\right] = (\lambda \circ \mu)(m).$$

Therefore

$$(\lambda \circ \mu)(m) = \bigvee_{m=m'r} [\lambda(m'r) \wedge \mu(r)] = \bigvee_{m=m'r} [\lambda(m) \wedge \mu(r)] = \lambda(m) \wedge \left(\bigvee_{m=m'r} \mu(r)\right).$$

We can show easily that

$$\begin{aligned}(\lambda\mu)(m) &= \lambda(m)\wedge\left(\bigvee_{m=m'r}\mu(r)\right)\\ &= \lambda(m)\wedge\left(\bigvee_{m=\sum\limits_{i=1}^{n}m_ir_i}\left[\bigwedge_{i=1}^{n}\mu(r_i)\right]\right)\\ &= \lambda(m)\wedge(\mathscr{M}\circ\mu)(m)=(\lambda\wedge\mathscr{M}\circ\mu)(m).\end{aligned}$$

Thus M is fuzzy normal over R.

3.2 Fuzzy Prime Subsemimodules of a Semimodule over a Semiring

In this section, we will define fuzzy prime subsemimodules of a right R-semimodule as an extension of the notion of fuzzy prime ideals of a semiring. We also characterize those semirings for which each fuzzy ideal is prime and also semirings for which each fuzzy right ideal is prime.

We now give the definition of a *fuzzy point* of a semiring R.

Definition 3.7. Let R be a semiring and $a \in R$. For any $t \in (0,1]$ the fuzzy subset of R defined by

$$a_t(x) = \begin{cases} t & \text{if } x = a \\ 0 & \text{if } x \neq a \end{cases}$$

is called a *fuzzy point* of R.

If λ is a fuzzy subset of R and $a_t \leq \lambda$, then we say that $a_t \in \lambda$. It is then clear that for any fuzzy subset λ, $\lambda = \bigvee\limits_{a_t\in\lambda} a_t$.

Definition 3.8. If λ is a fuzzy subset of R. Then the fuzzy left (right) ideal of R generated by λ is the smallest fuzzy left (right) ideal of R containing λ.

Lemma 3.4. *Let a_t be a fuzzy point of a semiring R. Then the fuzzy left (right) ideal of R generated by a_t is l_{a_t} (ξ_{a_t}) defined by*

$$l_{a_t}(x) = \begin{cases} t & \text{if } x \in Ra \\ 0 & \text{otherwise} \end{cases}$$

$$\xi_{a_t}(x) = \begin{cases} t & \text{if } x \in aR \\ 0 & \text{otherwise} \end{cases}$$

Proof. Let $x,y \in R$. If $x+y \in Ra$, then $l_{a_t}(x+y) = t \geq l_{a_t}(x) \wedge l_{a_t}(y)$. If $x+y \notin Ra$, then both x and y cannot be in Ra and so $l_{a_t}(x) \wedge l_{a_t}(y) = 0$. Thus $l_{a_t}(x+y) = 0 = l_{a_t}(x) \wedge l_{a_t}(y)$. Hence it follows that we always have $l_{a_t}(x+y) \geq l_{a_t}(x) \wedge l_{a_t}(y)$. If

$xy \in Ra$, then $l_{a_t}(xy) = t \geq l_{a_t}(y)$. If $xy \notin Ra$, then $y \notin Ra$ and then $l_{a_t}(xy) = 0 = l_{a_t}(y)$. Hence we always have $l_{a_t}(xy) \geq l_{a_t}(y)$. Thus l_{a_t} is a fuzzy left ideal of R.

By the definition of l_{a_t}, we observe that $a_t \leq l_{a_t}$. If λ is a fuzzy left ideal of R containing a_t, then if $x \in Ra$, since $t = a_t(a) \leq \lambda(a)$, it follows that $t \leq \lambda(a) \leq \lambda(ra)$ so $\lambda(x) \geq t = l_{a_t}(x)$. If $x \notin Ra$, then $l_{a_t}(x) = 0 \leq \lambda(x)$, and this implies that $l_{a_t} \leq \lambda$. Thus l_{a_t} is a fuzzy left ideal of R generated by a_t.

Corollary 3.3. *$l_{a_t} \circ \chi_R$ and $\chi_R \circ \xi_{a_t}$ are fuzzy ideals of R generated by a_t.*

Proof. By Lemma 3.1, $l_{a_t} \circ \chi_R$ is a fuzzy ideal of R.

Since

$$(l_{a_t} \circ \chi_R)(x) = \bigvee_{x = \sum_{i=1}^{p} y_i z_i} \left[\bigwedge_{i=1}^{p} [l_{a_t}(y_i) \wedge \chi_R(z_i)] \right] = \bigvee_{x = \sum_{i=1}^{p} y_i z_i} \left[\bigwedge_{i=1}^{p} l_{a_t}(y_i) \right]$$

because $\chi_R(z_i) = 1$.

Thus it follows that,

$$(l_{a_t} \circ \chi_R)(x) = \begin{cases} t & \text{if } x \in RaR \\ 0 & \text{otherwise.} \end{cases}$$

It is then clear that $a_t \leq (l_{a_t} \circ \chi_R)$.
If μ is a fuzzy ideal of R containing a_t, then $l_{a_t} \leq \mu$ and this implies that $l_{a_t} \circ \chi_R \leq \mu \circ \chi_R \leq \mu$. Therefore $l_{a_t} \circ \chi_R$ is a fuzzy ideal of R generated by a_t.

Similarly $\chi_R \circ \xi_{a_t}$ is a fuzzy ideal of R generated by a_t.

Corollary 3.4. *If λ is a fuzzy left (right) ideal of a semiring R and a_t, b_s are fuzzy points of R, such that $a_t \circ (\chi_R \circ b_s) \leq \lambda$ then $l_{a_t} \circ l_{b_s} \leq \lambda$ ($\xi_{a_t} \circ \xi_{b_s} \leq \lambda$).*

Proof. Let $a_t \circ (\chi_R \circ b_s) \leq \lambda$. Then $\chi_R \circ (a_t \circ (\chi_R \circ b_s)) \leq \chi_R \circ \lambda \leq \lambda$.

Now

$$\chi_R \circ a_t(x) = \bigvee_{x = \sum_{i=1}^{n} y_i z_i} \left[\bigwedge_{i=1}^{n} [\chi_R(y_i) \wedge a_t(z_i)] \right] = \bigvee_{x = \sum_{i=1}^{n} y_i z_i} \left[\bigwedge_{i=1}^{n} a_t(z_i) \right]$$

because $\chi_R(y_i) = 1$.

Thus it follows that

$$\chi_R \circ a_t(x) = \begin{cases} t & \text{if } x \in Ra \\ 0 & \text{otherwise.} \end{cases}$$

Similarly

$$\chi_R \circ b_s(x) = \begin{cases} s & \text{if } x \in Rb \\ 0 & \text{otherwise.} \end{cases}$$

Then we have

$$(\chi_R \circ a_t) \circ (\chi_R \circ b_s)(x) = \begin{cases} t \wedge s & \text{if } x \in RaRb \\ 0 & \text{otherwise.} \end{cases}$$

On the other hand

$$l_{a_t} \circ l_{b_s}(x) = \bigvee_{x = \sum_{i=1}^{n} y_i z_i} \left[\bigwedge_{i=1}^{n} [l_{a_t}(y_i) \wedge l_{b_s}(z_i)] \right].$$

Thus

$$l_{a_t} \circ l_{b_s}(x) = \begin{cases} t \wedge s & \text{if } x \in RaRb \\ 0 & \text{otherwise.} \end{cases}$$

Since $\chi_R \circ (a_t \circ (\chi_R \circ b_s)) \leq \lambda$. Thus $l_{a_t} \circ l_{b_s} \leq \lambda$.

Definition 3.9. A fuzzy ideal λ of a semiring R is called a *fuzzy prime* ideal of R if for any fuzzy ideals μ and ν of R, $\mu \circ \nu \leq \lambda$ implies that $\mu \leq \lambda$ or $\nu \leq \lambda$.

Proposition 3.4. *Let λ be a fuzzy ideal of a semiring R. Then the following are equivalent*
(i) λ is a fuzzy prime ideal of R.
(ii) $a_t \circ (\chi_R \circ b_s) \leq \lambda$ if and only if $a_t \in \lambda$ or $b_s \in \lambda$ for every fuzzy points a_t, b_s of R.
(iii) If a_t and b_s are fuzzy points of R such that $\tau_{a_t} \circ \tau_{b_s} \leq \lambda$, then either $a_t \in \lambda$ or $b_s \in \lambda$, where τ_{a_t} is the fuzzy ideal of R generated by a_t.

Proof. $(i) \Rightarrow (ii)$ Let us suppose that $a_t \circ (\chi_R \circ b_s) \leq \lambda$. Then by Corollary 3.4, $l_{a_t} \circ l_{b_s} \leq \lambda$. Now $l_{a_t} \circ \chi_R \circ l_{b_s} \circ \chi_R = l_{a_t} \circ (\chi_R \circ l_{b_s}) \circ \chi_R \leq l_{a_t} \circ l_{b_s} \circ \chi_R \leq \lambda \circ \chi_R \leq \lambda$. Therefore $l_{a_t} \circ \chi_R \leq \lambda$ or $l_{b_s} \circ \chi_R \leq \lambda$. That is either $a_t \in \lambda$ or $b_s \in \lambda$.

Conversely, suppose either $a_t \in \lambda$ or $b_s \in \lambda$. Then $a_t \circ \chi_R \circ b_s \leq \lambda$.

$(ii) \Rightarrow (iii)$ As $a_t \in \tau_{a_t}$ and $b_s \in \tau_{b_s} \Rightarrow a_t \circ \chi_R \leq \tau_{a_t} \circ \chi_R \leq \tau_{a_t}$. Thus $a_t \circ \chi_R \circ b_s \leq \tau_{a_t} \circ \tau_{bs}$. If $\tau_{a_t} \circ \tau_{bs} \leq \lambda$, then $a_t \circ \chi_R \circ b_s \leq \lambda$. Hence either $a_t \in \lambda$ or $b_s \in \lambda$.

$(iii) \Rightarrow (i)$ Let μ and ν be fuzzy ideals of R such that $\mu \circ \nu \leq \lambda$. Suppose $\mu \nleq \lambda$. Then there exists $x \in R$ such that $\mu(x) \nleq \lambda(x)$. Let $\mu(x) = t \in (0,1]$, then $x_t \in \mu$. Let $y_s \in \nu$. Then $\tau_{x_t} \circ \tau_{ys} \leq \mu \circ \nu \leq \lambda$. By (3) either $x_t \in \lambda$ or $y_s \in \lambda$. Since $x_t \notin \lambda$, it follows that $y_s \in \lambda$. Hence $\nu \leq \lambda$.

Definition 3.10. A fuzzy ideal λ of a semiring R is called a *fuzzy semiprime* ideal of R if for any fuzzy ideal μ of R, $\mu^2 \leq \lambda$ implies that $\mu \leq \lambda$. λ is called *fuzzy strongly irreducible* if for any fuzzy ideals μ and ν of R, $\mu \wedge \nu \leq \lambda$ implies that $\mu \leq \lambda$ or $\nu \leq \lambda$.

Proposition 3.5. *The following conditions on a fuzzy ideal λ of a semiring R are equivalent:*
(i) λ is a fuzzy semiprime ideal of R.
(ii) $a_t \circ \chi_R \circ a_t \leq \lambda$ if and only if $a_t \in \lambda$.

Proof. $(i) \Rightarrow (ii)$ Suppose $a_t \circ \chi_R \circ a_t \leq \lambda$. Then by Corollary 3.4, $l_{a_t} \circ l_{a_t} \leq \lambda$.
Now $l_{a_t} \circ (\chi_R \circ l_{a_t}) \circ \chi_R \leq \lambda$.

Therefore $l_{a_t} \circ \chi_R \leq \lambda$. But $l_{a_t} \circ \chi_R$ is a fuzzy ideal of R generated by a_t. So $a_t \in \lambda$.

If $a_t \in \lambda$, then the fuzzy ideal generated by a_t is contained in λ, and so $l_{a_t} \circ \chi_R \leq \lambda$ but $a_t \circ \chi_R \circ a_t \leq l_{a_t} \circ \chi_R \leq \lambda$.

$(ii) \Rightarrow (i)$ Let δ be a fuzzy ideal of R such that $\delta^2 \leq \lambda$. Let $a_t \in \delta$. Then $a_t \circ \chi_R \leq \delta \circ \chi_R \leq \delta$. Thus $a_t \circ \chi_R \circ a_t \leq \delta \circ a_t \leq \delta^2 \leq \lambda \Rightarrow a_t \in \lambda$ and therefore $\delta = \bigvee_{a_t \in \delta} a_t \leq \lambda$. Hence λ is a fuzzy semiprime ideal of R.

Proposition 3.6. *A fuzzy ideal λ of a semiring R is fuzzy prime if and only if it is fuzzy semiprime and fuzzy strongly irreducible.*

Proof. Let λ be a fuzzy prime ideal of R. Then λ is a fuzzy semiprime ideal of R. Let μ and ν be fuzzy ideals of R such that $\mu \wedge \nu \leq \lambda$. Since $\mu \circ \nu \leq \mu \wedge \nu$, it follows that $\mu \circ \nu \leq \lambda$ and as λ is a fuzzy prime ideal, either $\mu \leq \lambda$ or $\nu \leq \lambda$. Hence λ is a fuzzy strongly irreducible.

Conversely, assume that λ is a fuzzy ideal of R which is both fuzzy semiprime and fuzzy irreducible. If μ, ν are fuzzy ideals of R such that $\mu \circ \nu \leq \lambda$, then $(\mu \wedge \nu)^2 \leq \mu \circ \nu \leq \lambda$. Since λ is fuzzy semiprime, we have $\mu \wedge \nu \leq \lambda$ and also since λ is a fuzzy strongly irreducible, it follows that $\mu \leq \lambda$ or $\nu \leq \lambda$. Thus λ is a fuzzy prime ideal.

3.3 Fully Fuzzy Prime Semirings

A semiring R is called *fully fuzzy prime* (semiprime) if each of its fuzzy ideal is prime (semiprime). We call a semiring R *fully idempotent* if each ideal of R is idempotent.

We are now ready to prove the following characterization Theorem.

Theorem 3.6. *The following assertions on a semiring R are equivalent:*

(i) R is fully idempotent.

(ii) Each fuzzy ideal of R is idempotent.

(iii) For each pair of fuzzy ideals λ, μ of R, $\lambda \wedge \mu = \lambda \circ \mu$.

(iv) The set of all fuzzy ideals of R is a distributive lattice under the sum and product of fuzzy ideals.

(v) Each proper fuzzy ideal of R is the intersection of fuzzy prime ideals of R which contain it.

(vi) Each fuzzy ideal of R is a fuzzy semiprime ideal.

Proof. $(i) \Leftrightarrow (ii) \Leftrightarrow (iii)$ (See [9], Theorem 2.1)

$(i) \Leftrightarrow (iv)$ (See [9], Theorem 2.2)

$(i) \Leftrightarrow (v)$ (See [9], Theorem 2.6)

$(ii) \Rightarrow (vi)$ Let λ, δ be fuzzy ideals of R such that $\delta^2 \leq \lambda$. Now by condition (ii) $\delta = \delta^2$, so $\delta \leq \lambda$. Hence λ is a fuzzy semiprime ideal.

$(vi) \Rightarrow (ii)$ Let λ be any fuzzy ideal of R. Then λ^2 is also a fuzzy ideal of R. Since $\lambda^2 \leq \lambda^2$, it follows from the condition (vi), $\lambda \leq \lambda^2$. Note that $\lambda^2 \leq \lambda$, therefore $\lambda = \lambda^2$.

Theorem 3.7. *A semiring R is fully fuzzy prime semiring if and only if R is fully idempotent and the set of fuzzy ideals of R is totally ordered.*

Proof. First suppose R is a fully fuzzy prime semiring. Let λ be any fuzzy ideal of R. Clearly then λ^2 is also a fuzzy ideal of R. Since $\lambda^2 \leq \lambda^2$, we have $\lambda \leq \lambda^2$. Note on the other hand, that $\lambda^2 \leq \lambda$. Hence $\lambda^2 = \lambda$. Therefore it follows that every fuzzy ideal of R is idempotent. Hence by Theorem 3.6, R is fully idempotent. Let μ, ν be fuzzy ideals of R. Then $\mu \wedge \nu$ is a fuzzy ideal and hence a fuzzy prime ideal of R. On the other hand, $\mu \circ \nu \leq \mu \wedge \nu$, so either $\mu \leq \mu \wedge \nu$ or $\nu \leq \mu \wedge \nu$. That is, either $\mu \leq \nu$ or $\nu \leq \mu$.

Conversely, let us assume that R is a fully idempotent semiring and the set of fuzzy ideals of R is totally ordered. Let λ, μ, ν be fuzzy ideals of R such that $\lambda \circ \mu \leq \nu$. Since the set of fuzzy ideals of R is totally ordered, we have $\lambda \leq \mu$ or $\mu \leq \lambda$. Assume that $\mu \leq \lambda$. Then $\mu = \mu^2 \leq \mu \circ \lambda \leq \nu$. Thus ν is a fuzzy prime ideal of R and hence R is fully fuzzy prime.

In an analogous manner, we may also define fuzzy prime right ideals of a semiring and then by employing the arguments similar to those used in proving the above results we can also characterize those semirings for which each fuzzy right ideal is prime. Following this line of investigation, we first give a definition.

Definition 3.11. A fuzzy right ideal μ of a semiring R is called a *fuzzy prime (semiprime) right ideal* of R if for any fuzzy right ideals λ, δ of R, $\lambda \circ \delta \leq \mu$ implies $\lambda \leq \mu$ or $\delta \leq \mu$ ($\lambda^2 \leq \mu \Rightarrow \lambda \leq \mu$); μ is called a fuzzy irreducible (strongly irreducible) if for any fuzzy right ideals λ, δ of R, $\lambda \wedge \delta = \mu \Rightarrow \lambda = \mu$ or $\delta = \mu$ ($\lambda \wedge \delta \leq \mu \Rightarrow \lambda \leq \mu$ or $\delta \leq \mu$).

Now we state a couple of propositions on fuzzy prime right ideals and refer to the paper [11] for their proofs.

Proposition 3.7. *Let λ be a fuzzy right ideal of a semiring R. Then the following conditions are equivalent:*

(i) *λ is a fuzzy prime right ideal of R.*

(ii) *For any fuzzy points a_t, b_s of R, $a_t \circ \chi_R \circ b_s \leq \lambda \Rightarrow a_t \in \lambda$ or $b_s \in \lambda$.*

(iii) *For any fuzzy points a_t, b_s of R, $\xi_{a_t} \circ \xi_{b_s} \leq \lambda \Rightarrow a_t \in \lambda$ or $b_s \in \lambda$.*

Proposition 3.8. *Let λ be a fuzzy right ideal of a semiring R. Then the following conditions are equivalent:*

(i) *λ is a fuzzy semiprime right ideal of R.*

(ii) *For any fuzzy point a_t of R, $a_t \circ \chi_R \circ a_t \leq \lambda \Rightarrow a_t \in \lambda$.*

(iii) *For any fuzzy point a_t of R such that $\xi_{a_t} \circ \xi_{a_t} \leq \lambda \Rightarrow a_t \in \lambda$.*

Proposition 3.9. *A fuzzy semiprime strongly irreducible right ideal of a semiring R is a fuzzy prime right ideal.*

Proposition 3.10. *Let λ be a fuzzy right ideal of a semiring R with $\lambda(a) = \alpha$, where a is any element of R and $\alpha \in (0,1]$. Then there exists a fuzzy irreducible right ideal δ of R such that $\lambda \leq \delta$ and $\delta(a) = \alpha$.*

Proposition 3.11. *Every fuzzy right ideal of a semiring R is the intersection of all fuzzy irreducible right ideals of R which contain it.*

We will now state the following theorems based on the propositions stated above. The details of the proofs of these theorems can be found in [11].

Theorem 3.8. *The following assertions on a semiring R are equivalent:*

(i) R ie right weakly regular (Recall that R is right weakly regular if for each $x \in R$, $x \in (xR)^2$).

(ii) Each fuzzy right ideal of R is idempotent.

(iii) For each fuzzy right ideal μ and for each fuzzy ideal λ of R, $\lambda \wedge \mu = \lambda \circ \mu$.

(iv) Each fuzzy right ideal of R is a fuzzy semiprime right ideal of R.

Theorem 3.9. *If every fuzzy right ideal of a semiring R is fuzzy prime right ideal, then R is right weakly regular and the set of fuzzy ideals of R is totally ordered.*

Theorem 3.10. *If R is right weakly regular semiring and the set of all fuzzy right ideals of R is totally ordered, then every fuzzy right ideal of R is a fuzzy prime right ideal of R.*

Theorem 3.11. *If the set of all fuzzy right ideals of a semiring R is totally ordered, then the following are equivalent:*

(i) R is right weakly regular.

(ii) Every fuzzy right ideal of R is a fuzzy prime right ideal.

3.4 Fuzzy Prime Subsemimodules

We begin with a proposition.

Proposition 3.12. *Let R be a semiring and M a right R-semimodule. If λ is a fuzyy subsemimodule of M, then $A_\lambda(M) = \bigvee\{a_t : a_t$ is a fuzzy point of R satisfying $\chi_M \circ a_t \leq \lambda\}$ is a fuzzy ideal of R.*

Proof. Let $X = \{a_t : a_t$ is a fuzzy point of R satisfying $\chi_M \circ a_t \leq \lambda\}$. Let $m \in M$. Then

$$\chi_M \circ a_t(m) = \bigvee_{m=\sum_{i=1}^{p} m_i r_i} \left[\bigwedge_{i=1}^{p} [\chi_M(m_i) \wedge a_t(r_i)] \right]$$

$$= \bigvee_{m=\sum_{i=1}^{p} m_i r_i} \left[\bigwedge_{i=1}^{p} a_t(r_i) \right] \quad \text{(because } \chi_M(m_i) = 1\text{).}$$

Now since

$$a_t(x) = \begin{cases} t & \text{if } x = a \\ 0 & \text{otherwise.} \end{cases}$$

Therefore

$$\chi_M \circ a_t(m) = \begin{cases} t & \text{if } m = m_1 a \text{ for some } m_1 \in M \\ 0 & \text{otherwise.} \end{cases}$$

Let $x, y \in R$ and suppose $A_\lambda(M)(x) = \alpha$ and $A_\lambda(M)(y) = \beta$. This implies that there exist $x_\alpha, y_\beta \in X$. Consider the fuzzy point $(x+y)_{\alpha\wedge\beta}$. We show that $(x+y)_{\alpha\wedge\beta} \in X$. Now

$$\chi_M \circ (x+y)_{\alpha\wedge\beta}(m) = \begin{cases} \alpha\wedge\beta & \text{if } m = m'(x+y) \\ 0 & \text{otherwise.} \end{cases}$$

We show that $\chi_M \circ (x+y)_{\alpha\wedge\beta} \leq \lambda$.

If $\chi_M \circ (x+y)_{\alpha\wedge\beta}(m) = 0 \leq \lambda(m)$, when $\chi_M(x+y)_{\alpha\wedge\beta}(m) = \alpha\wedge\beta$, then $m = m'(x+y) = m'x + m'y$ and

$$\begin{aligned} \lambda(m) &= \lambda(m'x + m'y) \\ &\geq \lambda(m'x) \wedge \lambda(m'y) \\ &\geq \chi_M(x)_\alpha(m'x) \wedge \chi_M(y)_\beta(m'y) = \alpha\wedge\beta. \end{aligned}$$

Hence $\chi_M \circ (x+y)_{\alpha\wedge\beta} \leq \lambda$, so $(x+y)_{\alpha\wedge\beta} \in X$. Thus $A_\lambda(M)(x+y) \geq \alpha\wedge\beta = A_\lambda(M)(x) \wedge A_\lambda(M)(y)$.

Let $x, r \in R$ and suppose $A_\lambda(M)(x) = \alpha \Rightarrow x_\alpha \in X$. Then $\chi_M(x)_\alpha \leq \lambda$. Now consider the fuzzy point $(rx)_\alpha$. We show that $(rx)_\alpha \in X$, and

$$\chi_M \circ (rx)_\alpha(m) = \begin{cases} \alpha & \text{if } m = m_1(rx) \\ 0 & \text{otherwise.} \end{cases}$$

If $m = m_1(rx)$, then $\lambda(m) = \lambda(m_1 rx) \geq \chi_M \circ (x)_\alpha(m_1(rx)) = \alpha = \chi_M \circ (rx)_\alpha(m)$. If $m \neq m_1(rx)$, then $\chi_M \circ (rx)_\alpha(m) = 0 \leq \lambda(m)$.
Hence $\lambda \geq \chi_M \circ (rx)_\alpha$, thus $(rx)_\alpha \in X$. Hence $A_\lambda(M)(rx) \geq \alpha = A_\lambda(M)(x)$.

Similarly define the fuzzy point $(xr)_\alpha$

$$\chi_M \circ (xr)_\alpha(m) = \begin{cases} \alpha & \text{if } m = m_1(xr) \\ 0 & \text{otherwise.} \end{cases}$$

If $m = m_2(xr) = (m_2 x)r$, then $\lambda(m) = \lambda((m_2 x)r) = \lambda(m_2 x) \geq \chi_M \circ (x)_\alpha(m_2 x) = \alpha = \chi_M \circ (xr)_\alpha(m)$. Thus it follows $\lambda \geq \chi_M(xr)_\alpha$, and consequently $(xr)_\alpha \in X$ and so $A_\lambda(M)(xr) \geq \alpha = A_\lambda(M)(x)$. Hence $A_\lambda(M)$ is a fuzzy ideal of R.

Corollary 3.5. *Let R be a semiring and λ a fuzzy right ideal of R. Then $A_\lambda(R) = \bigvee\{a_t : a_t$ is a fuzzy point of R satisfying $\chi_R \circ a_t \leq \lambda\}$ is the greatest fuzzy ideal of R contained in λ.*

Proof. By the above proposition, $A_\lambda(R)$ is a fuzzy ideal of R. Also $a_t \leq \chi_R \circ a_t \Rightarrow a_t \leq \lambda$. Hence $A_\lambda(R) = \bigvee a_t \leq \lambda$.

Also, if δ is a fuzzy ideal of R contained in λ, then for every fuzzy point $x_t \in \delta$, we have $\chi_R \circ x_t \leq \delta \leq \lambda \Rightarrow x_t \in A_\lambda(R) \Rightarrow \delta \leq A_\lambda(R)$.

Proposition 3.13. *Let λ be a fuzzy subsemimodule of an R-semimodule M. Then for each fuzzy point m_α, $A_\lambda(m_\alpha) = \bigvee\{a_t : a_t$ is a fuzzy point of R satisfying $m_\alpha a_t \in \lambda\}$ is a fuzzy right ideal of R.*

Proof. Let $X = \{a_t : a_t$ is a fuzzy point of R satisfying $m_\alpha a_t \in \lambda\}$.

$$m_\alpha \circ a_t(m') = \bigvee_{m' = \sum_{i=1}^{p} m_i r_i} \left[\bigwedge_{i=1}^{p} [m_\alpha(m_i) \wedge a_t(r_i)] \right].$$

Thus we have

$$m_\alpha \circ a_t(m') = \begin{cases} \alpha \wedge t & \text{if } m' = ma \\ 0 & \text{otherwise.} \end{cases}$$

Let $x, y \in R$ and suppose $A_\lambda(m_\alpha)(x) = \gamma$ and $A_\lambda(m_\alpha)(y) = \beta$. Then $x_\gamma, y_\beta \in X$. We now consider the fuzzy point $(x+y)_{\gamma\wedge\beta}$ and show that $(x+y)_{\gamma\wedge\beta} \in X$.

$$m_\alpha \circ (x+y)_{\gamma\wedge\beta}(m') = \begin{cases} \alpha \wedge (\gamma \wedge \beta) & \text{if } m' = m(x+y) \\ 0 & \text{otherwise.} \end{cases}$$

If $m' = m(x+y) = mx + my$, then $\lambda(m') = \lambda(mx+my) \geq \lambda(mx) \wedge \lambda(my)$ $\geq m_\alpha \circ x_\gamma(mx) \wedge m_\alpha \circ y_\beta(my) = (\alpha \wedge \gamma) \wedge (\alpha \wedge \beta) = \alpha \wedge (\gamma \wedge \beta) = m_a \circ (x+y)_{\gamma\wedge\beta}(m')$.

If $m' \neq m(x+y)$ then $m_a \circ (x+y)_{\gamma\wedge\beta}(m') = 0 \leq \lambda(m')$. So $m_\alpha \circ (x+y)_{\gamma\wedge\beta} \leq \lambda$ and hence $(x+y)_{\gamma\wedge\beta} \in X$. Thus
$A_\lambda(m_\alpha)(x+y) \geq \alpha \wedge \beta = A_\lambda(m_\alpha)(x) \wedge A_\lambda(m_\alpha)(y)$.

Let $x, r \in R$, and suppose $A_\lambda(m_\alpha)(x) = \beta$. Then $x_\beta \in X$. Thus $m_\alpha \circ x_\beta \leq \lambda$. Consider the fuzzy point $(xr)_\beta$. We show that $(xr)_\beta \in X$.

$$m_\alpha (xr)_\beta(m') = \begin{cases} \alpha \wedge \beta & \text{if } m' = m(xr) \\ 0 & \text{otherwise.} \end{cases}$$

If $m' = mxr$, then $\lambda(m') = \lambda(mxr) \geq \lambda(mx) = m_\alpha \circ x_\beta(mx) = \alpha \wedge \beta = m_\alpha \circ (xr)_\beta(m')$.

If $m' \neq mxr$, then $m_\alpha (xr)_\beta(m') = 0 \leq \lambda(m')$.

Hence $(xr)_\beta \in X$. Thus $A_\lambda(m_\alpha)(xr) \geq \beta = A_\lambda(m_\alpha)(x)$. Therefore $A_\lambda(m_\alpha)$ is a fuzzy right ideal of R.

Corollary 3.6. *Let λ be a fuzzy right ideal of a semiring R and a_t be any fuzzy point of R. Then $A_\lambda(a_t)$ is a fuzzy right ideal of R.*

Lemma 3.5. *Let λ be a fuzzy subsemimodule of an R-semimodule M. Then* $A_\lambda(M) = \bigwedge_{m_r \leq \mathcal{M}} A_\lambda(m_r)$.

Definition 3.12. A fuzzy subsemimodule λ of a right R-semimodule M is called *fuzzy prime* if for any fuzzy point v_α of M and any fuzzy point a_t of R, $v_\alpha \circ \chi_R \circ a_t \leq \lambda \Rightarrow v_\alpha \in \lambda$ or $a_t \in A_\lambda(M)$, λ is called *fuzzy semiprime* if for any fuzzy point v_α of M and any fuzzy point a_t of R, $v_\alpha \circ a_t \circ \chi_R \circ a_t \leq \lambda \Rightarrow v_\alpha \circ a_t \in \lambda$.

Next we prove a couple of propositions.

Proposition 3.14. *A fuzzy right ideal λ of a semiring R is fuzzy prime if and only if λ is fuzzy prime as a fuzzy subsemimodule of R_R.*

Proof. Let λ be a fuzzy prime right ideal of R. Let a_t, b_r be any fuzzy points of R such that $a_t \circ \chi_R \circ b_r \leq \lambda$. By Corollary 3.4, $\xi_{a_t} \circ \xi_{b_r} \leq \lambda$.

Now $\xi_{a_t} \circ (\chi_R \circ \xi_{b_r}) = (\xi_{a_t} \circ \chi_R) \circ \xi_{b_r} \leq \xi_{a_t} \circ \xi_{b_r} \leq \lambda \Rightarrow \xi_{a_t} \leq \lambda$ or $\chi_R \circ \xi_{b_r} \leq \lambda$.

If $\chi_R \circ \xi_{b_r} \leq \eta$, then $b_r \leq \xi_{b_r} \Rightarrow \chi_R \circ b_r \leq \chi_R \circ \xi_{b_r} \leq \lambda \Rightarrow b_r \in A_\lambda(R)$. Hence λ is a fuzzy prime subsemimodule of R.

Conversely, assume that λ is a fuzzy prime subsemimodule of R_R. We show that λ is a fuzzy right ideal of R which is fuzzy prime. Let a_t, b_r be any fuzzy points of R such that $a_t \circ \chi_R \circ b_r \leq \lambda \rightarrow a_t \in \lambda$ or $b_r \in A_\lambda(R)$. If $b_r \in A_\lambda(R)$ then $\chi_R \circ b_r \leq \lambda$.

$$\chi_R \circ b_r(x) = \bigvee_{x=\sum_{i=1}^{p} y_i z_i} \left[\bigwedge_{i=1}^{p} [\chi_R(y_i) \wedge b_r(z_i)] \right]$$

$$= \bigvee_{x=\sum_{i=1}^{p} y_i z_i} \left[\bigwedge_{i=1}^{p} b_r(z_i) \right].$$

Thus

$$\chi_R \circ b_r(x) = \begin{cases} r & \text{if } x \in Rb \\ 0 & \text{otherwise} \end{cases}$$

$\Rightarrow b_r \leq \chi_R \circ b_r$. Thus $b_r \in \lambda$ if $\chi_R \circ b_r \leq \lambda$.

Hence either $a_t \in \lambda$ or $b_r \in \lambda$. So by Proposition 3.7, λ is a fuzzy prime right ideal of R.

Proposition 3.15. *A fuzzy right ideal λ of a semiring R is fuzzy semiprime right ideal if and only if λ is fuzzy semiprime as a fuzzy subsemimodule of R_R.*

Proof. Let λ be a fuzzy semiprime right ideal of R. Let a_t, b_r be any fuzzy points of R such that $b_r \circ a_t \circ \chi_R \circ a_t \leq \lambda$, as $\chi_R \circ b_r \leq \chi_R$. Thus

$$b_r \circ a_t \circ \chi_R \circ b_r \circ a_t \leq (b_r \circ a_t) \circ \chi_R \circ (b_r \circ a_t) \leq \lambda \Rightarrow (b_r \circ a_t) \leq \lambda.$$

Thus λ is a fuzzy semiprime fuzzysubsemimodule of R_R.

Conversely, assume that λ is a fuzzy prime subsemimodule of R_R. We show that λ is a fuzzy prime right ideal of R. Let a_t, b_r be any fuzzy points of R such that $a_t \circ \chi_R \circ b_r \leq \lambda \Rightarrow a_t \in \lambda$ or $b_r \in A_\lambda(R)$.

If $b_r \in A_\lambda(R)$ then $\chi_R \circ b_r \leq \lambda$.

$$\chi_R \circ b_r(x) = \bigvee_{x=\sum_{i=1}^{p} y_i z_i} \left[\bigwedge_{i=1}^{p} [\chi_R(y_i) \wedge b_r(z_i)] \right]$$

$$= \bigvee_{x=\sum_{i=1}^{p} y_i z_i} \left[\bigwedge_{i=1}^{p} b_r(z_i) \right].$$

Thus

$$\chi_R \circ b_r(x) = \begin{cases} r & \text{if } x \in Rb \\ 0 & \text{otherwise.} \end{cases}$$

This implies $b_r \leq \chi_R \circ b_r$. Thus $b_r \in \lambda$ if $\chi_R \circ b_r \leq \lambda$. Hence either $a_t \in \lambda$ or $b_r \in \lambda$.

Proposition 3.16. *If a fuzzy subsemimodule λ of a right R-semimodule M_R is fuzzy prime then $A_\lambda(M)$ is a fuzzy prime ideal of R.*

Proof. By Proposition 3.12, $A_\lambda(M)$ is a fuzzy ideal of R. Let a_t, b_r be any fuzzy points of R such that $a_t \circ \chi_R \circ b_r \leq A_\lambda(M)$.

Assume that $a_t \notin A_\lambda(M)$. Then $\chi_M \circ a_t \nleq \lambda$. Thus there exists $v \in M$ such that $\chi_M \circ a_t(v) \nleq \lambda(v)$.
Now

$$\chi_M \circ a_t(v) = \begin{cases} t & \text{if } v \in Ma \\ 0 & \text{otherwise.} \end{cases}$$

Suppose that $v = ma$ and consider the fuzzy point m_1 of M.

Now

$$m_1 \circ a_t(m') = \begin{cases} 1 \wedge t = t & \text{if } m' = ma \\ 0 & \text{otherwise.} \end{cases}$$

In particular $m_1 \circ a_t(m') = t$ so $m_1 \circ a_t \nleq \lambda$. Thus $m_1 \circ a_t \circ \chi_R \circ b_r \leq \chi_M \circ a_t \circ \chi_R \circ b_r \leq \lambda$.

Since λ is a fuzzy prime subsemimodule of M, $m_1 a_t$ or $b_r \in A_\lambda(M)$, but $m_1 \circ a_t \nleq \lambda$. Hence $b_r \in A_\lambda(M)$.

Example 3.1. Let S be a nonempty set. Define a binary operation $*$ on S as follows

$$x * y = y \quad \text{for all } x, y \text{ in } S.$$

Then $(S, *)$ is a semigroup. Adjoining an identity 1 to S in the usual manner and let $S^1 = S \cup \{1\}$. Then $(S^1, *)$ is a monoid. Now let $R^1 = S^1 \cup \{\infty\}$, where $\{\infty\}$ is a ring with a single element ∞. On the set R^1, define two binary operations $+$ and $\cdot$ as follows

$$s_1 + s_2 = \infty = s_1 + \infty = \infty + s_2 \text{ for } s_1, s_2 \text{ in } S^1, \text{ and}$$

$$s_1 \cdot s_2 = \text{ product in } S^1 \text{ if } s_1, s_2 \in S^1, \text{ and}$$

$$s_1 \cdot \infty = \infty = \infty \cdot s_2.$$

Now adjoin an absorbing zero $0 \notin R^1$ and let $R = R^1 \cup \{0\}$. Then $(R, +, \cdot)$ is a semiring with an absorbing zero 0 and identity 1. The (crisp) ideals of the semiring R are $\{0\}, \{0\} \cup \{\infty\}$, and $S \cup \{0\} \cup \{\infty\}$ and R, all of which are idempotent. The set of ideals is totally ordered by inclusion. We now examine the fuzzy ideals of R.

First we observe the following facts.

Fact 1. A fuzzy subset $\lambda : R \to [0,1]$ is a fuzzy ideal of R if and only if the following conditions hold:

(a) $\lambda(0) \geq \lambda(\infty)$,
(b) $\lambda(\infty) \geq \lambda(x)$ for all $x(\neq 0) \in R$, and
(c) $\lambda(x) = \lambda(y)$, for all x, y in S.

Proof. Suppose $\lambda : R \to [0,1]$ is a fuzzy ideal of R. We certainly have $\lambda(0) \geq \lambda(\infty)$. Further $\lambda(\infty) \geq \lambda(x \cdot \infty) \geq \lambda(x)$ for all $x(\neq 0) \in R$. Moreover, for $x, y \in S$, we have $\lambda(x) = \lambda(y \cdot x) \geq \lambda(y)$ and $\lambda(y) = \lambda(x \cdot y) \geq \lambda(x)$. Thus $\lambda(x) = \lambda(y)$.

Conversely, suppose λ is a fuzzy subset of R satisfying conditions $(a), (b), (c)$. We show that λ is a fuzzy ideal of R. Let $x, y \in R$ with both $x, y \neq 0$. Then $\lambda(x+y) = \lambda(\infty) \geq \lambda(x) \wedge \lambda(y)$, since $\lambda(\infty) \geq \lambda(x)$ for all $x(\neq 0) \in R$. In the case, one of x and y is zero, say $x = 0$, then $x + y = y$ and $\lambda(x+y) = \lambda(y)$ and $\lambda(x) \wedge \lambda(y) = \lambda(y)$, since $\lambda(0) \geq \lambda(a)$ for all $a \in R$. Thus, in any case $\lambda(x+y) \geq \lambda(x) \wedge \lambda(y)$.

Finally we show that $\lambda(xy) \geq \lambda(x)$ and $\lambda(xy) \geq \lambda(y)$ for all $x, y \in R$. If $x, y \in S$ then $\lambda(xy) = \lambda(y) = \lambda(x)$. Moreover, for all possible pairs x, y of elements of the set $\{0, 1, \infty\}$, it is easily verified that $\lambda(xy) \geq \lambda(x)$ and $\lambda(xy) \geq \lambda(y)$. Thus in all cases, $\lambda(xy) \geq \lambda(x)$ and $\lambda(xy) \geq \lambda(y)$ for $x, y \in R$, which proves that λ is a fuzzy ideal of R.

Fact 2. All fuzzy ideals of the semiring R constructed in the above example are fuzzy idempotent.

Proof. Let $\lambda : R \to [0,1]$ be a fuzzy ideal of R. Then

$$\lambda^2(0) = \bigvee_{0 = \sum_{i=1}^{n} y_i z_i} \left[\bigwedge_{i=1}^{n} [\lambda(y_i) \wedge \lambda(z_i)] \right]$$

One expression form for 0 is $0 = 0 \cdot 0 + 0 \cdot 0$ for which $\bigwedge_{i=1}^{n} [\lambda(y_i) \wedge \lambda(z_i)] = \lambda(0)$. Hence $\lambda^2(0) = \lambda(0)$. We now compute $\lambda^2(x)$ for all $x \in S$. Clearly no expression form of x involves only 0 and ∞. Thus for any expression form $x = \sum_{i=1}^{n} y_i z_i$ for an element x, we have $\bigwedge_{i=1}^{n} [\lambda(y_i) \wedge \lambda(z_i)] \neq \lambda(0), \lambda(\infty)$.

Note that $x = x \cdot 1$ and $x = x \cdot x$ are among the possible expression forms of x and therefore $\bigvee_{x = \sum_{i=1}^{n} y_i z_i} \left[\bigwedge_{i=1}^{n} [\lambda(y_i) \wedge \lambda(z_i)] \right] = \lambda(x)$, since $\lambda(x) = \lambda(y)$ and $\lambda(x) \geq \lambda(1)$ for $x, y \in S$.

Thus we have $\lambda^2(x) = \lambda(x)$ for all $x \in S$. We now compute $\lambda^2(\infty)$. First we note that no expression form of ∞ involves 0 only. Thus for any expression form of ∞, say $\infty = \sum_{i=1}^{m} y_i z_i$, we have $\bigwedge_{i=1}^{m} [\lambda(y_i) \wedge \lambda(z_i)] \neq \lambda(0)$.

Now one expression form for ∞ is $\infty = \infty \cdot \infty + \infty \cdot \infty$ for which we have $\bigwedge_{i=1}^{m} [\lambda(y_i) \wedge \lambda(z_i)] = \lambda(\infty)$ and therefore

$$\lambda^2(\infty) = \bigvee_{\infty = \sum_{i=1}^{n} y_i z_i} \left[\bigwedge_{i=1}^{n} [\lambda(y_i) \wedge \lambda(z_i)] \right] = \lambda(\infty),$$ since $\lambda(\infty) \geq \lambda(x)$ for all $x (\neq 0) \in R$.

Hence for all $x \in R$, we have $\lambda^2(x) = \lambda(x)$ showing λ is idempotent. Thus every fuzzy ideal is semiprime.

By Fact 1, $\lambda(0) = 1, \lambda(\infty) = .8$ and $\lambda(x) = .7$ for all $x \in S^1$ and $\mu(0) = 1, \mu(\infty) = .75$ and $\mu(x) = .72$ for all $x \in S^1$ are fuzzy ideals of R but neither $\lambda \nleq \mu$ nor $\mu \nleq \lambda$, that is the set of fuzzy ideals of R is not totally ordered. Let $\nu(0) = 1, \nu(\infty) = .77$ and $\nu(x) = .71$ for all $x \in S^1$. Then ν is a fuzzy ideal of R. $\lambda \circ \mu \leq \nu$ because $\lambda \circ \mu(0) = 1, \lambda \circ \mu(\infty) = .75$ and $\lambda \circ \mu(x) = .7$ but neither $\lambda \nleq \nu$ nor $\mu \nleq \nu$. Thus ν is not fuzzy prime.

The main source of the results of this section is [11].

Chapter 4
Fuzzy k-Ideals of Semirings

This chapter consists of eight sections. In section 1 we prove that the k-sum and k-product of fuzzy k-ideals of a semiring is a fuzzy k-ideal. section 2 is devoted to characterizing k-regular semirings in terms of fuzzy left (right) k-ideals. Section 3 contains various characterizations of right k-weakly regular semirings by fuzzy right k-ideals. In section 4, we define fuzzy prime and semiprime right k-ideals and it is shown that R is a right k-weakly regular semiring if and only if each fuzzy right k-ideal of R is semiprime. In section 5, we characterize semirings in which each fuzzy k-ideal is idempotent and section 6 presents some results on prime and semiprime fuzzy k-ideals. We then study "k-semirings" and present some results related to these semirings. Finally, Section 8 is devoted to a study of fuzzy congruences of semirings.

Recall that a left (right) ideal I of a semiring R is called a *left* (right) k-ideal of R if for any $a,b \in I$, $x \in R$ and $x+a=b$, it follows that $x \in I$.

If A is a subset of a semiring R, then the k-closure of A denoted by $\overline{A}$ is defined by

$$\overline{A} = \{x \in R : x+a=b \text{ for some } a,b \in A\}.$$

If A is a left (right) k-ideal of R, then $\overline{A} = A$.

A fuzzy subset λ of a semiring R is called a fuzzy left (right) ideal of R if for all $a,b \in R$ we have

(1) $\lambda(a+b) \geq \lambda(a) \wedge \lambda(b)$,
(2) $\lambda(ab) \geq \lambda(b)$, $(\lambda(ab) \geq \lambda(a))$.

A fuzzy left (right) ideal λ of a semiring R is called a *fuzzy left (right) k-ideal* if $x+y=z \Rightarrow \lambda(x) \geq \lambda(y) \wedge \lambda(z)$ for all $x,y,z \in R$.

Throughout this chapter R denotes a semiring with zero 0.

Proposition 4.1. *Let A be a nonempty subset of a semiring R. Then a fuzzy set* $\lambda_A^{s,t}$ *defined by*

$$\lambda_A^{s,t}(x) = \begin{cases} t & \text{if } x \in A \\ s & \text{otherwise} \end{cases}$$

where $0 \leq s < t \leq 1$*, is a fuzzy left (right) k-ideal of R if and only if A is a left (right) k-ideal of R.*

J. Ahsan et al.: Fuzzy Semirings with Applications, STUDFUZZ 278, pp. 53–82.
springerlink.com

Proof. Straightforward

Proposition 4.2. *A fuzzy subset λ of a semiring R is a fuzzy left (right) k-ideal of R if and only if each nonempty level subset of λ is a left (right) k-ideal of R.*

Proof. Suppose λ is a fuzzy left k-ideal of R and $t \in (0,1]$ be such that $U(\lambda;t) \neq \emptyset$. Let $a,b \in U(\lambda;t)$. Then $\lambda(a) \geq t$ and $\lambda(b) \geq t$. As $\lambda(a+b) \geq \lambda(a) \wedge \lambda(b)$, we have $\lambda(a+b) \geq t$. Hence $a+b \in U(\lambda;t)$. For $r \in R$, $\lambda(ra) \geq \lambda(a)$ so $\lambda(ra) \geq t$. This implies $ra \in U(\lambda;t)$. Hence $U(\lambda;t)$ is a left ideal of R. Now let $x+a=b$ for some $a,b \in U(\lambda;t)$. Then $\lambda(a) \geq t$ and $\lambda(b) \geq t$. Since $\lambda(x) \geq \lambda(a) \wedge \lambda(b)$, we have $\lambda(x) \geq t$. Hence $x \in U(\lambda;t)$. Thus $U(\lambda;t)$ is a left k-ideal of R.

Conversely, assume that each nonempty subset $U(\lambda;t)$ of R is a left k-ideal of R. Let $a,b \in R$ be such that $\lambda(a+b) < \lambda(a) \wedge \lambda(b)$. Take $t \in (0,1]$ such that $\lambda(a+b) < t \leq \lambda(a) \wedge \lambda(b)$. Then $a,b \in U(\lambda;t)$ but $a+b \notin U(\lambda;t)$, a contradiction. Hence $\lambda(a+b) \geq \lambda(a) \wedge \lambda(b)$.

Similarly we can show that $\lambda(ab) \geq \lambda(b)$.

Let $x,y,z \in R$ be such that $x+y=z$. If possible let $\lambda(x) < \lambda(y) \wedge \lambda(z)$. Take $t \in (0,1]$ such that $\lambda(x) < t \leq \lambda(y) \wedge \lambda(z)$. Then $y,z \in U(\lambda;t)$ but $x \notin U(\lambda;t)$, a contradiction. Hence $\lambda(x) \geq \lambda(y) \wedge \lambda(z)$. Thus λ is a fuzzy left k-ideal of R.

Example 4.1. The set $R = \{0,1,2,3\}$ with operations addition and multiplication given by the following Cayley tables

+	0	1	2	3
0	0	1	2	3
1	1	1	2	3
2	2	2	2	3
3	3	3	3	2

·	0	1	2	3
0	0	0	0	0
1	0	1	1	1
2	0	1	1	1
3	0	1	1	1

is a semiring. The ideals in R are $\{0\},\{0,1\},\{0,1,2\},\{0,1,2,3\}$. All ideals are k-ideals. Let $t_1,t_2,t_3,t_4 \in (0,1]$ be such that $t_1 \geq t_2 \geq t_3 \geq t_4$.

Define $\lambda : R \longrightarrow [0,1]$ by

$$\lambda(0) = t_1$$
$$\lambda(1) = t_2$$
$$\lambda(2) = t_3$$
$$\lambda(3) = t_4$$

Then

$$U(\lambda;t) = \begin{cases} \{0,1,2,3\} & \text{if } \quad t \leq t_4 \\ \{0,1,2\} & \text{if } t_4 < t \leq t_3 \\ \{0,1\} & \text{if } t_3 < t \leq t_2 \\ \{0\} & \text{if } t_2 < t \leq t_1 \\ \emptyset & \text{if } \quad t > t_1 \end{cases}$$

Thus by Proposition 4.2, λ is a fuzzy k-ideal of R.

4.1 k-Product and k-Sum of Fuzzy Subsets

In general, the sum and product of fuzzy k-ideals, as defined in Chapter 2, is not a fuzzy k-ideal. We now define the k-product and k-sum of fuzzy subsets of a semiring R and prove that k-product and k-sum yield a fuzzy k-ideal. Most of the results of this section are taken from [132].

Definition 4.1. The *k-product* of two fuzzy subsets μ and ν of R is defined by

$$(\mu \odot_k \nu)(x) = \bigvee_{x+\sum_{i=1}^{m} a_i b_i = \sum_{j=1}^{n} a'_j b'_j} \left[\begin{array}{c} \left[\bigwedge_{i=1}^{m} \mu(a_i)\right] \wedge \left[\bigwedge_{i=1}^{m} \nu(b_i)\right] \wedge \\ \left[\bigwedge_{j=1}^{n} \mu(a'_j)\right] \wedge \left[\bigwedge_{j=1}^{n} \nu(b'_j)\right] \end{array} \right]$$

and $(\mu \odot_k \nu)(x) = 0$ if x can not be expressed as $x + \sum_{i=1}^{m} a_i b_i = \sum_{j=1}^{n} a'_j b'_j$.

By direct calculations we obtain the following result.

Proposition 4.3. *Let $\mu, \nu, \omega, \lambda$ be fuzzy subsets of R. Then $\mu \le \omega$ and $\nu \le \lambda \Rightarrow \mu \odot_k \nu \le \omega \odot_k \lambda$.*

For any subset A of a semiring R, recall that χ_A denotes the characteristic function of A.

Lemma 4.1. *Let R be a semiring and $A, B \subseteq R$. Then we have*
(i) $A \subseteq B$ if and only if $\chi_A \le \chi_B$.
(ii) $\chi_A \bigwedge \chi_B = \chi_{A \cap B}$.
(iii) $\chi_A \odot_k \chi_B = \chi_{\overline{AB}}$.

Proof. (i) and (ii) are obvious. For (iii) let $x \in R$. If $x \in \overline{AB}$, then $\chi_{\overline{AB}}(x) = 1$ and $x + \sum_{i=1}^{m} p_i q_i = \sum_{j=1}^{n} p'_j q'_j$ for some $p_i, p'_j \in A$ and $q_i, q'_j \in B$. Thus we have

$$(\chi_A \odot_k \chi_B)(x) = \bigvee_{x+\sum_{i=1}^{m} a_i b_i = \sum_{j=1}^{n} a'_j b'_j} \left[\begin{array}{c} [\bigwedge_{i=1}^{m} \chi_A(a_i)] \wedge [\bigwedge_{i=1}^{m} \chi_B(b_i)] \wedge \\ [\bigwedge_{j=1}^{n} \chi_A(a'_j)] \wedge [\bigwedge_{j=1}^{n} \chi_B(b'_j)] \end{array} \right]$$

$$\ge \left[\begin{array}{c} [\bigwedge_{i=1}^{m} \chi_A(p_i)] \wedge [\bigwedge_{i=1}^{m} \chi_B(q_i)] \wedge \\ [\bigwedge_{j=1}^{n} \chi_A(p'_j)] \wedge [\bigwedge_{j=1}^{n} \chi_B(q'_j)] \end{array} \right] = 1$$

and so

$$(\chi_A \odot_k \chi_B)(x) = 1 = \chi_{\overline{AB}}$$

If $x \notin \overline{AB}$, then $\chi_{\overline{AB}} = 0$. Now if possible, let $(\chi_A \odot_k \chi_B)(x) \neq 0$. Then

$$(\chi_A \odot_k \chi_B)(x) = \bigvee_{x+\sum_{i=1}^{m} a_i b_i = \sum_{j=1}^{n} a_j' b_j'} \begin{bmatrix} [\bigwedge_{i=1}^{m} \chi_A(a_i)] \wedge [\bigwedge_{i=1}^{m} \chi_B(b_i)] \wedge \\ [\bigwedge_{j=1}^{n} \chi_A(a_j')] \wedge [\bigwedge_{j=1}^{n} \chi_B(b_j')] \end{bmatrix} \neq 0.$$

Hence there exist $p_i, q_i, p_j', q_j' \in R$ such that

$$x + \sum_{i=1}^{m} p_i q_i = \sum_{j=1}^{n} p_j' q_j'$$

and

$$\begin{bmatrix} [\bigwedge_{i=1}^{m} \chi_A(p_i)] \wedge [\bigwedge_{i=1}^{m} \chi_B(q_i)] \wedge \\ [\bigwedge_{j=1}^{n} \chi_A(p_j')] \wedge [\bigwedge_{j=1}^{n} \chi_B(q_j')] \end{bmatrix} \neq 0,$$

that is

$$\chi_A(p_i) = \chi_A(p_j') = \chi_B(q_i) = \chi_B(q_j') = 1.$$

Hence $p_i, p_j' \in A$ and $q_i, q_j' \in B$, and so $x \in \overline{AB}$ which is a contradiction. Thus we have $(\chi_A \odot_k \chi_B)(x) = 0 = \chi_{\overline{AB}}(x)$.

Thus in any case, we have $(\chi_A \odot_k \chi_B)(x) = \chi_{\overline{AB}}(x)$.

Theorem 4.1. *(i) If λ, μ are fuzzy k-ideals of R, then $\lambda \odot_k \mu$ is a fuzzy k-ideal of R.*

(ii) If λ is a fuzzy right k-ideal and μ is a fuzzy left k-ideal of R, then $\lambda \odot_k \mu \leq \lambda \wedge \mu$.

Proof. (i) Let λ, μ be fuzzy k-ideals of R and $x, y \in R$. Then

$$(\lambda \odot_k \mu)(x) = \bigvee_{x+\sum_{i=1}^{m} a_i b_i = \sum_{j=1}^{n} a_j' b_j'} \begin{bmatrix} \left[\bigwedge_{i=1}^{m} \lambda(a_i)\right] \wedge \left[\bigwedge_{i=1}^{m} \mu(b_i)\right] \wedge \\ \left[\bigwedge_{j=1}^{n} \lambda(a_j')\right] \wedge \left[\bigwedge_{j=1}^{n} \mu(b_j')\right] \end{bmatrix}$$

and

$$(\lambda \odot_k \mu)(y) = \bigvee_{y+\sum_{k=1}^{p} c_k d_k = \sum_{l=1}^{q} c_l' d_l'} \begin{bmatrix} \left[\bigwedge_{k=1}^{p} \lambda(c_k)\right] \wedge \left[\bigwedge_{k=1}^{p} \mu(d_k)\right] \wedge \\ \left[\bigwedge_{l=1}^{q} \lambda(c_l')\right] \wedge \left[\bigwedge_{l=1}^{q} \mu(d_l')\right] \end{bmatrix}.$$

Thus

$$(\lambda \odot_k \mu)(x) \wedge (\lambda \odot_k \mu)(y) = \left[\bigvee_{x+\sum_{i=1}^m a_ib_i=\sum_{j=1}^n a_j'b_j'} \left[\begin{matrix} \left[\bigwedge_{i=1}^m \lambda(a_i)\right] \wedge \left[\bigwedge_{i=1}^m \mu(b_i)\right] \wedge \\ \left[\bigwedge_{j=1}^n \lambda(a_j')\right] \wedge \left[\bigwedge_{j=1}^n \mu(b_j')\right] \end{matrix} \right] \right]$$
$$\wedge \left[\bigvee_{y+\sum_{k=1}^p c_kd_k=\sum_{l=1}^q c_l'd_l'} \left[\begin{matrix} \left[\bigwedge_{k=1}^p \lambda(c_k)\right] \wedge \left[\bigwedge_{k=1}^p \mu(d_k)\right] \wedge \\ \left[\bigwedge_{l=1}^q \lambda(c_l')\right] \wedge \left[\bigwedge_{l=1}^q \mu(d_l')\right] \end{matrix} \right] \right]$$
$$= \bigvee_{x+\sum_{i=1}^m a_ib_i=\sum_{j=1}^n a_j'b_j'} \left[\bigvee_{y+\sum_{k=1}^p c_kd_k=\sum_{l=1}^q c_l'd_l'} \left[\begin{matrix} \left[\bigwedge_{i=1}^m \lambda(a_i)\right] \wedge \left[\bigwedge_{i=1}^m \mu(b_i)\right] \wedge \\ \left[\bigwedge_{j=1}^n \lambda(a_j')\right] \wedge \left[\bigwedge_{j=1}^n \mu(b_j')\right] \wedge \\ \left[\bigwedge_{k=1}^p \lambda(c_k)\right] \wedge \left[\bigwedge_{k=1}^p \mu(d_k)\right] \wedge \\ \left[\bigwedge_{l=1}^q \lambda(c_l')\right] \wedge \left[\bigwedge_{l=1}^q \mu(d_l')\right] \end{matrix} \right] \right].$$

Since for each expression $x+\sum_{i=1}^m a_ib_i = \sum_{j=1}^n a_j'b_j'$ and $y+\sum_{k=1}^p c_kd_k = \sum_{l-1}^q c_l'd_l'$, we have $x+y+\sum_{i=1}^m a_ib_i + \sum_{k=1}^p c_kd_k = \sum_{j=1}^n a_j'b_j' + \sum_{l=1}^q c_l'd_l'$, so we have

$$(\lambda \odot_k \mu)(x) \wedge (\lambda \odot_k \mu)(y) \leq \bigvee_{x+y+\sum_{s=1}^u e_sf_s=\sum_{t=1}^v e_t'f_t'} \left[\begin{matrix} \left[\bigwedge_{s=1}^u \lambda(e_s)\right] \wedge \left[\bigwedge_{s=1}^u \mu(f_s)\right] \wedge \\ \left[\bigwedge_{t=1}^v \lambda(e_t')\right] \wedge \left[\bigwedge_{t=1}^v \mu(f_t')\right] \end{matrix} \right]$$
$$= (\lambda \odot_k \mu)(x+y).$$

Similarly,

$$(\lambda \odot_k \mu)(x) = \bigvee_{x+\sum_{i=1}^m a_ib_i=\sum_{j=1}^n a_j'b_j'} \left[\begin{matrix} \left[\bigwedge_{i=1}^m \lambda(a_i)\right] \wedge \left[\bigwedge_{i=1}^m \mu(b_i)\right] \wedge \\ \left[\bigwedge_{j=1}^n \lambda(a_j')\right] \wedge \left[\bigwedge_{j=1}^n \mu(b_j')\right] \end{matrix} \right]$$
$$\leq \bigvee_{x+\sum_{i=1}^m a_ib_i=\sum_{j=1}^n a_j'b_j'} \left[\begin{matrix} \left[\bigwedge_{i=1}^m \lambda(a_i)\right] \wedge \left[\bigwedge_{i=1}^m \mu(b_ir)\right] \wedge \\ \left[\bigwedge_{j=1}^n \lambda(a_j')\right] \wedge \left[\bigwedge_{j=1}^n \mu(b_j'r)\right] \end{matrix} \right]$$
$$\leq \bigvee_{xr+\sum_{k=1}^p g_kh_k=\sum_{l=1}^q g_l'h_l'} \left[\begin{matrix} \left[\bigwedge_{k=1}^p \lambda(g_k)\right] \wedge \left[\bigwedge_{k=1}^p \mu(h_k)\right] \wedge \\ \left[\bigwedge_{l=1}^q \lambda(g_l')\right] \wedge \left[\bigwedge_{l=1}^q \mu(h_l')\right] \end{matrix} \right]$$
$$= (\lambda \odot_k \mu)(xr).$$

Analogously we can verify that $(\lambda \odot_k \mu)(rx) \geq (\lambda \odot_k \mu)(x)$ for all $r,x \in R$. This means that $\lambda \odot_k \mu$ is a fuzzy ideal of R.

To prove that $x+a=b$ implies $(\lambda \odot_k \mu)(x) \geq (\lambda \odot_k \mu)(a) \wedge (\lambda \odot_k \mu)(b)$, observe that

$$a+\sum_{i=1}^{m} a_i b_i = \sum_{j=1}^{n} a'_j b'_j \text{ and } b+\sum_{k=1}^{l} c_k d_k = \sum_{q=1}^{p} c'_q d'_q \tag{4.1}$$

together with $x+a=b$, and this gives $x+a+\sum\limits_{i=1}^{m} a_i b_i = b+\sum\limits_{i=1}^{m} a_i b_i$. Thus

$$x+\sum_{j=1}^{n} a'_j b'_j = b+\sum_{i=1}^{m} a_i b_i .$$

Consequently,

$$\begin{aligned} x+\sum_{j=1}^{n} a'_j b'_j+\sum_{k=1}^{l} c_k d_k &= b+\sum_{k=1}^{l} c_k d_k+\sum_{i=1}^{m} a_i b_i \\ &= \sum_{q=1}^{p} c'_q d'_q+\sum_{i=1}^{m} a_i b_i \\ &= \sum_{i=1}^{m} a_i b_i+\sum_{q=1}^{p} c'_q d'_q . \end{aligned}$$

Therefore

$$x+\sum_{j=1}^{n} a'_j b'_j+\sum_{k=1}^{l} c_k d_k = \sum_{i=1}^{m} a_i b_i+\sum_{q=1}^{p} c'_q d'_q . \tag{4.2}$$

Now, we have

$$\begin{aligned} (\lambda \odot_k \mu)(a) \wedge (\lambda \odot_k \mu)(b) &= \left[\bigvee_{a+\sum\limits_{i=1}^{m} a_i b_i=\sum\limits_{j=1}^{n} a'_j b'_j} \left[\begin{array}{l} \left[\bigwedge\limits_{i=1}^{m} \lambda(a_i)\right] \wedge \left[\bigwedge\limits_{i=1}^{m} \mu(b_i)\right] \wedge \\ \left[\bigwedge\limits_{j=1}^{n} \lambda(a'_j)\right] \wedge \left[\bigwedge\limits_{j=1}^{n} \mu(b'_j)\right] \end{array} \right] \right] \\ &\quad \wedge \left[\bigvee_{b+\sum\limits_{k=1}^{l} c_k d_k=\sum\limits_{q=1}^{p} c'_q d'_q} \left[\begin{array}{l} \left[\bigwedge\limits_{k=1}^{p} \lambda(c_k)\right] \wedge \left[\bigwedge\limits_{k=1}^{p} \mu(d_k)\right] \wedge \\ \left[\bigwedge\limits_{l=1}^{q} \lambda(c'_l)\right] \wedge \left[\bigwedge\limits_{l=1}^{q} \mu(d'_l)\right] \end{array} \right] \right] \\ &= \bigvee_{a+\sum\limits_{i=1}^{m} a_i b_i=\sum\limits_{j=1}^{n} a'_j b'_j} \left(\bigvee_{b+\sum\limits_{k=1}^{p} c_k d_k=\sum\limits_{l=1}^{q} c'_l d'_l} \left(\begin{array}{l} \left[\bigwedge\limits_{i=1}^{m} \lambda(a_i)\right] \wedge \left[\bigwedge\limits_{i=1}^{m} \mu(b_i)\right] \wedge \\ \left[\bigwedge\limits_{j=1}^{n} \lambda(a'_j)\right] \wedge \left[\bigwedge\limits_{j=1}^{n} \mu(b'_j)\right] \wedge \\ \left[\bigwedge\limits_{k=1}^{p} \lambda(c_k)\right] \wedge \left[\bigwedge\limits_{k=1}^{p} \mu(d_k)\right] \wedge \\ \left[\bigwedge\limits_{l=1}^{q} \lambda(c'_l)\right] \wedge \left[\bigwedge\limits_{l=1}^{q} \mu(d'_l)\right] \end{array} \right) \right) \end{aligned}$$

$$\leq \bigvee_{x+\sum_{s=1}^{u} g_s h_s = \sum_{t=1}^{w} g_t' h_t'} \left(\begin{array}{c} \left[\bigwedge_{s=1}^{u} \lambda(g_s)\right] \wedge \left[\bigwedge_{s=1}^{u} \mu(h_s)\right] \wedge \\ \left[\bigwedge_{t=1}^{w} \lambda(g_t')\right] \wedge \left[\bigwedge_{t=1}^{w} \mu(h_t')\right] \end{array} \right)$$
$$= (\lambda \odot_k \mu)(x).$$

Thus $(\lambda \odot_k \mu)(a) \wedge (\lambda \odot_k \mu)(b) \leq (\lambda \odot_k \mu)(x)$. Hence $(\lambda \odot_k \mu)$ is a fuzzy k-ideal of R.

(ii) By simple calculations we can prove that $\lambda \odot_k \mu \leq \lambda \wedge \mu$.

Definition 4.2. The *k-sum* $\lambda +_k \mu$ of fuzzy subsets λ and μ of R is defined by

$$(\lambda +_k \mu)(x) = \sup_{x+(a_1+b_1)=(a_2+b_2)} [\lambda(a_1) \wedge \lambda(a_2) \wedge \mu(b_1) \wedge \mu(b_2)],$$

where $x, a_1, b_1, a_2, b_2 \in R$.

Theorem 4.2. *The k-sum of fuzzy k-ideals of R is also a fuzzy k-ideal of R.*

Proof. Let λ, μ be fuzzy k-ideals of R. Then for $x, y, r \in R$, we have

$$(\lambda +_k \mu)(x) \wedge (\lambda +_k \mu)(y) = \left[\bigvee_{x+(a_1+b_1)=(a_2+b_2)} [\lambda(a_1) \wedge \lambda(a_2) \wedge \mu(b_1) \wedge \mu(b_2)] \right] \wedge$$
$$\left[\bigvee_{y+(a_1'+b_1')=(a_2'+b_2')} [\lambda(a_1') \wedge \lambda(a_2') \wedge \mu(b_1') \wedge \mu(b_2')] \right]$$
$$= \bigvee_{\substack{x+(a_1+b_1)=(a_2+b_2) \\ y+(a_1'+b_1')=(a_2'+b_2')}} \left(\begin{array}{c} \lambda(a_1) \wedge \lambda(a_2) \wedge \mu(b_1) \wedge \mu(b_2) \wedge \\ \lambda(a_1') \wedge \lambda(a_2') \wedge \mu(b_1') \wedge \mu(b_2') \end{array} \right)$$
$$\leq \bigvee_{\substack{x+(a_1+b_1)=(a_2+b_2) \\ y+(a_1'+b_1')=(a_2'+b_2')}} \left(\begin{array}{c} \lambda(a_1+a_1') \wedge \lambda(a_2+a_2') \wedge \\ \mu(b_1+b_1') \wedge \mu(b_2+b_2') \end{array} \right)$$
$$\leq \bigvee_{(x+y)+(c_1+d_1)=(c_2+d_2)} [\lambda(c_1) \wedge \lambda(c_2) \wedge \mu(d_1) \wedge \mu(d_2)]$$
$$= (\lambda +_k \mu)(x+y).$$

Now,

$$\begin{aligned}(\lambda +_k \mu)(x) &= \bigvee_{x+(a_1+b_1)=(a_2+b_2)} [\lambda(a_1)\wedge\lambda(a_2)\wedge\mu(b_1)\wedge\mu(b_2)] \\ &\leq \bigvee_{x+(a_1+b_1)=(a_2+b_2)} [\lambda(ra_1)\wedge\lambda(ra_2)\wedge\mu(rb_1)\wedge\mu(rb_2)] \\ &\leq \bigvee_{rx+(a_1''+b_1'')=(a_2''+b_2'')} [\lambda(a_1'')\wedge\lambda(a_2'')\wedge\mu(b_1'')\wedge\mu(b_2'')] \\ &= (\lambda +_k \mu)(rx).\end{aligned}$$

Similarly $(\lambda +_k \mu)(x) \leq (\lambda +_k \mu)(xr)$. This proves that $(\lambda +_k \mu)$ is a fuzzy ideal of R.

Now we show that $x+a=b$ implies $(\lambda +_k \mu)(x) \geq (\lambda +_k \mu)(a) \wedge (\lambda +_k \mu)(b)$. For this let $a+(a_1+b_1)=(a_2+b_2)$ and $b+(c_1+d_1)=(c_2+d_2)$ Then,

$$x+a+(c_1+d_1)=(c_2+d_2)$$

whence

$$x+a+(c_1+d_1)+(a_1+b_1)=(c_2+d_2)+(a_1+b_1)$$

and

$$x+(a+a_1+b_1)+(c_1+d_1)=(c_2+d_2)+(a_1+b_1).$$

Hence

$$x+(a_2+b_2)+(c_1+d_1)=(c_2+d_2)+(a_1+b_1).$$

Thus

$$x+(a_2+c_1)+(b_2+d_1)=(a_1+c_2)+(b_1+d_2).$$

Therefore

$$\begin{aligned}(\lambda +_k \mu)(a) \wedge (\lambda +_k \mu)(b) &= \left[\bigvee_{a+(a_1+b_1)=(a_2+b_2)} [\lambda(a_1)\wedge\lambda(a_2)\wedge\mu(b_1)\wedge\mu(b_2)]\right] \wedge \\ &\quad \left[\bigvee_{b+(c_1+d_1)=(c_2+d_2)} [\lambda(c_1)\wedge\lambda(c_2)\wedge\mu(d_1)\wedge\mu(d_2)]\right] \\ &= \bigvee_{\substack{a+(a_1+b_1)=(a_2+b_2)\\ b+(c_1+d_1)=(c_2+d_2)}} \begin{pmatrix}\lambda(a_1)\wedge\lambda(a_2)\wedge\mu(b_1)\wedge\mu(b_2)\wedge \\ \lambda(c_1)\wedge\lambda(c_2)\wedge\mu(d_1)\wedge\mu(d_2)\end{pmatrix} \\ &\leq \bigvee_{\substack{a+(a_1+b_1)=(a_2+b_2)\\ b+(c_1+d_1)=(c_2+d_2)}} \begin{pmatrix}\lambda(a_2+c_1)\wedge\lambda(a_1+c_2)\wedge \\ \mu(b_2+d_1)\wedge\mu(b_1+d_2)\end{pmatrix} \\ &\leq \bigvee_{x+(a'+b')=(a''+b'')} [\lambda(a')\wedge\lambda(a'')\wedge\mu(b')\wedge\mu(b'')] \\ &= (\lambda +_k \mu)(x).\end{aligned}$$

Thus $\lambda +_k \mu$ is a fuzzy k-ideal of R.

Theorem 4.3. *If μ is a fuzzy subset of a semiring R, then the following are equivalent:*

(*a*) μ *satisfies* (1) $\mu(x+y) \geq \min\{\mu(x), \mu(y)\}$ *and*
(2) $x+a=b \Rightarrow \mu(x) \geq \min\{\mu(a), \mu(b)\}$,

(*b*) $\mu +_k \mu \leq \mu$.

Proof. (*a*) $\Rightarrow$ (*b*) Let $x \in R$. Then

$$\begin{aligned}(\mu +_k \mu)(x) &= \bigvee_{x+(a_1+b_1)=(a_2+b_2)} [\mu(a_1) \wedge \mu(a_2) \wedge \mu(b_1) \wedge \mu(b_2)] \\ &\leq \bigvee_{x+(a_1+b_1)=(a_2+b_2)} [\mu(a_1+a_2) \wedge \mu(b_1+b_2)] \qquad \text{by (1)} \\ &\leq \mu(x) \qquad \text{by (2)}\end{aligned}$$

Thus $\mu +_k \mu \leq \mu$.

(*b*) $\Rightarrow$ (*a*) First we show that $\mu(0) \geq \mu(x)$ for all $x \in R$.

$$\begin{aligned}\mu(0) &\geq (\mu +_k \mu)(0) \\ &= \bigvee_{0+(a_1+b_1)=(a_2+b_2)} [\mu(a_1) \wedge \mu(a_2) \wedge \mu(b_1) \wedge \mu(b_2)] \\ &\geq \mu(x) \wedge \mu(x) \wedge \mu(x) \wedge \mu(x) \quad \text{because } 0+x+x=x+x \\ &= \mu(x).\end{aligned}$$

Thus $\mu(0) \geq \mu(x)$ for all $x \in R$.

Now

$$\begin{aligned}\mu(x+y) &\geq (\mu +_k \mu)(x+y) \\ &= \bigvee_{x+y+(a_1+b_1)=(a_2+b_2)} [\mu(a_1) \wedge \mu(a_2) \wedge \mu(b_1) \wedge \mu(b_2)] \\ &\geq \mu(0) \wedge \mu(0) \wedge \mu(x) \wedge \mu(y) \quad \text{because } x+y+0+0=x+y \\ &= \mu(x) \wedge \mu(y) \quad (\text{because } \mu(0) \geq \mu(x) \text{ for all } x \in R).\end{aligned}$$

Again

$$\begin{aligned}\mu(x) &\geq (\mu +_k \mu)(x) \\ &= \bigvee_{x+(a_1+b_1)=(a_2+b_2)} [\mu(a_1) \wedge \mu(a_2) \wedge \mu(b_1) \wedge \mu(b_2)]\end{aligned}$$

If $x+a=b$ then $x+a+0=b+0$ and so

$$\mu(x) \geq \mu(a) \wedge \mu(0) \wedge \mu(b) \wedge \mu(0) = \mu(a) \wedge \mu(b) \quad \left(\begin{array}{c}\text{because } \mu(0) \geq \mu(x) \\ \text{for all } x \in R\end{array}\right).$$

Theorem 4.4. *A fuzzy subset μ in a semiring R is a fuzzy left (right) k-ideal if and only if*
(i) $\mu +_k \mu \leq \mu$
(ii) $\chi_R \odot_k \mu \leq \mu \quad (\mu \odot_k \chi_R \leq \mu)$.

Proof. Let μ be a fuzzy left k-ideal of R. By Theorem 4.3, μ satisfies (i). Now we prove condition (ii). Let $x \in R$. If $(\chi_R \odot_k \mu)(x) = 0$, then $(\chi_R \odot_k \mu)(x) \leq (\mu)(x)$. Otherwise, there exist elements $a_i, b_i, a'_j, b'_j \in R$ such that $x + \sum_{i=1}^{m} a_i b_i = \sum_{j=1}^{n} a'_j b'_j$. Then we have

$$
\begin{aligned}
(\chi_R \odot_k \mu)(x) &= \bigvee_{x+\sum_{i=1}^{m} a_i b_i = \sum_{j=1}^{n} a'_j b'_j} \left[\begin{array}{c} \left[\bigwedge_{i=1}^{m} \chi_R(a_i)\right] \wedge \left[\bigwedge_{i=1}^{m} \mu(b_i)\right] \wedge \\ \left[\bigwedge_{j=1}^{n} \chi_R(a'_j)\right] \wedge \left[\bigwedge_{j=1}^{n} \mu(b'_j)\right] \end{array} \right] \\
&\leq \bigvee_{x+\sum_{i=1}^{m} a_i b_i = \sum_{j=1}^{n} a'_j b'_j} \left[[\bigwedge_{i=1}^{m} \mu(a_i b_i)] \wedge [\bigwedge_{j=1}^{n} \mu(a'_j b'_j)] \right] \\
&\leq \bigvee_{x+\sum_{i=1}^{m} a_i b_i = \sum_{j=1}^{n} a'_j b'_j} \left[\mu(\sum_{i=1}^{m} a_i b_i) \wedge \mu(\sum_{j=1}^{n} a'_j b'_j) \right] \\
&\leq \bigvee_{x+\sum_{i=1}^{m} a_i b_i = \sum_{j=1}^{n} a'_j b'_j} \mu(x) = \mu(x).
\end{aligned}
$$

This implies that $\chi_R \odot_k \mu \leq \mu$.

Conversely, assume that the given conditions hold. In order to show that μ is a fuzzy left k-ideal of R it is sufficient to show that the condition $\mu(xy) \geq \mu(y)$ holds. Let $x, y \in R$. Then we have

$$
\begin{aligned}
\mu(xy) \geq (\chi_R \odot_k \mu)(xy) &= \bigvee_{xy+\sum_{i=1}^{m} a_i b_i = \sum_{j=1}^{n} a'_j b'_j} \left[\begin{array}{c} \left[\bigwedge_{i=1}^{m} \chi_R(a_i)\right] \wedge \left[\bigwedge_{i=1}^{m} \mu(b_i)\right] \wedge \\ \left[\bigwedge_{j=1}^{n} \chi_R(a'_j)\right] \wedge \left[\bigwedge_{j=1}^{n} \mu(b'_j)\right] \end{array} \right] \\
&= \bigvee_{xy+\sum_{i=1}^{m} a_i b_i = \sum_{j=1}^{n} a'_j b'_j} \left[\left[\bigwedge_{i=1}^{m} \mu(b_i)\right] \wedge \left[\bigwedge_{j=1}^{n} \mu(b'_j)\right] \right]
\end{aligned}
$$

Since $xy + 0y = xy$,

we have $\mu(xy) \geq \mu(y)$ and μ is a fuzzy left k-ideal of R.

4.2 k-Regular Semirings

Recall that an element a of a semiring R is called *regular* if there exists $x \in R$ such that $a = axa$. A semiring R is called *regular* if each element of R is regular. Generalizing the concept of regularity, k-regular semirings are defined as:

Definition 4.3. A semiring R is said to be *k-regular* if for each $a \in R$, there exist $x, y \in R$ such that $a + axa = aya$.

Obviously, every regular semiring is k-regular but the converse is not true. If R is a ring then the concepts of regular and k-regular coincide.

It is shown in [79]

Theorem 4.5. *A semiring R is k-regular if and only if $A \cap B = \overline{AB}$ for every right k-ideal A and left k-ideal B of R.*

We now prove its fuzzy version:

Theorem 4.6. *A semiring R is k-regular if and only if for any fuzzy right k-ideal μ and any fuzzy left k-ideal ν of R we have $\mu \odot_k \nu = \mu \wedge \nu$.*

Proof. Let R be a k-regular semiring and μ, ν be fuzzy right k-ideal and fuzzy left k-ideal of R, respectively. Then by Theorem 4.1, we have $\mu \odot_k \nu \leq \mu \wedge \nu$. To show the reverse inclusion, let $x \in R$. Since R is k-regular, there exist $a, a' \in R$ such that $x + xax = xa'x$. Then we have

$$(\mu \odot_k \nu)(x) = \bigvee_{x + \sum_{i=1}^{m} a_i b_i = \sum_{j=1}^{n} a'_j b'_j} \left(\begin{bmatrix} \bigwedge_{i=1}^{m} \mu(a_i) \end{bmatrix} \wedge \begin{bmatrix} \bigwedge_{i=1}^{m} \nu(b_i) \end{bmatrix} \wedge \begin{bmatrix} \bigwedge_{j=1}^{n} \mu(a'_j) \end{bmatrix} \wedge \begin{bmatrix} \bigwedge_{j=1}^{n} \nu(b'_j) \end{bmatrix} \right)$$

$$\geq \min\left\{\mu(xa), \mu(xa'), \nu(x)\right\} \geq \min\{\mu(x), \nu(x)\} = (\mu \wedge \nu)(x).$$

This implies that $\mu \odot_k \nu \geq \mu \wedge \nu$. Therefore $\mu \odot_k \nu = \mu \wedge \nu$.

Conversely, let C, D be any right k-ideal and any left k-ideal of R, respectively. Then the characteristic functions χ_C, χ_D of C, D are fuzzy right k-ideal and fuzzy left k-ideal of R, respectively. Now, by assumption and Lemma 4.1, we have

$$\chi_{\overline{CD}} = \chi_C \odot_k \chi_D = \chi_C \wedge \chi_D = \chi_{C \cap D}.$$

Thus, $\overline{CD} = C \cap D$. Hence by Theorem 4.5, R is k-regular semiring.

4.3 Right k-Weakly Regular Semirings

Generalizing the concept of k-regular semiring in [133], a right k-weakly regular semiring is defined as:

Definition 4.4. A semiring R is called right (left) *k-weakly regular* semiring if for each $x \in R$, $x \in \overline{(xR)^2}$ $\left(\text{res. } x \in \overline{(Rx)^2}\right)$.

That is, for each $x \in R$ we have $r_i, s_i, t_j, p_j \in R$ such that $x + \sum_{i=1}^{n} xr_ixs_i = \sum_{j=1}^{m} xt_jxp_j$ $(x + \sum_{i=1}^{n} r_ixs_ix = \sum_{j=1}^{m} t_jxp_jx)$. Thus each k-regular semiring with 1 is right k-weakly regular but the converse is not true. However for a commutative semiring both the concepts coincide.

The following characterizations of right k-weakly regular semirings are given in [133].

Proposition 4.4. *The following statements are equivalent for a semiring R with identity :*

(i) R is right k-weakly regular semiring.

(ii) All right k-ideals of R are k-idempotent (A right k-ideal B of R is k-idempotent if $\overline{B^2} = B$).

(iii) $\overline{BA} = B \cap A$ for all right k-ideals B and two-sided k-ideals A of R.

Theorem 4.7. *The collection of all k-ideals of a right k-weakly regular semiring R forms a complete distributive lattice.*

Now we give the fuzzy version of the above results.

Theorem 4.8. *For a semiring R with* 1, *the following assertions are equivalent:*

(i) R is right k-weakly regular semiring.

(ii) All fuzzy right k-ideals of R are k-idempotent (A fuzzy right k-ideal λ of R is k-idempotent if $\lambda \odot_k \lambda = \lambda$).

(iii) $\lambda \odot_k \mu = \lambda \wedge \mu$ for all fuzzy right k-ideals λ and all fuzzy two-sided k-ideals μ of R.

Proof. (i) $\Longrightarrow$ (ii) Let λ be a fuzzy right k-ideal of R. Then we have $\lambda \odot_k \lambda \leq \lambda$.

For the reverse inclusion, let$x \in R$. Since R is right k-weakly regular, there exist $s_i, t_i, s'_j, t'_j \in R$ such that

$$x + \sum_{i=1}^{m} xs_ixt_i = \sum_{j=1}^{n} xs'_jxt'_j.$$

Hence

$$\lambda(x) = \lambda(x) \wedge \lambda(x) \leq \bigwedge_{i=1}^{m} (\lambda(xs_i) \wedge \lambda(xt_i)).$$

Also

$$\lambda(x) = \lambda(x) \wedge \lambda(x) \leq \bigwedge_{j=1}^{n} \left(\lambda(xs'_j) \wedge \lambda(xt'_j)\right).$$

Therefore

$$\begin{aligned}
\lambda(x) &\leq \bigwedge_{i=1}^{m} (\lambda(xs_i) \wedge \lambda(xt_i)) \wedge \bigwedge_{j=1}^{n} \left(\lambda(xs'_j) \wedge \lambda(xt'_j)\right) \\
&\leq \bigvee_{x+\sum_{i=1}^{m} xs_i xt_i = \sum_{j=1}^{n} xs'_j xt'_j} \left[\bigwedge_{i=1}^{m} (\lambda(xs_i) \wedge \lambda(xt_i)) \wedge \bigwedge_{j=1}^{n} \left(\lambda(xs'_j) \wedge \lambda(xt'_j)\right)\right] \\
&= (\lambda \odot_k \lambda)(x).
\end{aligned}$$

Hence $\lambda \leq \lambda \odot_k \lambda$, which proves $\lambda \odot_k \lambda = \lambda$.

$(ii) \Longrightarrow (iii)$ Let λ and μ be fuzzy right and two sided k-ideals of R, respectively. Then $\lambda \wedge \mu$ is a fuzzy right k-ideal of R. By Theorem 4.1 $\lambda \odot_k \mu \leq \lambda \wedge \mu$. By hypothesis,

$$(\lambda \wedge \mu) = (\lambda \wedge \mu) \odot_k (\lambda \wedge \mu) \leq \lambda \odot_k \mu.$$

Hence $\lambda \odot_k \mu = \lambda \wedge \mu$.

$(iii) \Longrightarrow (i)$ Let B be a right k-ideal and A be a two-sided k-ideal of R. Then the characteristic functions χ_B and χ_A of B and A are fuzzy right and fuzzy two-sided k-ideals of R, respectively. Hence by hypothesis and Lemma 4.1, we have

$$\chi_B \odot_k \chi_A = \chi_B \wedge \chi_A \implies \chi_{\overline{BA}} = \chi_{B \cap A} \implies \overline{BA} = B \cap A.$$

Thus by Proposition 4.4, R is right k-weakly regular semiring.

Combining Proposition 4.4 and Theorem 4.8 we have the following Theorem.

Theorem 4.9. *For a semiring R with 1, the following assertions are equivalent:*

(i) R is right k-weakly regular semiring.

(ii) All right k-ideals of R are k-idempotent.

(iii) $\overline{BA} = B \cap A$ for all right k-ideals B and two-sided k-ideal A of R.

(iv) All fuzzy right k-ideals of R are k-idempotent.

(v) $\lambda \odot_k \mu = \lambda \wedge \mu$ for all fuzzy right k-ideals λ and all fuzzy two-sided k-ideals μ of R.

If R is commutative, then the above assertions are equivalent to

(vi) R is k-regular.

Theorem 4.10. *If R is a right k-weakly regular semiring, then the set $\Im_R$ of all fuzzy k-ideals of R (ordered by $\leq$) is a distributive lattice.*

Proof. The set $\Im_R$ of all fuzzy k-ideals of R (ordered by $\leq$) is clearly a lattice under the k-sum and intersection of fuzzy k-ideals. Now we show that $\Im_R$ is a distributive lattice, that is for any fuzzy k-ideals λ, μ, δ of R we have $(\lambda \wedge \delta) + \mu = (\lambda + \mu) \wedge (\delta + \mu)$.

For any $x \in R$

$$\begin{aligned}
[(\lambda\wedge\delta)+\mu](x) &= \bigvee_{x+(a_1+b_1)=(a_2+b_2)} \begin{bmatrix} (\lambda\wedge\delta)(a_1)\wedge(\lambda\wedge\delta)(a_2)\wedge \\ (\mu)(b_1)\wedge(\mu)(b_2) \end{bmatrix} \\
&= \bigvee_{x+(a_1+b_1)=(a_2+b_2)} \begin{bmatrix} \lambda(a_1)\wedge\lambda(a_2)\wedge\mu(b_1)\wedge \\ \mu(b_2)\wedge\delta(a_1)\wedge\delta(a_2) \end{bmatrix} \\
&= \bigvee_{x+(a_1+b_1)=(a_2+b_2)} \begin{bmatrix} [\lambda(a_1)\wedge\lambda(a_2)\wedge\mu(b_1)\wedge\mu(b_2)]\wedge \\ [\delta(a_1)\wedge\delta(a_2)\wedge\mu(b_1)\wedge\mu(b_2)] \end{bmatrix} \\
&= \left(\bigvee_{x+(a_1+b_1)=(a_2+b_2)} [\lambda(a_1)\wedge\lambda(a_2)\wedge\mu(b_1)\wedge\mu(b_2)] \right) \\
&\quad \wedge \left(\bigvee_{x+(a_1+b_1)=(a_2+b_2)} [\delta(a_1)\wedge\delta(a_2)\wedge\mu(b_1)\wedge\mu(b_2)] \right) \\
&= (\lambda+\mu)(x)\wedge(\delta+\mu)(x) \\
&= [(\lambda+\mu)\wedge(\delta+\mu)](x).
\end{aligned}$$

4.4 Prime and Fuzzy Prime Right k-Ideals

Recall that a right k-ideal P of a semiring R is called k-prime (k-semiprime) right k-ideal of R if for any right k-ideals A,B of R,

$$AB \subseteq P \Longrightarrow A \subseteq P \text{ or } B \subseteq P \;\; \left(A^2 \subseteq P \Longrightarrow A \subseteq P\right).$$

The following results are proved in [133].

Theorem 4.11. *The following assertions for a semiring R with* 1 *are equivalent:*
(*i*) *R is right k-weakly regular semiring.*
(*ii*) *Each right k-ideal of R is k-semiprime right k-ideal of R.*

Theorem 4.12. *If every right k-ideal of a semiring R is k-prime right k-ideal, then R is a right k-weakly regular semiring and the set of k-ideals of R is totally ordered.*

Theorem 4.13. *If R is right k-weakly regular semiring and the set of all right k-ideals of R is totally ordered then every right k-ideal of R is a k-prime right k-ideal of R.*

Now we give the fuzzy version of above results.

Definition 4.5. A fuzzy right k-ideal μ of a semiring R is called a *fuzzy k-prime (k-semiprime) right k-ideal* of R if for any fuzzy k-right ideals λ,δ of R,

$$\lambda \odot_k \delta \leq \mu \Rightarrow \lambda \leq \mu \;\text{ or }\; \delta \leq \mu \;\; (\lambda \odot_k \lambda \leq \mu \Rightarrow \lambda \leq \mu).$$

μ is called a fuzzy k-irreducible (k-strongly irreducible) if for any fuzzy right k-ideals λ,δ of R,

$$\lambda \wedge \delta = \mu \Rightarrow \lambda = \mu \text{ or } \delta = \mu \ (\lambda \wedge \delta \leq \mu \Rightarrow \lambda \leq \mu \text{ or } \delta \leq \mu).$$

Lemma 4.2. *The intersection of fuzzy k-prime right k-ideals of R is a fuzzy k-semiprime right k-ideal of R.*

Proof. Straightforward.

Proposition 4.5. *Let R be a right k-weakly regular semiring. If λ is a fuzzy right k-ideal of R with $\lambda(a) = \alpha$, where a is any element of R and $\alpha \in (0,1]$, then there exists a fuzzy k-irreducible right k-ideal δ of R such that $\lambda \leq \delta$ and $\delta(a) = \alpha$.*

Proof. Let $X = \{\mu : \mu$ is a fuzzy right k-ideal of R, $\mu(a) = \alpha$ and $\lambda \leq \mu\}$. Then $X \neq \emptyset$, since $\lambda \in X$. Let $\mathfrak{F}$ be a totally ordered subset of X, say $\mathfrak{F} = \{\lambda_i : i \in I\}$. We claim that $\bigvee_{i \in I} \lambda_i$ is a fuzzy right k-ideal of R. For any $x, r \in R$, we have

$$\left(\bigvee_i \lambda_i\right)(x) = \bigvee_i (\lambda_i(x)) \leq \bigvee_i (\lambda_i(xr)) = \left(\bigvee_i \lambda_i\right)(xr).$$

Let $x, y \in R$. Consider

$$\begin{aligned}
\left(\bigvee_i \lambda_i\right)(x) \wedge \left(\bigvee_i \lambda_i\right)(y) &= \left(\bigvee_i (\lambda_i(x))\right) \wedge \left(\bigvee_j (\lambda_j(y))\right) \\
&= \bigvee_j \left[\bigvee_i (\lambda_i(x)) \wedge \lambda_j(y)\right] \\
&= \bigvee_j \left[\bigvee_i (\lambda_i(x) \wedge \lambda_j(y))\right] \\
&\leq \bigvee_j \left[\bigvee_i \left(\lambda_i^j(x) \wedge \lambda_i^j(y)\right)\right] \\
\text{where } \lambda_i^j &= \max\{\lambda_i, \lambda_j\}, \text{ note that } \lambda_i^j \in \{\lambda_i : i \in I\} \\
&\leq \bigvee_j \left[\bigvee_i \left[\lambda_i^j(x+y)\right]\right] \\
&= \bigvee_{i,j} \left[\lambda_i^j(x+y)\right] \\
&\leq \bigvee_i [\lambda_i(x+y)] \\
&= \left(\bigvee_i \lambda_i\right)(x+y).
\end{aligned}$$

Now, let $x + a = b$ where $a, b \in R$. Then

$$\begin{aligned}
\left(\bigvee_i \lambda_i\right)(a)\wedge\left(\bigvee_i \lambda_i\right)(b) &= \left(\bigvee_i (\lambda_i(a))\right)\wedge\left(\bigvee_j (\lambda_j(b))\right)\\
&= \bigvee_j\left[\left(\bigvee_i (\lambda_i(a))\right)\wedge\lambda_j(b)\right]\\
&= \bigvee_j\left[\bigvee_i (\lambda_i(a)\wedge\lambda_j(b))\right]\\
&\leq \bigvee_j\left[\bigvee_i \left(\lambda_i^j(a)\wedge\lambda_i^j(b)\right)\right]\\
&\text{where } \lambda_i^j = \max\{\lambda_i,\lambda_j\},\text{ note that } \lambda_i^j\in\{\lambda_i : i\in I\}\\
&\leq \bigvee_j\left[\bigvee_i \left(\lambda_i^j(x)\right)\right] \quad \text{because } \lambda_i^j \text{ is a fuzzy } k\text{-ideal}\\
&= \bigvee_{i,j}\left[\lambda_i^j(x)\right]\\
&\leq \bigvee_i [\lambda_i(x)] = \left(\bigvee_i \lambda_i\right)(x).
\end{aligned}$$

Thus $\bigvee_i \lambda_i$ is a fuzzy right k-ideal of R. Clearly $\lambda \leq \bigvee_i \lambda_i$ and $\bigvee_i \lambda_i(a) = \bigvee_i (\lambda_i(a)) = \alpha$. Thus $\bigvee_i \lambda_i$ is the l.u.b of $\mathfrak{F}$. Hence by Zorn's lemma there exists a fuzzy right k-ideal δ of R which is maximal with respect to the property that $\lambda \leq \delta$ and $\delta(a) = \alpha$.

We will show that δ is fuzzy k-irreducible right k-ideal of R. Let $\delta = \delta_1 \wedge \delta_2$, where δ_1, δ_2 are fuzzy right k-ideals of R. Thus $\delta \leq \delta_1$ and $\delta \leq \delta_2$. We claim that either $\delta = \delta_1$ or $\delta = \delta_2$. Suppose $\delta \neq \delta_1$ and $\delta \neq \delta_2$. Since δ is maximal with respect to the property that $\delta(a) = \alpha$ and since $\delta \lneqq \delta_1$ and $\delta \lneqq \delta_2$, we have $\delta_1(a) \neq \alpha$ and $\delta_2(a) \neq \alpha$. Hence $\alpha = \delta(a) = (\delta_1 \wedge \delta_2)(a) = (\delta_1)(a) \wedge (\delta_2)(a) \neq \alpha$, which is impossible. Thus $\delta = \delta_1$ or $\delta = \delta_2$. Hence δ is fuzzy k-irreducible right k-ideal of R.

Theorem 4.14. *Every fuzzy right k-ideal of a semiring R is the intersection of all fuzzy k-irreducible right k-ideals of R which contain it.*

Proof. Let λ be a fuzzy right k-ideal of R and let $\{\lambda_\alpha : \alpha \in \Lambda\}$ be the family of all fuzzy k-irreducible right k-ideals of R which contain λ. Obviosly $\lambda \leq \bigwedge_{\alpha\in\Lambda} \lambda_\alpha$. We show that $\bigwedge_{\alpha\in\Lambda} \lambda_\alpha \leq \lambda$. Let a be any element of R. Then by Proposition 4.5, there exists a fuzzy k-irreducible right k-ideal λ_β such that $\lambda \leq \lambda_\beta$ and $\lambda(a) = \lambda_\beta(a)$. Hence $\lambda_\beta \in \{\lambda_\alpha : \alpha \in \Lambda\}$. Thus $\bigwedge_{\alpha\in\Lambda} \lambda_\alpha \leq \lambda_\beta$, so $\bigwedge_{\alpha\in\Lambda} \lambda_\alpha(a) \leq \lambda_\beta(a) = \lambda(a) \Longrightarrow \bigwedge_{\alpha\in\Lambda} \lambda_\alpha \leq \lambda$. Hence $\bigwedge_{\alpha\in\Lambda} \lambda_\alpha = \lambda$.

Theorem 4.15. *The following assertions for a semiring R are equivalent:*

(i) *R is right k-weakly regular semiring.*

(ii) *All fuzzy right k-ideals of R are k-idempotent (A fuzzy right k-ideal λ of R is idempotent if $\lambda \odot_k \lambda = \lambda$).*

(iii) *$\lambda \odot_k \mu = \lambda \wedge \mu$ for all fuzzy right k-ideals λ and all fuzzy two-sided k-ideals μ of R.*

(iv) *Each fuzzy right k-ideal of R is a fuzzy k-semiprime right k-ideal of R.*

Proof. $(i) \Longleftrightarrow (ii) \Longleftrightarrow (iii)$ by Theorem 4.8

$(ii) \Rightarrow (iv)$ Let δ be any fuzzy right k-ideal of R and $\lambda \odot_k \lambda \leq \delta$, where λ is a fuzzy right k-ideal of R. By (ii) $\lambda \odot_k \lambda = \lambda$, so $\lambda \leq \delta$. Thus δ is a fuzzy k-semiprime right k-ideal of R.

$(iv) \Rightarrow (ii)$ Let δ be any fuzzy right k-ideal of R. Then $\delta \odot_k \delta$ is also a fuzzy right k-ideal of R and so by (iv) $\delta \odot_k \delta$ is a fuzzy k-semiprime right k-ideal of R. As $\delta \odot_k \delta \leq \delta \odot_k \delta \Rightarrow \delta \leq \delta \odot_k \delta$ but $\delta \odot_k \delta \leq \delta$ always holds. So $\delta \odot_k \delta = \delta$.

Theorem 4.16. *If every fuzzy right k-ideal of a semiring R is fuzzy k-prime right k-ideal, then R is right k-weakly regular semiring and the set of fuzzy k-ideals of R is totally ordered.*

Proof. Suppose R is a semiring in which each fuzzy right k-ideal is fuzzy k-prime right k-ideal. Let λ be a fuzzy right k-ideal of R. Then $\lambda \odot_k \lambda$ is also a fuzzy right k-ideal of R. As $\lambda \odot_k \lambda \leq \lambda \odot_k \lambda \Longrightarrow \lambda \leq \lambda \odot_k \lambda$. But $\lambda \odot_k \lambda \leq \lambda$ always. Hence $\lambda = \lambda \odot_k \lambda$. Thus R is right k-weakly regular hemiring.

Let λ, μ be any fuzzy k-ideals of R. Then $\lambda \odot_k \mu \leq \lambda \wedge \mu$. As $\lambda \wedge \mu$ is a fuzzy k-ideal of R so a fuzzy k-prime right k-ideal. Thus either $\lambda \leq \lambda \wedge \mu$ or $\mu \leq \lambda \wedge \mu$. That is either $\lambda \leq \mu$ or $\mu \leq \lambda$.

Theorem 4.17. *If R is right k-weakly regular semiring and the set of all fuzzy right k-ideals of R is totally ordered, then every fuzzy right k-ideal of R is a fuzzy k-prime right k-ideal of R.*

Proof. Let λ, μ, ν be fuzzy right k-ideals of R such that $\lambda \odot_k \mu \leq \nu$. Since the set of all fuzzy right k-ideals of R is totally ordered, we have $\lambda \leq \mu$ or $\mu \leq \lambda$. If $\lambda \leq \mu$, then $\lambda = \lambda \odot_k \lambda \leq \lambda \odot_k \mu \leq \nu$. If $\mu \leq \lambda$, then $\mu = \mu \odot_k \mu \leq \lambda \odot_k \mu \leq \nu$. Thus ν is a fuzzy k-prime right k-ideal.

4.5 Idempotent k-Ideals

In this section we characterize those semirings in which each fuzzy k-ideal is idempotent. The following results are proved in [132].

Proposition 4.6. *The following statements are equivalent for a semiring R:*

(i) *Each k-ideal of R is idempotent.*

(ii) *$A \cap B = \overline{AB}$ for each pair of k-ideals A, B of R.*

(iii) *$x \in \overline{RxRxR}$ for every $x \in R$.*

(*iv*) *$X \subseteq \overline{RXRXR}$ for every nonempty subset X of R.*
(*v*) *$A = \overline{RARAR}$ for every k-ideal A of R.*
If R is commutative, then the above assertions are equivalent to
(*vi*) *R is k-regular.*

Proposition 4.7. *The following statements are equivalent for a semiring R.*
(*i*) *Each fuzzy k-ideal of R is idempotent.*
(*ii*) *$\lambda \odot_k \mu = \lambda \wedge \mu$ for all fuzzy k-ideals of R.*
If R is commutative, then the above assertions are equivalent to
(*iii*) *R is k-regular.*

Proof. $(i) \Rightarrow (ii)$ Let λ and μ be fuzzy k-ideals of R. Since $\lambda \wedge \mu$ is a fuzzy k-ideal of R, so by hypothesis $\lambda \wedge \mu$ is idempotent. Thus $\lambda \wedge \mu = (\lambda \wedge \mu) \odot_k (\lambda \wedge \mu) \leq \lambda \odot_k \mu$. By Theorem 4.1, $\lambda \odot_k \mu \leq \lambda \wedge \mu$. Hence $\lambda \odot_k \mu = \lambda \wedge \mu$.
$(ii) \Rightarrow (i)$ Obvious.
If R is commutative then by Theorem 4.6, $(ii) \Leftrightarrow (iii)$.

Theorem 4.18. *Let R be a semiring with identity* 1. *Then the following assertions are equivalent.*
(*i*) *Each k-ideal of R is idempotent.*
(*ii*) *$A \cap B = \overline{AB}$ for each pair of k-ideals A, B of R.*
(*iii*) *Each fuzzy k-ideal of R is idempotent.*
(*iv*) *$\lambda \odot_k \mu = \lambda \wedge \mu$ for all fuzzy k-ideals λ, μ of R.*

Proof. $(i) \Leftrightarrow (ii)$ By Proposition 4.6.
$(iii) \Leftrightarrow (iv)$ By Proposition 4.7.
$(i) \Rightarrow (iii)$ Let $x \in R$. The smallest k-ideal of R containing x has the form $\overline{RxR}$. By hypothesis, we have $\overline{RxR} = \overline{(\overline{RxR})(\overline{RxR})} = \overline{RxRRxR}$. Thus $x \in \overline{RxR} = \overline{RxRRxR}$, this implies

$$x + \sum_{i=1}^{m} r_i x s_i u_i x t_i = \sum_{j=1}^{n} r'_j x s'_j u'_j t'_j$$

for some $r_i, s_i, u_i, t_i, r'_j, s'_j, u'_j, t'_j \in R$.
As $\lambda(x) \leq \lambda(r_i x s_i)$ and $\lambda(x) \leq \lambda(u_i x t_i)$ for each $i \in \{1, 2, ...m\}$, so

$$\lambda(x) \leq \bigwedge_{i=1}^{m} \lambda(r_i x s_i) \text{ and } \lambda(x) \leq \bigwedge_{i=1}^{m} (u_i x t_i).$$

Therefore $\lambda(x) \leq \left[\bigwedge_{i=1}^{m} \lambda(r_i x s_i)\right] \wedge \left[\bigwedge_{i=1}^{m} (u_i x t_i)\right]$.
Similarly

$$\lambda(x) \leq \left[\bigwedge_{j=1}^{n} \lambda(r'_j x s'_j)\right] \wedge \left[\bigwedge_{j=1}^{n} \lambda(u'_j x t'_j)\right].$$

Therefore

$$\lambda(x) \leq \left[\bigwedge_{i=1}^{m} \lambda(r_i x s_i)\right] \wedge \left[\bigwedge_{i=1}^{m} (u_i x t_i)\right] \wedge \left[\bigwedge_{j=1}^{n} \lambda(r_j' x s_j')\right] \wedge \left[\bigwedge_{j=1}^{n} \lambda(u_j' x t_j')\right]$$

$$\leq \bigvee_{x+\sum_{i=1}^{m} r_i x s_i u_i x t_i = \sum_{j=1}^{n} r_j' x s_j' u_j' t_j'} \left(\begin{array}{l} \left[\bigwedge_{i=1}^{m} \lambda(r_i x s_i)\right] \wedge \left[\bigwedge_{i=1}^{m} (u_i x t_i)\right] \wedge \\ \left[\bigwedge_{j=1}^{n} \lambda(r_j' x s_j')\right] \wedge \left[\bigwedge_{j=1}^{n} \lambda(u_j' x t_j')\right] \end{array} \right)$$

$$= (\lambda \odot_k \lambda)(x).$$

Hence $\lambda \leq \lambda \odot_k \lambda$. By Theorem 4.1, $\lambda \odot_k \lambda \leq \lambda$. Thus $\lambda \odot_k \lambda = \lambda$.

$(iii) \Rightarrow (i)$ Let A be a k-ideal of R, then the characteristic function χ_A of A is a fuzzy k-ideal of R. Hence by hypothesis $\chi_A = \chi_A \odot_k \chi_A = \chi_{\overline{AA}}$. Thus $A = \overline{AA}$.

Theorem 4.19. *Each fuzzy k-ideal of R is idempotent if and only if the set of all fuzzy k-ideals of R (ordered by $\leq$) forms a distributive lattice under the k-sum and k-product of fuzzy k-ideals with $\lambda \odot_k \mu = \lambda \wedge \mu$.*

Proof. Suppose that each fuzzy k-ideal of R is idempotent. Then by Proposition 4.7, $\lambda \odot_k \mu = \lambda \wedge \mu$. Let $\mathscr{FL}_R$ be the collection of all fuzzy k-ideals of R. Then $\mathscr{FL}_R$ is a lattice (ordered by $\leq$) under the k-sum and k-product of fuzzy k-ideals.

We show that $(\lambda \odot_k \delta) +_k \mu = (\lambda +_k \mu) \odot_k (\delta +_k \mu)$ for all $\lambda, \mu, \delta \in \mathscr{FL}_R$. Let $x \in R$. Then

$$((\lambda \odot_k \delta) +_k \mu)(x) = \bigvee_{x+(a_1+b_1)=(a_2+b_2)} [(\lambda \wedge \delta)(a_1) \wedge (\lambda \wedge \delta)(a_2) \wedge \mu(b_1) \wedge \mu(b_2)]$$

$$= \bigvee_{x+(a_1+b_1)=(a_2+b_2)} [\lambda(a_1) \wedge \lambda(a_2) \wedge \mu(b_1) \wedge \mu(b_2) \wedge \delta(a_1) \wedge \delta(a_2)]$$

$$= \left[\bigvee_{x+(a_1+b_1)=(a_2+b_2)} [\lambda(a_1) \wedge \lambda(a_2) \wedge \mu(b_1) \wedge \mu(b_2)]\right] \wedge$$

$$\left[\bigvee_{x+(a_1+b_1)=(a_2+b_2)} [\delta(a_1) \wedge \delta(a_2) \wedge \mu(b_1) \wedge \mu(b_2)]\right]$$

$$= (\lambda +_k \mu)(x) \wedge (\delta +_k \mu)(x)$$

$$= [(\lambda +_k \mu) \wedge (\delta +_k \mu)](x)$$

$$= ((\lambda +_k \mu) \odot_k (\delta +_k \mu))(x).$$

So, $\mathscr{FL}_R$ is a distributive lattice.

The converse is obvious.

4.6 Prime and Semiprime Fuzzy k-Ideals

A proper k-ideal P of R is called prime (semiprime) if for any k-ideals A, B of R, $AB \subseteq P$ implies $A \subseteq P$ or $B \subseteq P$ ($A^2 \subseteq P$ implies $A \subseteq P$). A proper k-ideal P of R is called *irreducible* if for any k-ideals A, B of R, $A \cap B = P$ implies $A = P$ or $B = P$. By analogy a fuzzy k-ideal δ of R is called *k-prime* (k-semiprime) if for any fuzzy k-ideals λ, μ of R, $\lambda \odot_k \mu \leq \delta$ implies $\lambda \leq \delta$ or $\mu \leq \delta$ ($\lambda \odot_k \lambda \leq \delta$ implies $\lambda \leq \delta$), and *irreducible* if $\lambda \wedge \mu = \delta$ implies $\lambda = \delta$ or $\mu = \delta$.

Theorem 4.20. *Let R be a semiring in which all fuzzy k-ideals are idempotent. Then a fuzzy k-ideal of R is irreducible if and only if it is k-prime.*

Proof. Assume that all fuzzy k-ideals of R are idempotent and let δ be an arbitrary irreducible fuzzy k-ideal of R. We prove that it is k-prime. If $\lambda \odot_k \mu \leq \delta$ for some fuzzy k-ideals λ, μ of R, then also $\lambda \wedge \mu \leq \delta$. Since the set $\mathscr{FL}_R$ of all fuzzy k-ideals of R is a distributive lattice, we have $\delta = (\lambda \wedge \mu) +_k \delta = (\lambda +_k \delta) \wedge (\mu +_k \delta)$. Thus $\lambda +_k \delta = \delta$ or $\mu +_k \delta = \delta$. Hence $\lambda \leq \delta$ or $\mu \leq \delta$. This proves that δ is k-prime.

Conversely, if δ is a k-prime fuzzy k-ideal of R and $\lambda \wedge \mu = \delta$ for some $\lambda, \mu \in \mathscr{FL}_R$, then $\lambda \odot_k \mu = \delta$, which implies $\lambda \leq \delta$ or $\mu \leq \delta$. Since $\delta = \lambda \wedge \mu$, we have also $\delta \leq \lambda$ and $\delta \leq \mu$. Thus $\lambda = \delta$ or $\mu = \delta$. So, δ is irreducible.

Lemma 4.3. *Let R be a semiring in which each fuzzy k-ideal is idempotent. If λ is a fuzzy k-ideal of R with $\lambda(a) = \alpha$, where a is any element of R and $\alpha \in [0,1]$, then there exists an irreducible k-prime fuzzy k-ideal δ of R such that $\lambda \leq \delta$ and $\delta(a) = \alpha$.*

Proof. Let λ be an arbitrary fuzzy k-ideal of R and $a \in R$ be fixed. Consider the following collection of fuzzy k-ideals of R

$$\mathscr{B} = \{\mu \mid \mu(a) = \lambda(a),\ \lambda \leq \mu\}.$$

$\mathscr{B}$ is nonempty since $\lambda \in \mathscr{B}$. Let $\mathscr{F}$ be a totally ordered subset of $\mathscr{B}$ containing λ, say $\mathscr{F} = \{\lambda_i \mid i \in I\}$.

We claim that $\bigvee_{i \in I} \lambda_i$ is a fuzzy k-ideal of R.

For any $x, y \in R$, we have

$$\begin{aligned}
\left(\bigvee_{i\in I} \lambda_i\right)(x) \wedge \left(\bigvee_{i\in I} \lambda_i\right)(y) &= \left(\bigvee_{i\in I} \lambda_i(x)\right) \wedge \left(\bigvee_{j\in I} \lambda_j(y)\right) \\
&= \bigvee_{i,j\in I} (\lambda_i(x) \wedge \lambda_j(y)) \\
&\leq \bigvee_{i,j\in I} ((\lambda_i(x) \vee \lambda_j(x)) \wedge (\lambda_i(y) \vee \lambda_j(y))) \\
&\leq \bigvee_{i,j\in I} (\lambda_i(x+y) \vee \lambda_j(x+y)) \\
&\leq \bigvee_{i\in I} \lambda_i(x+y) = \left(\bigvee_{i\in I} \lambda_i\right)(x+y).
\end{aligned}$$

Similarly

$$\left(\bigvee_{i\in I}\lambda_i\right)(x)=\bigvee_{i\in I}\lambda_i(x)\leq\bigvee_{i\in I}\lambda_i(xr)=\left(\bigvee_{i\in I}\lambda_i\right)(xr)$$

and

$$\left(\bigvee_{i\in I}\lambda_i\right)(x)\leq\left(\bigvee_{i\in I}\lambda_i\right)(rx)$$

for all $x,r\in R$. Thus $\bigvee_{i\in I}\lambda_i$ is a fuzzy ideal.

Now, let $x+a=b$, where $a,b\in R$. Then

$$\begin{aligned}\left(\bigvee_{i\in I}\lambda_i\right)(a)\wedge\left(\bigvee_{i\in I}\lambda_i\right)(b)&=\left(\bigvee_{i\in I}\lambda_i(a)\right)\wedge\left(\bigvee_{j\in I}(\lambda_j(b)\right)\\&=\bigvee_{i,j\in I}(\lambda_i(a)\wedge\lambda_j(b))\\&\leq\bigvee_{i,j\in I}(\lambda_i(a)\vee\lambda_j(a))\wedge(\lambda_i(b)\vee\lambda_j(b))\\&\leq\bigvee_{i,j}(\lambda_i(x)\vee\lambda_j(x))\leq\bigvee_{i\in i}\lambda_i(x)=\left(\bigvee_{i\in I}\lambda_i\right)(x).\end{aligned}$$

Hence $\bigvee_{i\in I}\lambda_i$ is a fuzzy k-ideal of R. Clearly $\lambda\leq\bigvee_{i\in I}\lambda_i$ and $(\bigvee_{i\in I}\lambda_i)(a)=\lambda(a)=\alpha$. Thus $\bigvee_{i\in I}\lambda_i$ is the least upper bound of $\mathscr{F}$. Hence by Zorn's lemma there exists a fuzzy k-ideal δ of R which is maximal with respect to the property that $\lambda\leq\delta$ and $\delta(a)=\alpha$.

We will show that δ is an irreducible fuzzy k-ideal of R. Let $\delta=\delta_1\wedge\delta_2$, where δ_1,δ_2 are fuzzy k-ideals of R. Then $\delta\leq\delta_1$ and $\delta\leq\delta_2$. We claim that either $\delta=\delta_1$ or $\delta=\delta_2$. Suppose $\delta\neq\delta_1$ and $\delta\neq\delta_2$. Since δ is maximal with respect to the property that $\delta(a)=\alpha$ and since $\delta\lneqq\delta_1$ and $\delta\lneqq\delta_2$, so $\delta_1(a)\neq\alpha$ and $\delta_2(a)\neq\alpha$. Thus $\alpha=\delta(a)=(\delta_1\wedge\delta_2)(a)=\delta_1(a)\wedge\delta_2(a)\neq\alpha$, which is impossible. Hence $\delta=\delta_1$ or $\delta=\delta_2$. Thus δ is an irreducible fuzzy k-ideal of R. By Theorem 4.20, δ is k-prime.

Theorem 4.21. *Each fuzzy k-ideal of R is idempotent if and only if each fuzzy k-ideal of R is the intersection of those k-prime fuzzy k-ideals of R which contain it.*

Proof. Suppose each fuzzy k-ideal of R is idempotent. Let λ be a fuzzy k-ideal of R and let $\{\lambda_\alpha\,|\,\alpha\in\Lambda\}$ be the family of all k-prime fuzzy k-ideals of R which contain λ. Obviously $\lambda\leq\bigwedge_{\alpha\in\Lambda}\lambda_\alpha$. We now show that $\bigwedge_{\alpha\in\Lambda}\lambda_\alpha\leq\lambda$. Let a be an arbitrary element of R. Then, by Lemma 4.3, there exists an irreducible k-prime fuzzy k-ideal δ such that $\lambda\leq\delta$ and $\lambda(a)=\delta(a)$. Hence $\delta\in\{\lambda_\alpha\,|\,\alpha\in\Lambda\}$ and $\bigwedge_{\alpha\in\Lambda}\lambda_\alpha\leq\delta$. Thus, $\bigwedge_{\alpha\in\Lambda}\lambda_\alpha(a)\leq\delta(a)=\lambda(a)$. Thus $\bigwedge_{\alpha\in\Lambda}\lambda_\alpha\leq\lambda$. Therefore $\bigwedge_{\alpha\in\Lambda}\lambda_\alpha=\lambda$.

Conversely, assume that each fuzzy k-ideal of R is the intersection of those k-prime fuzzy k-ideals of R which contain it. Let λ be a fuzzy k-ideal of R. Then $\lambda \odot_k \lambda$ is also a fuzzy k-ideal of R, so

$\lambda \odot_k \lambda = \bigwedge_{\alpha \in \Lambda} \lambda_\alpha$ where λ_α are k-prime fuzzy k-ideals of R. Thus each λ_α contains $\lambda \odot_k \lambda$, and hence λ. So $\lambda \leq \bigwedge_{\alpha \in \Lambda} \lambda_\alpha = \lambda \odot \lambda$, but $\lambda \odot_k \lambda \leq \lambda$ always. Hence $\lambda = \lambda \odot_k \lambda$.

Theorem 4.22. *Each fuzzy k-ideal of R is idempotent if and only if each fuzzy k-ideal of R is semiprime.*

Proof. For any fuzzy k-ideal λ of R we have $\lambda \odot_h \lambda \leq \lambda$. If each fuzzy k-ideal of R is semiprime, then $\lambda \odot_k \lambda \leq \lambda \odot_k \lambda$ implies $\lambda \leq \lambda \odot_k \lambda$. Hence $\lambda \odot_k \lambda = \lambda$.

The converse is obvious.

Every fuzzy k-prime k-ideal is fuzzy k-semiprime k-ideal but the converse is not true.

Example 4.2. Consider the semiring $R = \{0, a, b, c\}$ defined by the following operation tables:

$+$	0	a	b	c
0	0	a	b	c
a	a	b	c	a
b	b	c	a	b
c	c	a	b	c

$\cdot$	0	a	b	c
0	0	0	0	0
a	0	a	b	c
b	0	b	b	c
c	0	c	c	c

This semiring has two k-ideals $\{0, c\}$ and R. Obviously these k-ideals are idempotent.

For any fuzzy ideal λ of R and any $x \in R$, we have $\lambda(0) \geq \lambda(x) \geq \lambda(a)$. Indeed, $\lambda(0) = \lambda(0x) \geq \lambda(x) = \lambda(xa) \geq \lambda(a)$. This together with $\lambda(a) = \lambda(b+b) \geq \lambda(b) \wedge \lambda(b) = \lambda(b)$ implies $\lambda(a) = \lambda(b)$. Consequently, $\lambda(c) = \lambda(a+b) \geq \lambda(a) \wedge \lambda(b) = \lambda(b)$. Therefore $\lambda(0) \geq \lambda(c) \geq \lambda(b) = \lambda(a)$ for every fuzzy k-ideal of this semiring.

Now we prove that each fuzzy k-ideal of R is idempotent. Since $\lambda \odot_k \lambda \leq \lambda$ always holds, we have to show that $\lambda \odot_k \lambda \geq \lambda$. Obviously, for every $x \in R$ we have

$$\begin{aligned}(\lambda \odot_k \lambda)(x) &= \sup_{x + \sum_{i=1}^{m} a_i b_i = \sum_{j=1}^{n} a'_j b'_j} \left[\bigwedge_{i=1}^{n} (\lambda(a_i) \wedge \lambda(b_i)) \wedge \bigwedge_{j=1}^{m} \left(\lambda(a'_j) \wedge \lambda(b'_j) \right) \right] \\ &\geq \sup_{x+cd=c'd'} [\lambda(c) \wedge \lambda(d) \wedge \lambda(c') \wedge \lambda(d')] \\ &= \lambda(c) \wedge \lambda(d) \wedge \lambda(c') \wedge \lambda(d').\end{aligned}$$

So, $x + cd = c'd'$implies $(\lambda \odot_k \lambda)(x) \geq \lambda(c) \wedge \lambda(d) \wedge \lambda(c') \wedge \lambda(d')$. Hence $0 + 00 = 00$ implies $(\lambda \odot_k \lambda)(0) \geq \lambda(0)$. Similarly $a + bb = bc$ implies $(\lambda \odot_k \lambda)(a) \geq \lambda(b) \wedge \lambda(c) = \lambda(b) = \lambda(a)$, $b + aa = bc$ implies $(\lambda \odot_k \lambda)(b) \geq \lambda(a) \wedge \lambda(b) \wedge \lambda(c) = \lambda(b)$. Analogously, from $c + 00 = cc$ it follows that $(\lambda \odot_k \lambda)(c) \geq \lambda(0) \wedge \lambda(c) = \lambda(c)$. This proves that $(\lambda \odot_k \lambda)(x) \geq \lambda(x)$ for every $x \in R$. Therefore $\lambda \odot_k \lambda = \lambda$ for every fuzzy k-ideal of R, which, by Theorem 4.22, means that each fuzzy k-ideal of R is semiprime.

Consider the following three fuzzy sets:

$$\lambda(0) = \lambda(c) = 0.8,\ \lambda(a) = \lambda(b) = 0.4,$$
$$\mu(0) = \mu(c) = 0.6,\ \mu(a) = \mu(b) = 0.5,$$
$$\delta(0) = \delta(c) = 0.7,\ \delta(a) = \delta(b) = 0.45.$$

These three fuzzy sets are idempotent fuzzy k-ideals. Since all fuzzy k-ideal of this hemiring are idempotent, by Proposition 4.7, we have $\lambda \odot_k \mu = \lambda \wedge \mu$. Thus $(\lambda \odot_k \mu)(0) = (\lambda \odot_k \mu)(c) = 0.6$ and $(\lambda \odot_k \mu)(a) = (\lambda \odot_k \mu)(b) = 0.4$. So, $\lambda \odot_k \mu \leq \delta$ but neither $\lambda \leq \delta$ nor $\mu \leq \delta$, that is δ is not a k-prime fuzzy k-ideal.

4.7 k-Fuzzy Ideals and k-Semirings

Definition 4.6. A semiring R with zero is called a *k-semiring* if for any $a, b \in R$ there exists a unique element c in R such that either $b + c = a$ or $a + c = b$ but not both.

Let R be a k-semiring and R' be a set of the same cardinality with $R - \{0\}$ such that $R \cap R' = \emptyset$. Let us denote the image of $a \in R - \{0\}$ under a given bijection by a'. Let $\oplus$ and $\odot$ denote addition and multiplication respectively on a set $\overline{R} = R \cup R'$ as follows:

$$a \oplus b = \begin{cases} a+b & \text{if } a, b \in R \\ (x+y)' & \text{if } a = x', b = y' \in R' \\ c & \text{if } a \in R, b = y' \in R', a = y + c \\ c' & \text{if } a \in R, b = y' \in R', a + c = y \end{cases}$$

where c is the unique element in R such that either $a = y + c$ or $a + c = y$ but not both, and

$$a \odot b = \begin{cases} ab & \text{if } a, b \in R \\ xy & \text{if } a = x', b = y' \in R' \\ (ay)' & \text{if } a \in R, b = y' \in R' \\ (xb)' & \text{if } a = x' \in R', b \in R \end{cases}$$

It is shown in [36], that these operations are well defined.

Theorem 4.23. *If R is a k-semiring, then $(R', \oplus, \odot)$ is a ring, called the extension ring of R.*

Definition 4.7. A fuzzy ideal λ of a semiring R is called a *k-fuzzy ideal* of R if $\lambda(x+y) = \lambda(0)$ and $\lambda(y) = \lambda(0)$ imply $\lambda(x) = \lambda(0)$.

The following results are taken from [88].

Theorem 4.24. *Let I be a k-ideal of a semiring R. Then the characteristic function χ_I is a k-fuzzy ideal of R.*

Proof. Straightforward.

Theorem 4.25. *Let λ be a fuzzy ideal of a semiring R. If λ_t is a k-ideal of R for each $t\,(\leq \lambda(0))$, then λ is a k-fuzzy ideal of R.*

Proof. Let $x,y \in R$ be such that $\lambda(x+y) = \lambda(0)$ and $\lambda(y) = \lambda(0)$. Then $x+y \in \lambda_0$, $y \in \lambda_0$. Since λ_t is a k-ideal of R, we have λ_0 is a k-ideal of R. Thus $x \in \lambda_0$, that is $\lambda(x) = \lambda(0)$. Hence λ is a k-fuzzy ideal of R.

The following example shows that the converse of the above Theorem is not true.

Example 4.3. Let $R = Z^*$, the set of nonnegative integers. Define a fuzzy subset λ of R by

$$\lambda(x) = \begin{cases} 1 & \text{if } x \in (2), \\ \frac{1}{2} \text{ if } x \in (2,3) - (2), \\ 0 \text{ if } x \in Z^* - (2,3). \end{cases}$$

Then λ is a k-fuzzy ideal but $\lambda_{\frac{1}{2}} = \left\{x \in Z^* \mid \lambda(x) \geq \frac{1}{2}\right\} = (2,3)$ is not a k-ideal of R.

Theorem 4.26. *Let $f\colon R \to R'$ be an epimorphism of semirings. Let μ be a fuzzy ideal of R'. Then μ is a k-fuzzy ideal of R' if and only if $\left(f^{-1}(\mu)\right)(x) = \mu(f(x))$ for all $x \in R$ is a k-fuzzy ideal of R.*

Proof. Suppose μ is a k-fuzzy ideal of R'. Let $x,y \in R$, then

$$\begin{aligned} f^{-1}(\mu)(x+y) &= \mu(f(x+y)) \\ &= \mu(f(x)+f(y)) \\ &\geq \mu(f(x)) \wedge \mu(f(y)) \\ &= f^{-1}(\mu)(x) \wedge f^{-1}(\mu)(y) \end{aligned}$$

and

$$f^{-1}(\mu)(xy) = \mu(f(xy)) = \mu(f(x)f(y)) \geq \mu(f(x)) = f^{-1}(\mu)(x).$$

Similarly

$$f^{-1}(\mu)(xy) = \mu(f(xy)) = \mu(f(x)f(y)) \geq \mu(f(y)) = f^{-1}(\mu)(y).$$

Let

$$f^{-1}(\mu)(x+y) = f^{-1}(\mu)(0) \text{ and } f^{-1}(\mu)(y) = f^{-1}(\mu)(0).$$

Then

$$\mu(f(x+y)) = \mu(f(x)+f(y)) = \mu(0) \text{ and } \mu(f(y)) = \mu(0).$$

Since μ is a k-fuzzy ideal of R', we have $\mu(f(x)) = \mu(0)$. That is, $f^{-1}(\mu)(x) = f^{-1}(\mu)(0)$. Thus $f^{-1}(\mu)$ is a k-fuzzy ideal of R.

Conversely, assume that $f^{-1}(\mu)$ is a k-fuzzy ideal of R. Let $x,y \in R'$. Then there exist $a,b \in R$ such that $f(a) = x$ and $f(b) = y$. Now

$$\begin{aligned}\mu(x+y) = \mu(f(a)+f(b)) = \mu(f(a+b)) &= f^{-1}(\mu)(a+b)\\ &\geq f^{-1}(\mu)(a) \wedge f^{-1}(\mu)(b)\\ &= \mu(f(a)) \wedge \mu(f(b)) = \mu(x) \wedge \mu(y).\end{aligned}$$

$$\begin{aligned}\mu(xy) &= \mu(f(a)f(b)) = \mu(f(ab)) = f^{-1}(\mu)(ab)\\ &\geq f^{-1}(\mu)(a)\\ &= \mu(f(a)) = \mu(x).\end{aligned}$$

Similarly $\mu(xy) \geq \mu(y)$.

If $\mu(x+y) = \mu(0)$ and $\mu(y) = \mu(0)$, then $f^{-1}(\mu)(a+b) = f^{-1}(\mu)(0)$ and $f^{-1}(\mu)(b) = f^{-1}(\mu)(0)$. Since $f^{-1}(\mu)$ is a k-fuzzy ideal of R, we have $f^{-1}(\mu)(a) = f^{-1}(\mu)(0)$. Thus μ is a k-fuzzy ideal of R'.

Definition 4.8. Let $f\colon R \to R'$ be a homomorphism of semirings. Let λ be a fuzzy subset of R. We define a fuzzy subset $f(\lambda)$ of R' by

$$f(\lambda(y)) = \begin{cases} Sup\{\lambda(t) \mid t \in R, f(t) = y\} & \text{if } f^{-1}(y) \neq \emptyset \\ 0 & \text{if } f^{-1}(y) = \emptyset. \end{cases}$$

Definition 4.9. Let R and R' be any sets and let $f\colon R \to R'$ be any function. A fuzzy subset λ of R is called *f-invariant* if $f(x) = f(y)$ implies $\lambda(x) = \lambda(y)$ for all $x, y \in R$.

Theorem 4.27. *Let $f\colon R \to R'$ be an epimorphism of semirings. Let λ be an f-invariant fuzzy ideal of R. Then λ is a k-fuzzy ideal of R if and only if $f(\lambda)$ is a k-fuzzy ideal of R'.*

Proof. Let $x, y \in R'$. Then there exist $a, b \in R$ such that $f(a) = x$ and $f(b) = y$. Then $f(a+b) = x+y$ and $f(ab) = xy$. Since λ is f-invariant,

$$f(\lambda)(x+y) = f(\lambda)f(a+b) = \lambda(a+b) \geq \lambda(a) \wedge \lambda(b)$$

and

$$f(\lambda)(xy) = f(\lambda)f(ab) = \lambda(ab) \geq \lambda(a) \vee \lambda(b).$$

Hence $f(\lambda)$ is a fuzzy ideal of R'.

Let $f(\lambda)(x+y) = f(\lambda)(0)$ and $f(\lambda)(y) = f(\lambda)(0)$. Then $f(\lambda)(x+y) = f(\lambda)f(a+b) = \lambda(a+b)$ and $f(\lambda)(y) = f(\lambda)f(b) = \lambda(b)$, since λ is f-invariant. Thus $\lambda(a+b) = \lambda(0)$ and $\lambda(b) = \lambda(0)$, since $f(\lambda)(0) = \lambda(0)$. Since λ is a k-fuzzy ideal of R, $\lambda(a) = \lambda(0)$. Thus $f(\lambda)(x) = f(\lambda)(f(a)) = f(\lambda)(0)$. Hence $f(\lambda)$ is a k-fuzzy ideal of R'.

Conversely, if $f(\lambda)$ is a k-fuzzy ideal, then for any $x \in R$
$f^{-1}(f(\lambda))(x) = f(\lambda)(f(x)) = \sup\{\lambda(t) : t \in R, f(t) = f(x)\}$
$= \sup\{\lambda(t) : t \in R, \lambda(t) = \lambda(x)\} = \lambda(x)$.
So $f^{-1}(f(\lambda)) = \lambda$. Thus λ is a k-fuzzy ideal by Theorem 4.26.

Let R be a commutative k-semiring, and $\overline{R}$ its extension ring. Let λ be a fuzzy ideal of R such that all its level subsets are k-ideals of R. Then $R = \underset{t \in Im\lambda}{\cup} \lambda_t$, $\overline{R} = \underset{t \in Im\lambda}{\cup} \overline{\lambda}_t$ and $s > t$ if and only if $\lambda_s \subset \lambda_t$ if and only if $\overline{\lambda}_s < \overline{\lambda}_t$.

The following results are taken from 89.

Theorem 4.28. *Let R be a commutative k-semiring, $\overline{R}$ its extension ring. Let λ be a fuzzy ideal of R such that all its level subsets are k-ideals of R. Define the fuzzy subset $\overline{\lambda}$ of $\overline{R}$ for all $x \in \overline{R}$, $\overline{\lambda}(x) = \sup\{t : x \in \overline{\lambda_t}, t \in Im\lambda\}$. Then $\overline{\lambda}$ is a fuzzy ideal of $\overline{R}$.*

Theorem 4.29. *Let $f : R \to S$ be an epimorphism of k-semirings R and S, and λ as in Theorem 4.28, and $\lambda_R \subseteq Ker f$. Then there exists a unique epimorphism ϕ from R/A onto S such that $f = \phi \circ g$ where $g(x) = x + \lambda$ for any $x \in R$.*

Where $x+\lambda$ is defined as $x+\lambda : R \to [0,1]$, $(x+\lambda)(z) = \overline{\lambda}(z \oplus x)$.

Proof. Define a map $\phi : R/A \to S$ by $\phi(x+\lambda) = f(x)$ for each $x \in R$. Then ϕ is well defined, since $x+\lambda = y+\lambda \Leftrightarrow \overline{\lambda}(x \oplus y') = \lambda(0) = \overline{\lambda}(0) \Leftrightarrow x \oplus y' \in \overline{\lambda}_{\overline{R}}$.

Since $\overline{\lambda}_{\overline{R}} = \overline{\lambda_R} \subseteq \overline{\ker f} = \ker \overline{f}$,

$\overline{f}(x \oplus y') = 0 \Rightarrow \overline{f}(x) = \overline{f}(y) \Rightarrow f(x) = f(y) \Rightarrow \phi(x+\lambda) = \phi(y+\lambda)$.

Further, since f is onto, ϕ is also onto. A routine computation will establish that ϕ is a homomorphism. On the other hand, $f(x) = \phi(x+\lambda) = \phi(g(x)\lambda) = (\phi \circ g)(x)$ for all $x \in R$. Finally we show that ϕ is unique. For this, suppose that $f = h \circ g$ for some $h : R/A \to S$. Then $\phi(x+\lambda) = f(x) = (h \circ g)(x) = h(x+\lambda)$ for all $x \in R$.

Corollary 4.1. *The induced homomorphism ϕ is an isomorphism if and only if λ is f-invariant.*

Proof. Assume that ϕ is an isomorphism and let $f(x) = f(y)$. Then

$\phi(x+\lambda) = \phi(y+\lambda) \Rightarrow x+\lambda = y+\lambda \Rightarrow (x+\lambda)(0) = (y+\lambda)(0)$

$\Rightarrow \overline{\lambda}(x \oplus 0') = \overline{\lambda}(y \oplus 0') \Rightarrow \lambda(x) = \lambda(y)$.

Thus λ is f-invariant.

Conversely, assume that λ is f-invariant and let $\phi(x+\lambda) = \phi(0+\lambda) = 0$. Then $f(x) = f(0)$ and so $\lambda(x) = \lambda(0)$, since λ is f-invariant. Thus $x+\lambda = 0+\lambda$ and thus ϕ is $1-1$. Hence ϕ is an isomorphism.

4.8 Fuzzy Congruences

In this section, we give the definition of fuzzy congruence on a semiring and study the relation between fuzzy congruences and fuzzy ideals of a semiring. We also study the quotient semiring of a semiring over a fuzzy congruence.

Definition 4.10. Let R be a semiring. A fuzzy subset ρ of $R \times R$ is called a *fuzzy relation* on R.

A fuzzy relation ρ on R is called a *fuzzy equivalence relation* if it satisfies the following conditions:

(1) $\rho(x,x) = \sup_{y,z \in R} \rho(y,z)$ for all $x \in R$. (fuzzy reflexive)

(2) $\rho(x,y) = \rho(y,x)$ for all $x,y \in R$. (fuzzy symmetric)

(3) $\rho(x,z) \geq \sup_{y \in R} \{\rho(x,y) \wedge \rho(y,z)\}$ for all $x,z \in R$. (fuzzy transitive)

A fuzzy equivalence relation ρ on a semiring R is called a *fuzzy congruence* if

$$\rho(x+y,z+t) \geq \rho(x,z) \wedge \rho(y,t)$$
$$\text{and } \rho(xy,zt) \geq \rho(x,z) \wedge \rho(y,t)$$

for all $x,y,z,t \in R$.

Example 4.4. Let $\mathbb{N}$ be the set of all nonnegative integers. Then $\mathbb{N}$ is a semiring with respect to the usual addition and multiplication of numbers. The fuzzy relation ρ on $\mathbb{N}$ defined by

$$\rho(x,y) = \begin{cases} 1 & \text{if } x = y \\ 0.5 & \text{if } x \neq y \text{ and both } x,y \text{ are either even or odd} \\ 0 & \text{otherwise} \end{cases}$$

is a fuzzy congruence on R.

Theorem 4.30. *Let σ be a nonempty relation on a semiring R and $t,r \in [0,1]$ be such that $t \geq r$. Define a fuzzy subset ρ of $R \times R$ by*

$$\rho(x,y) = \begin{cases} t & \text{if } (x,y) \in \sigma \\ r & \text{otherwise} \end{cases}$$

Then σ is a congruence on R if and only if ρ is a fuzzy congruence on R.

Proof. Let σ be a congruence on R. Then $\rho(x,x) = t = \sup_{y,z \in R} \rho(y,z)$ for all $x \in R$. If $(x,y) \in \sigma$, then $(y,x) \in \sigma$. So $\rho(x,y) = t = \rho(y,x)$. If $(x,y) \notin \sigma$, then $(y,x) \notin \sigma$. So $\rho(x,y) = r = \rho(y,x)$. Obviously $\rho(x,z) \geq \sup_{y \in R}\{\rho(x,y) \wedge \rho(y,z)\}$. Also $\rho(x+y,z+t) \geq \rho(x,z) \wedge \rho(y,t)$ and $\rho(xy,zt) \geq \rho(x,z) \wedge \rho(y,t)$ for all $x,y,z,t \in R$.

Conversely, assume that ρ is a fuzzy congruence on R. Since $\sigma \neq \emptyset$, $\rho(x,x) = \sup_{y,z \in R} \rho(y,z) = t$, that is $(x,x) \in \sigma$ for all $x \in R$. If $(x,y) \in \sigma$, then $\rho(y,x) = \rho(x,y) = t$. This implies that $(y,x) \in \sigma$. If $(x,y),(y,z) \in \sigma$, then $\rho(x,z) \geq \sup_{y \in R}\{\rho(x,y) \wedge \rho(y,z)\} = t$. Thus $\rho(x,z) = t$, that is $(x,z) \in \sigma$. Hence σ is an equivalence relation. If (x,y) and $(a,b) \in \sigma$, then $\rho(x,y) = t$ and $\rho(a,b) = t$. Since $\rho(x+a,y+b) \geq \rho(x,y) \wedge \rho(a,b)$ and $\rho(xa,yb) \geq \rho(x,y) \wedge \rho(a,b)$ for all $x,y,a,b \in R$, we have $\rho(x+a,y+b) = t$ and $\rho(xa,yb) = t$, that is $(x+a,y+b)$ and (xa,yb) are in σ. Hence σ is a congruence relation on R.

Corollary 4.2. *Let σ be a nonempty relation on a semiring R. Then σ is a congruence on R if and only if χ_σ, the characteristic function of σ, is a fuzzy congruence on R.*

Theorem 4.31. *A fuzzy relation ρ on a semiring R is a fuzzy congruence on R if and only if $\rho_t \neq \emptyset$ is a congruence on R for all $t \in [0,1]$.*

Proof. Let ρ be a fuzzy congruence on R and $x \in R$. Then $\rho(x,x) = \sup_{y,z \in R} \rho(y,z)$, so $(x,x) \in \rho_t$, if $\rho_t \neq \emptyset$. If $(x,y) \in \rho_t$, then $\rho(x,y) \geq t$. Since $\rho(x,y) = \rho(y,x)$, we have $\rho(y,x) \geq t$. Hence $(y,x) \in \rho_t$. If $(x,z), (z,y) \in \rho_t$, then $\rho(x,z) \geq t$ and $\rho(z,y) \geq t$. Since $\rho(x,y) \geq \sup_{z \in R} \{\rho(x,z) \wedge \rho(z,y)\}$, we have $\rho(x,y) \geq t \wedge t = t$. Thus $(x,y) \in \rho_t$. This shows that ρ_t is an equivalence relation on R. Now suppose $(x,y), (a,b) \in \rho_t$. Then $\rho(x,y) \geq t$ and $\rho(a,b) \geq t$. Since $\rho(x+a, y+b) \geq \rho(x,y) \wedge \rho(a,b)$ and $\rho(xa, yb) \geq \rho(x,y) \wedge \rho(a,b)$, we have $\rho(x+a, y+b) \geq t$ and $\rho(xa, yb) \geq t$. This implies $(x+a, y+b)$ and (xa, yb) are in ρ_t. Hence ρ_t is a congruence on R.

Conversely, assume that $\rho_t \neq \emptyset$ is a congruence on R for all $t \in [0,1]$. Let $x \in R$ be such that $\rho(x,x) \neq \sup_{y,z \in R} \rho(y,z)$. If $\rho(x,x) < \sup_{y,z \in R} \rho(y,z)$, then there exist $t \in [0,1]$ such that $\rho(x,x) < t \leq \sup_{y,z \in R} \rho(y,z)$. This implies $\rho_t \neq \emptyset$ and $(x,x) \notin \rho_t$, which is a contradiction. As $\rho(x,x) > \sup_{y,z \in R} \rho(y,z)$ is not possible, $\rho(x,x) = \sup_{y,z \in R} \rho(y,z)$.

Let $x, y \in R$ be such that $\rho(x,y) \neq \rho(y,x)$. Suppose $\rho(x,y) < \rho(y,x)$. Then there exist $t \in [0,1]$ such that $\rho(x,y) < t \leq \rho(y,x)$. This implies $(y,x) \in \rho_t$ but $(x,y) \notin \rho_t$, which is a contradiction. Hence $\rho(x,y) = \rho(y,x)$.

If there exist $x, y, z \in R$ such that $\rho(x,z) < \rho(x,y) \wedge \rho(y,z)$, then we can select $t \in [0,1]$ such that $\rho(x,z) < t \leq \rho(x,y) \wedge \rho(y,z)$. This implies that $(x,y), (y,z) \in \rho_t$ but $(x,z) \notin \rho_t$, which is a contradiction. Hence $\rho(x,z) \geq \sup_{y \in R} \{\rho(x,y) \wedge \rho(y,z)\}$ for all $x, z \in R$.

Now let $x, y, a, b \in R$ be such that $\rho(x+a, y+b) < \rho(x,y) \wedge \rho(a,b)$. Then we can find $t \in [0,1]$ such that $\rho(x+a, y+b) < t \leq \rho(x,y) \wedge \rho(a,b)$. This implies that $(x,y), (a,b) \in \rho_t$ but $(x+a, y+b) \notin \rho_t$. However, this is a contradiction. Hence $\rho(x+a, y+b) \geq \rho(x,y) \wedge \rho(a,b)$. Similarly we can show that $\rho(xa, yb) \geq \rho(x,y) \wedge \rho(a,b)$. Thus ρ is a fuzzy congruence on R.

Theorem 4.32. *Let ρ be a fuzzy congruence on a semiring R. Define a fuzzy subset λ_ρ of R as follows:*

$$\lambda_\rho(x) = \rho(x,0) \text{ for all } x \in R.$$

Then λ_ρ is a fuzzy k-ideal of R.

Proof. Let $x, y \in R$. Then

$$\lambda_\rho(x+y) = \rho(x+y, 0) \geq \rho(x,0) \wedge \rho(y,0) = \lambda_\rho(x) \wedge \lambda_\rho(y)$$

and

$$\lambda_\rho(xy) = \rho(xy, 0) \geq \rho(x,x) \wedge \rho(y,0) = \rho(y,0) = \lambda_\rho(y)$$
$$\lambda_\rho(xy) = \rho(xy, 0) \geq \rho(x,0) \wedge \rho(y,y) = \rho(x,0) = \lambda_\rho(x).$$

Let $a,b,c \in R$ be such that $a+b=c$. Then

$$\begin{aligned}\lambda_\rho(a) &= \rho(a,0)\\ &\geq \sup_{y\in R}\{\rho(a,y)\wedge\rho(y,0)\}\\ &\geq \rho(a,c)\wedge\rho(c,0)\\ &\geq \{\rho(a,a)\wedge\rho(0,b)\}\wedge\rho(c,0)\\ &= \rho(0,b)\wedge\rho(c,0) \quad (\text{ because } a+b=c)\\ &= \rho(b,0)\wedge\rho(c,0) = \lambda_\rho(b)\wedge\lambda_\rho(c).\end{aligned}$$

Hence λ_ρ is a fuzzy k-ideal of R.

Theorem 4.33. *Let λ be a fuzzy ideal of a semiring R. Define a fuzzy relation ρ_λ on R as follows*

$$\rho_\lambda(x,y) = \sup_{x+a=y+b}\{\lambda(a)\wedge\lambda(b)\}$$

for all $x,y,a,b \in R$. Then ρ_λ is a fuzzy congruence on R.

Proof. Let λ be a fuzzy ideal of R and $a \in R$. Then
$\rho_\lambda(a,a) = \sup_{a+x=a+y}\{\lambda(x)\wedge\lambda(y)\} \geq \lambda(0)\wedge\lambda(0)$
$\geq \lambda(x)\wedge\lambda(y)$ for any $x,y \in R$, since $\lambda(0) \geq \lambda(x)$.
Again $\rho_\lambda(x,y) = \sup_{x+a=y+b}\{\lambda(a)\wedge\lambda(b)\}$, so $\rho_\lambda(a,a) \geq \rho_\lambda(x,y)$ for any $x,y \in R$.
Hence $\rho_\lambda(a,a) = \sup_{x,y\in R}\rho_\lambda(x,y)$. Therefore ρ_λ is fuzzy reflexive. Obviously ρ_λ is fuzzy symmetric. Now

$$\begin{aligned}\rho_\lambda(x,y) &= \sup_{x+a=y+b}\{\lambda(a)\wedge\lambda(b)\}\\ &\geq \sup_{x+a=z+c}\sup_{z+c=y+b}[\{\lambda(a)\wedge\lambda(c)\}\wedge\{\lambda(c)\wedge\lambda(b)\}]\\ &= \min\left[\sup_{x+a=z+c}[\lambda(a)\wedge\lambda(c)],\ \sup_{z+c=y+b}[\lambda(c)\wedge\lambda(b)]\right]\\ &= \min[\rho_\lambda(x,z),\rho_\lambda(z,y)].\end{aligned}$$

Hence $\rho_\lambda(x,y) \geq \sup[\min[\rho_\lambda(x,z),\rho_\lambda(z,y)]]$. Thus ρ_λ is a fuzzy equivalence relation on R.

Similarly, we can show that ρ_λ is a fuzzy congruence on R.

Theorem 4.34. *Let ρ be a fuzzy congruence on a semiring R and $a \in R$. Define a fuzzy subset ρ_a of R by $\rho_a(x) = \rho(a,x)$ for all $x \in R$. Then $\rho_a = \rho_b$ if and only if $\rho(a,b) = \rho(b,a) = \rho(a,a) = \rho(b,b)$ for all $a,b \in R$.*

Proof. Suppose $\rho_a = \rho_b$. Then $\rho_a(x) = \rho_b(x)$ for all $x \in R$. In particular $\rho_a(a) = \rho_b(a)$. This implies $\rho(a,a) = \rho(b,a)$. Similarly $\rho(b,b) = \rho(a,b)$. Hence $\rho(a,b) = \rho(b,a) = \rho(a,a) = \rho(b,b)$.

Conversely, assume that $\rho(a,b) = \rho(b,a) = \rho(a,a) = \rho(b,b)$.
Now

$$\begin{aligned}\rho(a,x) &\geq \sup_{y\in R}\{\rho(a,y)\wedge\rho(y,x)\} \\ &\geq \rho(a,b)\wedge\rho(b,x) = \rho(b,x) \quad \left(\text{because } \rho(a,b) = \rho(a,a) = \sup_{y,z\in R}\rho(y,z)\right).\end{aligned}$$

Similarly, $\rho(b,x) \geq \rho(a,x)$. Hence $\rho(b,x) = \rho(a,x)$, that is $\rho_a = \rho_b$.

Definition 4.11. Let ρ be a fuzzy congruence on a semiring R and $a \in R$. Then the fuzzy subset ρ_a of R defined by $\rho_a(x) = \rho(a,x)$ for all $x \in R$ is called a *fuzzy congruence class*.

Theorem 4.35. *Let ρ be a fuzzy congruence on a semiring R. The set R/ρ of all fuzzy congruence classes is a semiring with respect to the binary operations defined by*

$$\rho_a + \rho_b = \rho_{a+b} \text{ and } \rho_a\rho_b = \rho_{ab}$$

for all $a,b \in R$.

Proof. First, we show that the given binary operations are well defined. Let $a,b,c,d \in R$ be such that $\rho_a = \rho_b$ and $\rho_c = \rho_d$. Then by Theorem 4.34, we have $\rho(a,b) = \rho(a,a)$ and $\rho(c,d) = \rho(a,a)$. Since $\rho(a+c,b+d) \geq \rho(a,b)\wedge\rho(c,d) = \rho(a,a)$, we have $\rho(a+c,b+d) = \rho(a,a)$. This implies $\rho_{a+c} = \rho_{b+d}$. Similarly $\rho_{ac} = \rho_{bd}$. Hence the binary operations are well defined. Now it is a routine matter to verify that the set R/ρ of all fuzzy congruence classes is a semiring with respect to the binary operations defined above.

Chapter 5
Fuzzy Quasi-ideals and Fuzzy Bi-ideals in Semirings

The object of this chapter is to study fuzzy quasi-ideals and fuzzy bi-ideals and Section 1 provides a study of these ideals. Section 2 presents various characterizations of regular semirings involving these fuzzy ideals. In Section 3, we examine and characterize regular and intra regular semirings in this context. Section 4 provides a study of fuzzy k-quasi-ideals and fuzzy k-bi-ideals of semirings and Section 5 contains characterizations of k-regular semirings by the ideals studied in Section 4. Section 6 contains characterizations of k-intra regular semirings by the ideals examined in the preceding section.

Throughout this chapter R will denote a semiring with zero.

5.1 Fuzzy Quasi-ideals and Bi-ideals

Recall that a nonempty subset Q of a semiring R is called a *quasi-ideal* of R if $(Q,+)$ is a subsemigroup of $(R,+)$ and $RQ \cap QR \subseteq Q$.

It is clear that every one-sided ideal of R is a quasi-ideal of R and every quasi-ideal of R is a subsemiring of R but the converse is not true.

Example 5.1. Let $R = \left\{ \begin{bmatrix} a & b \\ c & d \end{bmatrix} : a,b,c,d \text{ are non negative integers} \right\}$. Then R is a semiring under the usual addition and multiplication of matrices.

Let $Q = \left\{ \begin{bmatrix} x & 0 \\ 0 & 0 \end{bmatrix} : \text{x is a non negative integer} \right\}$. Then Q is a quasi-ideal of R but not an ideal of R.

A nonempty subset B of a semiring R is called a *bi-ideal* of R if B is closed under addition and multiplication and $BRB \subseteq B$.

Every quasi-ideal of a semiring is a bi-ideal but the converse is not true.

Example 5.2. Let $\mathbb{N}$ and $\mathbb{R}^+$ denote the sets of all positive integers and positive real numbers, respectively. Then the set R of all matrices of the form $\begin{pmatrix} a & 0 \\ b & c \end{pmatrix}$ $(a,b \in \mathbb{R}^+, c \in \mathbb{N})$ together with $\begin{pmatrix} 0 & 0 \\ 0 & 0 \end{pmatrix}$ is a semiring with respect to the usual addition and multiplication of matrices. Let A,B be the sets of all

J. Ahsan et al.: Fuzzy Semirings with Applications, STUDFUZZ 278, pp. 83–103.
springerlink.com

matrices $\begin{pmatrix} a & 0 \\ b & c \end{pmatrix}$ $(a,b \in \mathbb{R}^+, c \in \mathbb{N}, a < b)$ together with $\begin{pmatrix} 0 & 0 \\ 0 & 0 \end{pmatrix}$ and $\begin{pmatrix} p & 0 \\ q & d \end{pmatrix}$ $(p,q \in \mathbb{R}^+, d \in \mathbb{N}, 3 < q)$ together with $\begin{pmatrix} 0 & 0 \\ 0 & 0 \end{pmatrix}$, respectively. Then A and B are right and left ideals of R, respectively. Now the product AB is a bi-ideal of R but it is not a quasi-ideal of R. Indeed, the element

$$\begin{pmatrix} 6 & 0 \\ 9 & 1 \end{pmatrix} = \begin{pmatrix} 6 & 0 \\ 3 & 1 \end{pmatrix} \left(\begin{pmatrix} 1 & 0 \\ 2 & 1 \end{pmatrix} \begin{pmatrix} 1 & 0 \\ 4 & 1 \end{pmatrix} \right)$$
$$= \left(\begin{pmatrix} 1 & 0 \\ \frac{7}{6} & 1 \end{pmatrix} \begin{pmatrix} 24 & 0 \\ 4 & 1 \end{pmatrix} \right) \begin{pmatrix} \frac{1}{4} & 0 \\ 1 & 1 \end{pmatrix}$$

belongs to the intersection $R(AB) \cap (AB)R$, but it is not an element of AB.

This example also shows that product of quasi-ideals need not be a quasi-ideal.

Definition 5.1. A fuzzy subset λ of a semiring R is called a fuzzy *quasi-ideal* of R if

(i) $\lambda(x+y) \geq \lambda(x) \wedge \lambda(y)$ for all $x, y \in R$.
(ii) $\chi_R \circ \lambda \wedge \lambda \circ \chi_R \leq \lambda$.

Obviously every fuzzy quasi-ideal of R is a fuzzy subsemiring of R and every fuzzy left (right) ideal of R is a fuzzy quasi-ideal of R.

Theorem 5.1. *A fuzzy subset λ of a semiring R is a fuzzy quasi-ideal of R if and only if each nonempty level subset $U(\lambda;t)$ of λ is a quasi-ideal of R.*

Proof. Suppose λ is a fuzzy quasi-ideal of R and $t \in (0,1]$ be such that $U(\lambda;t) \neq \emptyset$. Let $a, b \in U(\lambda;t)$. Then $\lambda(a) \geq t$ and $\lambda(b) \geq t$. As $\lambda(a+b) \geq \lambda(a) \wedge \lambda(b)$, so $\lambda(a+b) \geq t$. Hence $a+b \in U(\lambda;t)$.

Let $x \in U(\lambda;t)R \cap RU(\lambda;t)$. Then $x = \sum_{i=1}^{m} u_i r_i$ and $x = \sum_{k=1}^{p} s_k v_k$ for some $u_i, v_k \in U(\lambda;t)$ and $r_i, s_k \in R$. Now

$$\begin{aligned}
\lambda(x) &\geq [(\chi_R \circ \lambda) \wedge (\lambda \circ \chi_R)](x) \\
&= (\chi_R \circ \lambda)(x) \wedge (\lambda \circ \chi_R)(x) \\
&= \bigvee_{x = \sum_{k=1}^{p} s_k v_k} \left(\bigwedge_{k=1}^{p} [\chi_R(s_k) \wedge \lambda(v_k)] \right) \wedge \bigvee_{x = \sum_{i=1}^{m} u_i r_i} \left(\bigwedge_{i=1}^{m} [\lambda(u_i) \wedge \chi_R(r_i)] \right) \\
&\geq t \wedge t = t.
\end{aligned}$$

So, $\lambda(x) \geq t$. Thus, $x \in U(\lambda;t)$. Hence $U(\lambda;t)R \cap RU(\lambda;t) \subseteq U(\lambda;t)$.

Conversely, assume that each nonempty subset $U(\lambda;t)$ of R is a quasi-ideal of R. Let $a, b \in R$ be such that $\lambda(a+b) < \lambda(a) \wedge \lambda(b)$. Take $t \in (0,1]$ such that $\lambda(a+b) < t \leq \lambda(a) \wedge \lambda(b)$. Then $a, b \in U(\lambda;t)$ but $a+b \notin U(\lambda;t)$, a contradiction. Hence $\lambda(a+b) \geq \lambda(a) \wedge \lambda(b)$.

Let $x \in R$. If possible let $\lambda(x) < [(\lambda \circ \chi_R) \wedge (\chi_R \circ \lambda)](x)$. Take $t \in (0,1]$ such that $\lambda(x) < t \leq [(\lambda \circ \chi_R) \wedge (\chi_R \circ \lambda)](x)$. If $[(\lambda \circ \chi_R) \wedge (\chi_R \circ \lambda)](x) \geq t$, then

$$(\chi_R \circ \lambda)(x) \wedge (\lambda \circ \chi_R)(x) = \left[\bigvee_{x=\sum_{k=1}^{p} s_k v_k} \left(\bigwedge_{k=1}^{p} [\chi_R(s_k) \wedge \lambda(v_k)] \right) \right] \wedge \left[\bigvee_{x=\sum_{i=1}^{m} u_i r_i} \left(\bigwedge_{i=1}^{m} [\lambda(u_i) \wedge \chi_R(r_i)] \right) \right]$$

Hence $\bigvee_{x=\sum_{k=1}^{p} s_k v_k} \left(\bigwedge_{k=1}^{p} [\chi_R(s_k) \wedge \lambda(v_k)] \right) \geq t$

and $\bigvee_{x=\sum_{i=1}^{m} u_i r_i} \left(\bigwedge_{i=1}^{m} [\lambda(u_i) \wedge \chi_R(r_i)] \right) \geq t$, so, $\lambda(u_i) \geq t, \lambda(v_k) \geq t$, that is, $u_i, v_k \in U(\lambda;t)$.

Thus $\sum_{i=1}^{m} u_i r_i \in U(\lambda;t)R$ and $\sum_{k=1}^{p} s_k v_k \in RU(\lambda;t)$. This implies $x \in U(\lambda;t)R \cap RU(\lambda;t) \subseteq U(\lambda;t)$, and hence $x \in U(\lambda;t)$, that is $\lambda(x) \geq t$, a contradiction. Hence $(\lambda \circ \chi_R) \wedge (\chi_R \circ \lambda) \leq \lambda$. Thus λ is a fuzzy quasi-ideal of R.

Corollary 5.1. *Let Q be a nonempty subset of a semiring R. Then Q is a quasi-ideal of R if and only if the characteristic function χ_Q of Q is a fuzzy quasi-ideal of R.*

Proposition 5.1. *The intersection of any two fuzzy quasi-ideals of a semiring R is a fuzzy quasi-ideal of R.*

Proof. Let μ, ν be fuzzy quasi-ideals of a semiring R and $x, y \in R$. Then

$$\begin{aligned}(\mu \wedge \nu)(x+y) &= \mu(x+y) \wedge \nu(x+y) \geq [\mu(x) \wedge \mu(y)] \wedge [\nu(x) \wedge \nu(y)] \\ &= [\mu(x) \wedge \nu(x)] \wedge [\mu(y) \wedge \nu(y)] = (\mu \wedge \nu)(x) \wedge (\mu \wedge \nu)(y).\end{aligned}$$

Also,

$$\begin{aligned}((\mu \wedge \nu) \circ \chi_R) \wedge (\chi_R \circ (\mu \wedge \nu)) &\leq (\mu \circ \chi_R) \wedge (\chi_R \circ \mu) \leq \mu \\ (\chi_R \circ (\mu \wedge \nu)) \wedge (\chi_R \circ (\mu \wedge \nu)) &\leq (\nu \circ \chi_R) \wedge (\chi_R \circ \nu) \leq \nu.\end{aligned}$$

Thus $((\mu \wedge \nu) \circ \chi_R) \wedge (\chi_R \circ (\mu \wedge \nu)) \leq \mu \wedge \nu$.

This completes the proof.

Corollary 5.2. *Let μ and ν be fuzzy right and fuzzy left ideals of a semiring R, respectively. Then $\mu \wedge \nu$ is a fuzzy quasi-ideal of R.*

Definition 5.2. A fuzzy subset λ of a semiring R is called a *fuzzy bi-ideal* of R if for all $x,y,z \in R$

1. $\lambda(x+y) \geq \lambda(x) \wedge \lambda(y)$
2. $\lambda(xy) \geq \lambda(x) \wedge \lambda(y)$
3. $\lambda(xyz) \geq \lambda(x) \wedge \lambda(z)$.

Theorem 5.2. *A fuzzy subset μ of a semiring R is a fuzzy bi-ideal of R if and only if*

(i) $\mu + \mu \leq \mu$
(ii) $\mu \circ \mu \leq \mu$
(iii) $\mu \circ \chi_R \circ \mu \leq \mu$.

Proof. Let μ be a fuzzy bi-ideal of R and $x \in R$. Then $(\mu+\mu)(x) = \bigvee\limits_{x=y+z} \{\mu(y) \wedge \mu(z)\} \leq \bigvee\limits_{x=y+z} \mu(y+z) = \mu(x)$.

Let $x \in R$. If $(\mu \circ \mu)(x) = 0$, then $\mu \circ \mu \leq \mu$. Otherwise, there exist $a_i, b_i \in R$ such that $x = \sum\limits_{i=1}^{m} a_i b_i$. Then we have

$$\begin{aligned}
(\mu \circ \mu)(x) &= \bigvee_{x=\sum_{i=1}^{m} a_i b_i} \left[\bigwedge_{i=1}^{m} [\mu(a_i) \wedge \mu(b_i)] \right] \\
&\leq \bigvee_{x=\sum_{i=1}^{m} a_i b_i} \left[\bigwedge_{i=1}^{m} [\mu(a_i b_i)] \right] \\
&\leq \bigvee_{x=\sum_{i=1}^{m} a_i b_i} \left[\mu\Big(\sum_{i=1}^{m} a_i b_i\Big) \right] \\
&\leq \mu(x).
\end{aligned}$$

Hence $\mu \circ \mu \leq \mu$.

Let $x \in R$. If $(\mu \circ \chi_R \circ \mu)(x) = 0$, then $\mu \circ \chi_R \circ \mu \leq \mu$. Otherwise, there exist $a_i, b_i \in R$ such that $x = \sum\limits_{i=1}^{m} a_i b_i$. Then we have

$$
\begin{aligned}
(\mu \circ \chi_R \circ \mu)(x) &= \bigvee_{x=\sum_{i=1}^{m} a_i b_i} \left[\bigwedge_{i=1}^{m} [(\mu \circ \chi_R)(a_i) \wedge \mu(b_i)] \right] \\
&= \bigvee_{x=\sum_{i=1}^{m} a_i b_i} \left(\bigwedge_{i=1}^{m} \left[\bigvee_{a_i=\sum_{k=1}^{p} c_k d_k} \left\{ \bigwedge_{k=1}^{p} \mu(c_k) \wedge \bigwedge_{q=1}^{r} \chi_R(d_q) \right\} \wedge \mu(b_i) \right] \right) \\
&= \bigvee_{x=\sum_{i=1}^{m} a_i b_i} \bigwedge_{i=1}^{m} \left[\bigvee_{a_i=\sum_{k=1}^{p} c_k d_k} \left\{ \bigwedge_{k-1}^{p} [\mu(c_k) \wedge \mu(b_i)] \right\} \right]
\end{aligned}
$$

$$
\begin{aligned}
(\mu \circ \chi_R \circ \mu)(x) &\leq \bigvee_{x=\sum_{i=1}^{m} a_i b_i} \bigwedge_{i=1}^{m} \left[\bigvee_{a_i=\sum_{k=1}^{p} c_k d_k} \mu \left(\sum_{k=1}^{p} c_k d_k b_i \right) \right] \\
&= \bigvee_{x=\sum_{i=1}^{m} a_i b_i} \bigwedge_{i=1}^{m} \mu(a_i b_i) \\
&\leq \bigvee_{x=\sum_{i=1}^{m} a_i b_i} \mu \left(\sum_{i=1}^{m} a_i b_i \right) \\
&= \mu(x).
\end{aligned}
$$

Hence,

$$(\mu \circ \chi_R \circ \mu)(x) \leq \mu(x).$$

Conversely, assume that μ satisfies (i), (ii), and (iii). Let $x, y, z \in R$. Then

$$
\begin{aligned}
\mu(x+y) &\geq (\mu + \mu)(x+y) \\
&= \bigvee_{x+y=a+b} \{\mu(a) \wedge \mu(b)\} \\
&\geq \mu(x) \wedge \mu(y).
\end{aligned}
$$

Also,

$$\begin{aligned}\mu(xy) &\geq (\mu \circ \mu)(xy) \\ &= \bigvee_{xy=\sum_{i=1}^{m} a_i b_i} \left[\bigwedge_{i=1}^{m} [\mu(a_i) \wedge \mu(b_i)] \right] \\ &\geq \mu(x) \wedge \mu(y).\end{aligned}$$

Similarly,

$$\mu(xyz) \geq (\mu(x) \wedge \mu(z).$$

Hence μ is a fuzzy bi-ideal of R.

From the above Theorem it follows that every fuzzy quasi-ideal of a semiring is a fuzzy bi-ideal of R.

Theorem 5.3. *A fuzzy subset λ of a semiring R is a fuzzy bi-ideal of R if and only if each nonempty level subset $U(\lambda;t)$ of λ is a bi-ideal of R.*

Proof. Suppose λ is a fuzzy bi-ideal of R and $t \in (0,1]$ be such that $U(\lambda;t) \neq \emptyset$. Let $a,b \in U(\lambda;t)$. Then $\lambda(a) \geq t$ and $\lambda(b) \geq t$. As $\lambda(a+b) \geq \lambda(a) \wedge \lambda(b)$, so $\lambda(a+b) \geq t$. Hence $a+b \in U(\lambda;t)$. Also, $\lambda(ab) \geq \lambda(a) \wedge \lambda(b)$ so $\lambda(ab) \geq t$. This implies that $ab \in U(\lambda;t)$.

Let $x \in U(\lambda;t)RU(\lambda;t)$. Then $x = \sum_{i=1}^{m} a_i r_i b_i$ for some $a_i, b_i \in U(\lambda;t)$ and $r_i \in R$. Since λ is a fuzzy bi-ideal of R, we have $\lambda(a_i r_i b_i) \geq \lambda(a_i) \wedge \lambda(b_i) \geq t$. As $\lambda\left(\sum_{i=1}^{m} a_i r_i b_i\right) \geq \bigwedge_{i=1}^{m} \lambda(a_i r_i b_i) \geq t$, so $\sum_{i=1}^{m} a_i r_i b_i \in U(\lambda;t)$. Thus $U(\lambda;t)RU(\lambda;t) \subseteq U(\lambda;t)$. Hence $U(\lambda;t)$ is a bi-ideal of R.

Conversely, assume that each nonempty subset $U(\lambda;t)$ of R is a bi-ideal of R. Let $a,b \in R$ be such that $\lambda(a+b) < \lambda(a) \wedge \lambda(b)$. Take $t \in (0,1]$ such that $\lambda(a+b) < t \leq \lambda(a) \wedge \lambda(b)$. Then $a,b \in U(\lambda;t)$ but $a+b \notin U(\lambda;t)$, a contradiction. Hence $\lambda(a+b) \geq \lambda(a) \wedge \lambda(b)$. Similarly we can show that $\lambda(ab) \geq \lambda(a) \wedge \lambda(b)$ and $\lambda(abc) \geq \lambda(a) \wedge \lambda(c)$. This shows that λ is a fuzzy bi-ideal of R.

Corollary 5.3. *Let B be a nonempty subset of a semiring R. Then B is a bi-ideal of R if and only if the characteristic function χ_B of B is a fuzzy bi-ideal of R.*

5.2 Regular Semirings in Terms of Fuzzy Quasi-ideals and Fuzzy Bi-ideals

In this section we characterize regular semirings by the properties of their fuzzy quasi-ideals and fuzzy bi-ideals.

Theorem 5.4. *The following assertions are equivalent for a semiring R.*

(i) R is regular.
(ii) $\mu = \mu \circ \chi_R \circ \mu$ for every fuzzy bi-ideal μ of R.
(iii) $\mu = \mu \circ \chi_R \circ \mu$ for every fuzzy quasi-ideal μ of R.

Proof. (*i*) $\Rightarrow$ (*ii*) Let R be a regular semiring and μ be any fuzzy bi-ideal of R. For $x \in R$, there exists $a \in R$ such that $x = xax$. Thus we have

$$\begin{aligned}
(\mu \circ \chi_R \circ \mu)(x) &= \bigvee_{x=\sum_{i=1}^{m} a_i b_i} \left[\bigwedge_{i=1}^{m} [(\mu \circ \chi_R)(a_i) \wedge \mu(b_i)] \right] \\
&\geq (\mu \circ \chi_R)(xa) \wedge \mu(x) \\
&= \left\{ \bigvee_{xa=\sum_{i=1}^{m} c_i d_i} \left[\bigwedge_{i=1}^{m} \mu(c_i) \wedge \chi_R(d_i) \right] \right\} \wedge \mu(x) \\
&\geq \mu(x) \wedge \mu(x) = \mu(x).
\end{aligned}$$

This implies that $\mu \leq \mu \circ \chi_R \circ \mu$. But by Theorem 5.2, $\mu \geq \mu \circ \chi_R \circ \mu$. Hence $\mu = \mu \circ \chi_R \circ \mu$.

(*ii*) $\Rightarrow$ (*iii*) Straightforward.

(*iii*) $\Rightarrow$ (*i*) Let λ, μ be fuzzy right and fuzzy left ideals of R, respectively. Then $\lambda \wedge \mu$ is a fuzzy quasi-ideal of R. Hence by hypothesis

$$\begin{aligned}
\lambda \wedge \mu &\leq (\lambda \wedge \mu) \circ \chi_R \circ (\lambda \wedge \mu) \\
&\leq \lambda \circ \chi_R \circ \mu \\
&\leq \lambda \circ \mu.
\end{aligned}$$

But $\lambda \circ \mu \leq \lambda \wedge \mu$ always hold. Hence $\lambda \circ \mu = \lambda \wedge \mu$. Thus by Theorem 2.8, R is a regular semiring.

Theorem 5.5. *The following assertions are equivalent for a semiring R:*

(i) R is regular.
(ii) $\mu \wedge \nu \leq \mu \circ \nu$ for every fuzzy bi-ideal μ and every fuzzy left ideal ν of R.
(iii) $\mu \wedge \nu \leq \mu \circ \nu$ for every fuzzy quasi-ideal μ and every fuzzy left ideal ν of R.
(iv) $\mu \wedge \nu \leq \mu \circ \nu$ for every fuzzy right ideal μ and every fuzzy bi-ideal ν of R.
(v) $\mu \wedge \nu \leq \mu \circ \nu$ for every fuzzy right ideal μ and every fuzzy quasi-ideal ν of R.
(vi) $\mu \wedge \nu \wedge \omega \leq \mu \circ \nu \circ \omega$ for every fuzzy right ideal μ, every fuzzy bi-ideal ν and every fuzzy left ideal ω of R.
(vii) $\mu \wedge \nu \wedge \omega \leq \mu \circ \nu \circ \omega$ for every fuzzy right ideal μ, every fuzzy quasi-ideal ν and every fuzzy left ideal ω of R.

Proof. (*i*) $\Rightarrow$ (*ii*) Let μ and ν be any fuzzy bi-ideal and fuzzy left ideal of R, respectively. For $x \in R$ there exists $a \in R$ such that $x = xax$. Thus

$$\begin{aligned}
(\mu \circ \nu)(x) &= \bigvee_{x=\sum_{i=1}^{m} a_i b_i} \left[\bigwedge_{i=1}^{m} [\mu(a_i) \wedge \nu(b_i)] \right] \\
&\geq \mu(x) \wedge \nu(ax) \geq \mu(x) \wedge \nu(x) = (\mu \wedge \nu)(x).
\end{aligned}$$

This implies that $\mu \wedge \nu \leq \mu \circ \nu$.

$(ii) \Rightarrow (iii)$ Straightforward.

$(iii) \Rightarrow (i)$ Let μ and λ be any fuzzy right ideal and fuzzy left ideal of R, respectively. Then μ is a fuzzy quasi-ideal of R. Thus by hypothesis $\mu \wedge \lambda \leq \mu \circ \lambda$. But $\mu \circ \lambda \leq \mu \wedge \lambda$ always holds. Thus $\mu \circ \lambda = \mu \wedge \lambda$. Therefore by Theorem2.8, R is a regular semiring.

Similarly we can prove $(i) \Leftrightarrow (iv) \Leftrightarrow (v)$.

$(i) \Rightarrow (vi)$ Let μ, ν, ω be any fuzzy right ideal, fuzzy bi-ideal and fuzzy left ideal of R, respectively. For $x \in R$ there exists $a \in R$ such that $x = xax$. Then

$$
\begin{aligned}
(\mu \circ \nu \circ \omega)(x) &= \bigvee_{x=\sum_{i=1}^{m} a_i b_i} \left[\bigwedge_{i=1}^{m} [(\mu \circ \nu)(a_i) \wedge \omega(b_i)] \right] \\
&\geq (\mu \circ \nu)(x) \wedge \omega(ax) \\
&= \left(\bigvee_{x=\sum_{i=1}^{m} a_i b_i} \left[\bigwedge_{i=1}^{m} [\mu(a_i) \wedge \nu(b_i)] \right] \right) \wedge \omega(ax) \\
&\geq \mu(xa) \wedge \nu(x) \wedge \omega(ax) \\
&\geq \mu(x) \wedge \nu(x) \wedge \omega(x) = (\mu \wedge \nu \wedge \omega)(x).
\end{aligned}
$$

So, $\mu \wedge \nu \wedge \omega \leq \mu \circ \nu \circ \omega$.

$(vi) \Rightarrow (vii)$ Straightforward.

$(vii) \Rightarrow (i)$ Let λ and μ be any fuzzy right and fuzzy left ideals of R, respectively. Since χ_R is a fuzzy quasi-ideal of R, by assumption we have

$$
\begin{aligned}
\lambda \wedge \chi_R \wedge \mu &\leq \lambda \circ \chi_R \circ \mu \\
&\Rightarrow \lambda \wedge \mu \leq \lambda \circ \mu.
\end{aligned}
$$

But $\lambda \circ \mu \leq \lambda \wedge \mu$ is always true. Therefore $\lambda \wedge \mu = \lambda \circ \mu$. Hence R is regular.

Lemma 5.1. *A semiring R is regular if and only if each right ideal and each left ideal of R is idempotent and the product of a right ideal and left ideal of R is a quasi-ideal of R.*

Next we give the fuzzy version of the above Lemma.

Theorem 5.6. *A semiring R is regular if and only if each fuzzy right ideal and fuzzy left ideal of R is idempotent and for any fuzzy right ideal μ and fuzzy left ideal ν of R, $\mu \circ \nu$ is a fuzzy quasi-ideal of R.*

Proof. Let R be a regular semiring and μ be a fuzzy right ideal of R. Then $\mu \circ \mu \leq \mu \circ \chi_R \leq \mu$. Let $x \in R$. Then there exists $a \in R$ such that $x = xax$. So we have

$$
\begin{aligned}
(\mu \circ \mu)(x) &= \bigvee_{x=\sum_{i=1}^{m} a_i b_i} \left[\bigwedge_{i=1}^{m} [\mu(a_i) \wedge \mu(b_i)] \right] \\
&\geq \mu(xa) \wedge \mu(x) \\
&\geq \mu(x).
\end{aligned}
$$

This implies that $\mu \circ \mu \geq \mu$. Hence $\mu = \mu \circ \mu$, so μ is idempotent. Similarly we can prove that every fuzzy left ideal of R is idempotent. Now let μ and ν be any fuzzy right ideal and fuzzy left ideal of R, respectively. By Theorem 2.8, we have $\mu \circ \nu = \mu \wedge \nu$ and it follows from Corollary 5.2, that $\mu \circ \nu$ is a fuzzy quasi-ideal of R.

Conversely, let A be a right ideal of R. Then χ_A, the characteristic function of A, is a fuzzy right ideal of R. By hypothesis $\chi_A = \chi_A \circ \chi_A = \chi_{A^2}$. This implies that $A = A^2$, that is, A is idempotent. Similarly we can show that each left ideal of R is idempotent. Now let A be a right ideal and B be a left ideal of R. Then $\chi_{AB} = \chi_A \circ \chi_B$ is a fuzzy quasi-ideal of R, that is, AB is a quasi-ideal of R. Therefore by Lemma 5.1, R is regular.

5.3 Intra-regular Semirings in Terms of Fuzzy Quasi-ideals and Fuzzy Bi-ideals

In this section we characterize intra-regular semirings and regular and intra-regular semirings by the properties of their fuzzy quasi-ideals and fuzzy bi-ideals.

Recall that a semiring R is *intra-regular* if for each $x \in R$, $x \in Rx^2R$, that is, $x = \sum_{i=1}^{n} r_i x^2 r_i'$ for $r_i, r_i' \in R$.

Theorem 5.7. *A semiring R is intra-regular if and only if $A \cap B \subseteq AB$ for all right ideals B and left ideals A of R.*

Now we give the fuzzy version of this Theorem.

Theorem 5.8. *The following assertions are equivalent for a semiring R.*

(i) R is intra-regular.
(ii) $\mu \wedge \nu \leq \mu \circ \nu$ for every fuzzy left ideal μ and every fuzzy right ideal ν of R.

Proof. $(i) \Rightarrow (ii)$ Let R be an intra-regular semiring and μ, ν be any fuzzy left ideal and fuzzy right ideal of R, respectively. For $x \in R$ there exist $a_i, b_i \in R$ such that

$$x = \sum_{i=1}^{m} a_i x^2 b_i.$$

Hence we have

$$
\begin{aligned}
(\mu \circ \nu)(x) &= \bigvee_{x=\sum_{i=1}^{m} a_i b_i} \left[\bigwedge_{i=1}^{m} [\mu(a_i) \wedge \nu(b_i)] \right] \\
&\geq \bigwedge_{i=1}^{m} [\mu(a_i x) \wedge \nu(x b_i)] \\
&\geq \mu(x) \wedge \nu(x) = (\mu \wedge \nu)(x).
\end{aligned}
$$

Thus $\mu \wedge \nu \leq \mu \circ \nu$.

$(ii) \Rightarrow (i)$ Let A, B be left and right ideals of R, respectively. Then χ_A and χ_B, the characteristic functions of A and B are fuzzy left and fuzzy right ideals of R, respectively. Now, by the hypothesis,

$$\chi_{A \cap B} = \chi_A \wedge \chi_B \leq \chi_A \circ \chi_B = \chi_{AB}.$$

Thus $A \cap B \subseteq AB$. So R is intra-regular.

Theorem 5.9. *The following assertions are equivalent for a semiring R.*

(i) R is both regular and intra-regular.
(ii) $B = BB$ for each bi-ideal B of R.
(iii) $Q = QQ$ for each quasi-ideal Q of R.

Next we prove the fuzzy version of this Theorem.

Theorem 5.10. *The following assertions are equivalent for a semiring R.*

(i) R is both regular and intra-regular.
(ii) $\mu \circ \mu = \mu$ for each fuzzy bi-ideal μ of R.
(iii) $\mu \circ \mu = \mu$ for each fuzzy quasi-ideal μ of R.

Proof. $(i) \Rightarrow (ii)$ Let R be both regular and intra-regular semiring and μ be a fuzzy bi-ideal of R. For $x \in R$ there exist elements $p, a_i, b_i \in R$ such that $x = xpx$ and $x = \sum_{i=1}^{m} a_i x^2 b_i$. Thus $x = xpx = xpxpx = xp\left(\sum_{i=1}^{m} a_i x^2 b_i\right)px = \sum_{i=1}^{m} (xpa_i x)(xb_i px)$. Hence we have

$$
\begin{aligned}
(\mu \circ \mu)(x) &= \bigvee_{x=\sum_{i=1}^{m} a_i b_i} \left[\bigwedge_{i=1}^{m} [\mu(a_i) \wedge \mu(b_i)] \right] && (5.1) \\
&\geq \bigwedge_{i=1}^{m} \{\mu(xpa_i x) \wedge \mu(xb_i px)\} && (5.2) \\
&\geq \bigwedge_{i=1}^{m} \{\mu(x) \wedge \mu(x)\} && (5.3) \\
&= \mu(x). && (5.4)
\end{aligned}
$$

This implies that $\mu \leq \mu \circ \mu$. But $\mu \circ \mu \leq \mu$ always holds.

Thus $\mu \circ \mu = \mu$.

$(ii) \Rightarrow (iii)$ Straightforward.

$(iii) \Rightarrow (i)$ Let Q be any quasi-ideal of R. Then χ_Q, the characteristic function of Q, is a fuzzy quasi-ideal of R. By the assumption

$$\chi_Q = \chi_Q \circ \chi_Q = \chi_{Q^2}$$

Thus $Q = Q^2$.

Hence by Theorem 5.9, R is both regular and intra-regular.

Theorem 5.11. *The following assertions are equivalent for a semiring R.*

(i) R is both regular and intra-regular.
(ii) $\mu \wedge \nu \leq \mu \circ \nu$ *for all fuzzy bi-ideals* μ, ν *of R.*
(iii) $\mu \wedge \nu \leq \mu \circ \nu$ *for all fuzzy bi-ideals* μ *and fuzzy quasi-ideals* ν *of R.*
(iv) $\mu \wedge \nu \leq \mu \circ \nu$ *for all fuzzy quasi-ideals* μ *and fuzzy bi-ideals* ν *of R.*
(v) $\mu \wedge \nu \leq \mu \circ \nu$ *for all fuzzy quasi-ideals* μ, ν *of R.*

Proof. $(i) \Rightarrow (ii)$ Let μ and ν be fuzzy bi-ideals of R and $x \in R$. Since R is both regular and intra-regular, there exist elements $p, a_i, b_i \in R$ such that $x = xpx$ and $x = \sum_{i=1}^{m} a_i x^2 b_i$. Thus $x = xpx = xpxpx = xp\left(\sum_{i=1}^{m} a_i x^2 b_i\right) px = \sum_{i=1}^{m} (xpa_i x)(xb_i px)$.

Thus we have

$$\begin{aligned}(\mu \circ \nu)(x) &= \bigvee_{x = \sum_{i=1}^{m} a_i b_i} \left[\bigwedge_{i=1}^{m} [\mu(a_i) \wedge \nu(b_i)]\right] \\ &\geq \bigwedge_{i=1}^{m} \{\mu(xpa_i x) \wedge \nu(xb_i px)\} \\ &\geq \bigwedge_{i=1}^{m} \{\mu(x) \wedge \nu(x)\} \\ &\geq (\mu \wedge \nu)(x).\end{aligned}$$

Hence $\mu \wedge \nu \leq \mu \circ \nu$.

$(ii) \Rightarrow (iii) \Rightarrow (v)$ and $(ii) \Rightarrow (iv) \Rightarrow (v)$, since every fuzzy quasi-ideal of R is a fuzzy bi-ideal of R.

$(v) \Rightarrow (i)$ Let Q be any quasi-ideal of R. Then χ_Q, the characteristic function of Q, is a fuzzy quasi-ideal of R. Now by assumption

$$\chi_Q = \chi_Q \wedge \chi_Q \leq \chi_Q \circ \chi_Q = \chi_{Q^2}$$

So $Q \subseteq Q^2$. Since $Q \supseteq Q^2$ always true, therefore $Q = Q^2$.

Hence R is both regular and intra-regular.

5.4 Fuzzy k-Bi-ideals and Fuzzy k-Quasi-ideals of Semirings

In this section, we introduce the concepts of fuzzy k-bi-ideal and fuzzy k-quasi-ideal of a semiring. We characterize different classes of semirings by the properties of fuzzy k-bi-ideals and fuzzy k-quasi-ideals.

Definition 5.3. A nonempty subset A of a semiring R is called a *k-quasi-ideal* of R if A is closed under addition, $\overline{RA} \cap \overline{AR} \subseteq A$ and $x + a = b$ implies $x \in A$ for all $x \in R$ and $a, b \in A$.

A nonempty subset A of a semiring R is called a *k-bi-ideal* of R if A is closed under addition and multiplication, $\overline{ARA} \subseteq A$ and $x + a = b$ implies $x \in A$ for all $x \in R$ and $a, b \in A$.

Every left (right) k-ideal of a semiring R is a k-quasi-ideal of R and every k-quasi-ideal of R is a k-bi-ideal of R. But the converse does not hold, as shown in the following examples.

Example 5.3. Let $R = \left\{ \begin{bmatrix} a & b \\ c & d \end{bmatrix} : a,b,c,d \text{ are non negative integers} \right\}$. Then R is a semiring under the usual addition and multiplication of matrices. Consider the set Q of all matrices of the form $\begin{pmatrix} 0 & 0 \\ 0 & a \end{pmatrix}$ ($a \in \mathbb{N}_\circ$). Evidently Q is a k-quasi-ideal of R but not a left (right) k-ideal of R.

Example 5.4. Let $\mathbb{N}$ and $\mathbb{R}^+$ denote the sets of all positive integers and positive real numbers, respectively. The set R of all matrices of the form $\begin{pmatrix} a & 0 \\ b & c \end{pmatrix}$ $(a, b \in \mathbb{R}^+, c \in \mathbb{N})$ together with $\begin{pmatrix} 0 & 0 \\ 0 & 0 \end{pmatrix}$ is a semiring under the usual addition and multiplication of matrices. Let A, B be the sets of all matrices $\begin{pmatrix} a & 0 \\ b & c \end{pmatrix}$ $(a, b \in \mathbb{R}^+, c \in \mathbb{N}, a < b)$ together with $\begin{pmatrix} 0 & 0 \\ 0 & 0 \end{pmatrix}$ and $\begin{pmatrix} p & 0 \\ q & d \end{pmatrix}$ $(a, b \in \mathbb{R}, d \in \mathbb{N}, 3 < q)$ together with $\begin{pmatrix} 0 & 0 \\ 0 & 0 \end{pmatrix}$, respectively. It is easy to show that A and B are right k -ideal and left k-ideal of R, respectively. The product AB is a k -bi-ideal of R but it is not a k-quasi-ideal of R. Indeed, the element

$$\begin{pmatrix} 6 & 0 \\ 9 & 1 \end{pmatrix} = \begin{pmatrix} 6 & 0 \\ 3 & 1 \end{pmatrix} \left(\begin{pmatrix} 1 & 0 \\ 2 & 1 \end{pmatrix} \begin{pmatrix} 1 & 0 \\ 4 & 1 \end{pmatrix} \right) = \left(\begin{pmatrix} 1 & 0 \\ \frac{7}{6} & 1 \end{pmatrix} \begin{pmatrix} 24 & 0 \\ 4 & 1 \end{pmatrix} \right) \begin{pmatrix} \frac{1}{4} & 0 \\ 1 & 1 \end{pmatrix}$$

belongs to the intersection $\overline{R(AB)} \cap \overline{(AB)R}$, but it is not an element of AB. Hence $\overline{R(AB)} \cap \overline{(AB)R} \nsubseteq AB$.

This example also shows that the product of k-quasi-ideals of a semiring R need not to be a k-quasi-ideal of R.

Definition 5.4. A fuzzy subset λ of a semiring R is called a *fuzzy k-bi-ideal* of R if for all $x, y, z \in R$ we have

1. $\lambda(x+y) \geq \lambda(x) \wedge \lambda(y)$
2. $\lambda(xy) \geq \lambda(x) \wedge \lambda(y)$
3. $\lambda(xyz) \geq \lambda(x) \wedge \lambda(z)$
4. $x+y=z \Longrightarrow \lambda(x) \geq \lambda(y) \wedge \lambda(z)$.

Theorem 5.12. *A fuzzy subset μ in a semiring R is a fuzzy k-bi-ideal of R if and only if*

(i) $\mu +_k \mu \leq \mu$
(ii) $\mu \odot_k \mu \leq \mu$
(iii) $\mu \odot_k \chi_R \odot_k \mu \leq \mu$.

Proof. Let μ be a fuzzy k-bi-ideal of R. By Theorem 4.3, μ satisfies (i).

Let $x \in R$. If $(\mu \odot_k \mu)(x) = 0$ then $\mu \odot_k \mu \leq \mu$. Otherwise, there exist $a_i, b_i, a_j^{'}, b_j^{'} \in R$ such that $x + \sum_{i=1}^{m} a_i b_i = \sum_{j=1}^{n} a_j^{'} b_j^{'}$. Then we have

$$\begin{aligned}
(\mu \odot_k \mu)(x) &= \bigvee_{x+\sum_{i=1}^{m} a_i b_i = \sum_{j=1}^{n} a_j^{'} b_j^{'}} \left[\bigwedge_{i=1}^{m} [\mu(a_i) \wedge \mu(b_i)] \wedge \bigwedge_{j=1}^{n} \left[\mu(a_j^{'}) \wedge \mu(b_j^{'})\right]\right] \\
&\leq \bigvee_{x+\sum_{i=1}^{m} a_i b_i = \sum_{j=1}^{n} a_j^{'} b_j^{'}} \left[\bigwedge_{i=1}^{m} [\mu(a_i b_i)] \wedge \bigwedge_{j=1}^{n} \left[\mu(a_j^{'} b_j^{'})\right]\right] \\
&\leq \bigvee_{x+\sum_{i=1}^{m} a_i b_i = \sum_{j=1}^{n} a_j^{'} b_j^{'}} \left[\mu\Big(\sum_{i=1}^{m} a_i b_i\Big) \wedge \mu\Big(\sum_{j=1}^{n} a_j^{'} b_j^{'}\Big)\right] \\
&\leq \mu(x).
\end{aligned}$$

Hence $\mu \odot_k \mu \leq \mu$.

Let $x \in R$. If $(\mu \odot_k \chi_R \odot_k \mu)(x) = 0$, then $\mu \odot_k \chi_R \odot_k \mu \leq \mu$. Otherwise, there exist $a_i, b_i, a_j^{'}, b_j^{'} \in R$ such that $x + \sum_{i=1}^{m} a_i b_i = \sum_{j=1}^{n} a_j^{'} b_j^{'}$. Then we have

$$\begin{aligned}
(\mu \odot_k \chi_R \odot_k \mu)(x) &= \bigvee_{x+\sum_{i=1}^{m} a_i b_i = \sum_{j=1}^{n} a_j^{'} b_j^{'}} \left[\begin{array}{l} \bigwedge_{i=1}^{m} [(\mu \odot_k \chi_R)(a_i) \wedge \mu(b_i)] \wedge \\ \bigwedge_{j=1}^{n} \left[(\mu \odot_k \chi_R)(a_j^{'}) \wedge \mu(b_j^{'})\right] \end{array}\right] \\
&= \bigvee_{x+\sum_{i=1}^{m} a_i b_i = \sum_{j=1}^{n} a_j^{'} b_j^{'}} \left[\begin{array}{l} \bigwedge_{i=1}^{m} \left[\bigvee_{a_i+\sum_{k=1}^{p} c_k d_k = \sum_{q=1}^{r} c_q^{'} d_q^{'}} \left\{\bigwedge_{k=1}^{p} \mu(c_k) \wedge \bigwedge_{q=1}^{r} \mu\left(c_q^{'}\right)\right\} \wedge \mu(b_i)\right] \wedge \\ \bigwedge_{j=1}^{n} \left[\bigvee_{a_j^{'}+\sum_{l=1}^{s} e_l f_l = \sum_{u=1}^{t} e_u^{'} f_u^{'}} \left\{\bigwedge_{l=1}^{s} \mu(e_l) \wedge \bigwedge_{u=1}^{t} \mu\left(e_u^{'}\right)\right\} \wedge \mu\left(b_j^{'}\right)\right] \end{array}\right]
\end{aligned}$$

$$= \bigvee_{x+\sum_{i=1}^{m} a_i b_i = \sum_{j=1}^{n} a'_j b'_j} \left[\begin{array}{l} \bigwedge_{i=1}^{m} \left[\bigvee_{a_i + \sum_{k=1}^{p} c_k d_k = \sum_{q=1}^{r} c'_q d'_q} \left\{ \begin{array}{l} \bigwedge_{k=1}^{p} [\mu(c_k) \wedge \mu(b_i)] \wedge \\ \bigwedge_{q=1}^{r} \left[\mu\left(c'_q\right) \wedge \mu(b_i)\right] \end{array} \right\} \right] \wedge \\ \bigwedge_{j=1}^{n} \left[\bigvee_{a'_j + \sum_{l=1}^{s} e_l f_l = \sum_{u=1}^{t} e'_u f'_u} \left\{ \begin{array}{l} \bigwedge_{l=1}^{s} \left[\mu(e_l) \wedge \mu\left(b'_j\right)\right] \wedge \\ \bigwedge_{u=1}^{t} \left[\mu\left(e'_u\right) \wedge \mu\left(b'_j\right)\right] \end{array} \right\} \right] \end{array} \right]$$

Since

$$a_i + \sum_{k=1}^{p} c_k d_k = \sum_{q=1}^{r} c'_q d'_q \text{ and } a'_j + \sum_{l=1}^{s} e_l f_l = \sum_{u=1}^{t} e'_u f'_u$$

so,

$$a_i b_i + \sum_{k=1}^{p} c_k d_k b_i = \sum_{q=1}^{r} c'_q d'_q b_i \text{ and } a'_j b'_j + \sum_{l=1}^{s} e_l f_l b'_j = \sum_{u=1}^{t} e'_u f'_u b'_j$$

$$(\mu \odot_k \chi_R \odot_k \mu)(x) \leq \bigvee_{x+\sum_{i=1}^{m} a_i b_i = \sum_{j=1}^{n} a'_j b'_j} \left[\begin{array}{l} \bigwedge_{i=1}^{m} \left[\bigvee_{a_i + \sum_{k=1}^{p} c_k d_k = \sum_{q=1}^{r} c'_q d'_q} \left\{ \mu\left(\sum_{k=1}^{p} c_k d_k b_i\right) \wedge \mu\left(\sum_{q=1}^{r} c'_q d'_q b_i\right) \right\} \right] \wedge \\ \bigwedge_{j=1}^{n} \left[\bigvee_{a'_j + \sum_{l=1}^{s} e_l f_l = \sum_{u=1}^{t} e'_u f'_u} \left\{ \mu\left(\sum_{l=1}^{s} e_l f_l b'_j\right) \wedge \mu\left(\sum_{u=1}^{t} e'_u f'_u b'_j\right) \right\} \right] \end{array} \right]$$

Since μ is a fuzzy k-bi-ideal of R, so
$\mu(a_i b_i) \geq \mu\left(\sum_{k=1}^{p} c_k d_k b_i\right) \wedge \mu\left(\sum_{q=1}^{r} c'_q d'_q b_i\right)$ and $\mu\left(a'_j b'_j\right) \geq \mu\left(\sum_{l=1}^{s} e_l f_l b'_j\right) \wedge$
$\mu\left(\sum_{u=1}^{t} e'_u f'_u b'_j\right)$

Hence,

$$\begin{aligned} (\mu \odot_k \chi_R \odot_k \mu)(x) &\leq \bigvee_{x+\sum_{i=1}^{m} a_i b_i = \sum_{j=1}^{n} a'_j b'_j} \left[\bigwedge_{i=1}^{m} [\mu(a_i b_i)] \wedge \bigwedge_{j=1}^{n} \left[\mu\left(a'_j b'_j\right)\right] \right] \\ &= \bigvee_{x+\sum_{i=1}^{m} a_i b_i = \sum_{j=1}^{n} a'_j b'_j} \left[\mu\left(\sum_{i=1}^{m} a_i b_i\right) \wedge \mu\left(\sum_{j=1}^{n} a'_j b'_j\right) \right] \\ &\leq \mu(x). \end{aligned}$$

Thus $\mu \odot_k \chi_R \odot_k \mu \leq \mu$.

Conversely, assume that μ satisfies the given conditions. Then by Theorem 4.3, $\mu(x+y) \geq \mu(x) \wedge \mu(y)$ and $x+y=z \Longrightarrow \mu(x) \geq \mu(y) \wedge \mu(z)$ for all $x, y, z \in R$. Let $x, y \in R$. Then

$$\begin{aligned}
\mu(xy) &\geq (\mu \odot_k \mu)(xy) \\
&= \bigvee_{xy+\sum_{i=1}^{m} a_i b_i = \sum_{j=1}^{n} a'_j b'_j} \left[\bigwedge_{i=1}^{m} [\mu(a_i) \wedge \mu(b_i)] \wedge \bigwedge_{j=1}^{n} \left[\mu(a'_j) \wedge \mu(b'_j)\right] \right] \\
&\geq \mu(0) \wedge \mu(x) \wedge \mu(y) \quad (\text{because } xy + 00 = xy\,) \\
&\geq \mu(x) \wedge \mu(y) \quad (\text{because } \mu(0) \geq \mu(x) \text{ for all } x \in R).
\end{aligned}$$

Similarly we can show that $\mu(xyz) \geq \mu(x) \wedge \mu(z)$.

Theorem 5.13. *A fuzzy subset λ of a semiring R is a fuzzy k-bi-ideal of R if and only if each nonempty level subset $U(\lambda;t)$ of λ is a k-bi-ideal of R.*

Proof. Suppose that λ is a fuzzy k-bi-ideal of R and $t \in (0,1]$ be such that $U(\lambda;t) \neq \emptyset$. Let $a,b \in U(\lambda;t)$. Then $\lambda(a) \geq t$ and $\lambda(b) \geq t$. As $\lambda(a+b) \geq \lambda(a) \wedge \lambda(b)$, so $\lambda(a+b) \geq t$. Hence $a+b \in U(\lambda;t)$. Also, $\lambda(ab) \geq \lambda(a) \wedge \lambda(b)$, so $\lambda(ab) \geq t$. This implies $ab \in U(\lambda;t)$.

Let $x \in \overline{U(\lambda;t)RU(\lambda;t)}$. Then $x + \sum_{i=1}^{m} a_i r_i b_i = \sum_{j=1}^{n} a'_j r'_j b'_j$, for some $a_i, b_i, a'_j, b'_j \in U(\lambda;t)$ and $r_i, r'_j \in R$. Since λ is a fuzzy k-bi-ideal of R, so $\lambda(a_i r_i b_i) \geq \lambda(a_i) \wedge \lambda(b_i) \geq t$. Hence $\sum_{i=1}^{m} a_i r_i b_i \in U(\lambda;t)$. Similarly $\sum_{j=1}^{n} a'_j r'_j b'_j \in U(\lambda;t)$. Hence $x \in U(\lambda;t)$. Thus $\overline{U(\lambda;t)RU(\lambda;t)} \subseteq U(\lambda;t)$.

Now let $x + a = b$ for some $a,b \in U(\lambda;t)$. Then $\lambda(a) \geq t$ and $\lambda(b) \geq t$. Since $\lambda(x) \geq \lambda(a) \wedge \lambda(b)$, so $\lambda(x) \geq t$. Hence $x \in U(\lambda;t)$. Thus $U(\lambda;t)$ is a k-bi-ideal of R.

Conversely, assume that each nonempty subset $U(\lambda;t)$ of R is a k-bi-ideal of R. Let $a,b \in R$ be such that $\lambda(a+b) < \lambda(a) \wedge \lambda(b)$. Take $t \in (0,1]$ such that $\lambda(a+b) < t \leq \lambda(a) \wedge \lambda(b)$. Then $a,b \in U(\lambda;t)$ but $a+b \notin U(\lambda;t)$, a contradiction. Hence $\lambda(a+b) \geq \lambda(a) \wedge \lambda(b)$. Similarly we can show that $\lambda(ab) \geq \lambda(a) \wedge \lambda(b)$ and $\lambda(abc) \geq \lambda(a) \wedge \lambda(c)$.

Let $x,y,z \in R$ be such that $x+y=z$. If possible let $\lambda(x) < \lambda(y) \wedge \lambda(z)$. Take $t \in (0,1]$ such that $\lambda(x) < t \leq \lambda(y) \wedge \lambda(z)$. Then $y,z \in U(\lambda;t)$ but $x \notin U(\lambda;t)$, a contradiction. Hence $\lambda(x) \geq \lambda(y) \wedge \lambda(z)$. Thus λ is a fuzzy k-bi-ideal of R.

Corollary 5.4. *Let B be a nonempty subset of a semiring R. Then B is a k-bi-ideal of R if and only if the characteristic function χ_B of B is a fuzzy k-bi-ideal of R.*

Definition 5.5. A fuzzy subset λ of a semiring R is called a *fuzzy k-quasi-ideal* of R if for all $x,y,z \in R$ we have

1. $\lambda(x+y) \geq \lambda(x) \wedge \lambda(y)$,
2. $(\lambda \odot_k \chi_R) \wedge (\chi_R \odot_k \lambda) \leq \lambda$,
3. $x+y=z \Longrightarrow \lambda(x) \geq \lambda(y) \wedge \lambda(z)$.

Obviously every fuzzy left (right) k-ideal of R is a fuzzy k-quasi-ideal of R and every fuzzy k-quasi-ideal of R is a fuzzy k-bi-ideal of R. But the converse does not hold.

Theorem 5.14. *A fuzzy subset λ of a semiring R is a fuzzy k-quasi-ideal of R if and only if each nonempty level subset $U(\lambda;t)$ of λ is a k-quasi-ideal of R.*

Proof. The proof is similar to the proof of Theorem 5.1.

Corollary 5.5. *Let Q be a nonempty subset of a semiring R. Then Q is a k-quasi-ideal of R if and only if the characteristic function χ_Q of Q is a fuzzy k-quasi-ideal of R.*

Proposition 5.2. *The intersection of fuzzy k-quasi-ideals of a hemiring R is a fuzzy k-quasi-ideal of R.*

Proof. The proof is similar to the proof of Proposition 5.1.

Corollary 5.6. *Let μ and ν be fuzzy right k-ideal and fuzzy left k-ideal of a semiring R, respectively. Then $\mu \wedge \nu$ is a fuzzy k-quasi-ideal of R.*

5.5 *k*-Regular Semirings

Recall the definition of *k*-regular semiring.

A semiring R is called a *k-regular semiring* if for each $x \in R$, there exist $a, a^{'} \in R$ such that $x + xax = xa^{'}x$.

The following is a well known characterization of *k*-regular semirings.

Theorem 5.15. *The following assertions are equivalent for a semiring R.*

(i) R is k-regular
(ii) $B = \overline{BRB}$ for every k-bi-ideal B of R
(iii) $Q = \overline{QRQ}$ for every k-quasi-ideal Q of R.

Now we give the fuzzy version of this Theorem.

Theorem 5.16. *The following assertions are equivalent for a semiring R.*

(i) R is k-regular.
(ii) $\mu = \mu \odot_k \chi_R \odot_k \mu$ for every fuzzy k-bi-ideal μ of R.
(iii) $\mu = \mu \odot_k \chi_R \odot_k \mu$ for every fuzzy k-quasi-ideal μ of R.

Proof. The proof is similar to the proof of Theorem 5.4.

Theorem 5.17. *The following assertions are equivalent for a semiring R.*

(i) R is k-regular.
(ii) $\mu \wedge \nu \leq \mu \odot_k \nu \odot_k \mu$ for every fuzzy k-bi-ideal μ and every fuzzy k-ideal ν of R.
(iii) $\mu \wedge \nu \leq \mu \odot_k \nu \odot_k \mu$ for every fuzzy k-quasi-ideal μ and every fuzzy k-ideal ν of R.

Proof. $(i) \Rightarrow (ii)$ Let μ and ν be any fuzzy k-bi-ideal and fuzzy k-ideal of R, respectively. For $x \in R$, there exist $a, a' \in R$ such that $x + xax = xa'x$. Thus we have

$$
\begin{aligned}
(\mu \odot_k \nu \odot_k \mu)(x) &= \bigvee_{x+\sum_{i=1}^{m} a_i b_i = \sum_{j=1}^{n} a'_j b'_j} \left[\bigwedge_{i=1}^{m} [(\mu \odot_k \nu)(a_i) \wedge \mu(b_i)] \wedge \bigwedge_{j=1}^{n} \left[(\mu \odot_k \nu)(a'_j) \wedge \mu(b'_j) \right] \right] \\
&\geq (\mu \odot_k \nu)(xa) \wedge (\mu \odot_k \nu)\left(xa'\right) \wedge \mu(x) \\
&= \left\{ \bigvee_{xa+\sum_{i=1}^{m} c_i d_i = \sum_{j=1}^{n} c'_j d'_j} \left[\bigwedge_{i=1}^{m} [\mu(c_i) \wedge \nu(d_i)] \wedge \bigwedge_{j=1}^{n} \left[\mu(c'_j) \wedge \nu(d'_j) \right] \right] \right\} \\
&\quad \wedge \left\{ \bigvee_{xa'+\sum_{i=1}^{m} l_i f_i = \sum_{j=1}^{n} l'_j f'_j} \left[\bigwedge_{i=1}^{m} [\mu(l_i) \wedge \nu(f_i)] \wedge \bigwedge_{j=1}^{n} \left[\mu(l'_j) \wedge \nu(f'_j) \right] \right] \right\} \wedge \mu(x) \\
&\geq \left[\mu(x) \wedge \nu(axa) \wedge \nu\left(a'xa\right) \right] \wedge \left[\mu(x) \wedge \nu\left(axa'\right) \wedge \nu\left(a'xa'\right) \right] \wedge \mu(x) \\
&\quad \left(\begin{array}{c} \text{because } x + xax = xa'x \text{ implies } xa + xaxa = xa'xa \\ \text{and } xa' + xaxa' = xa'xa' \end{array} \right) \\
&\geq \mu(x) \wedge \nu(x) \wedge \mu(x) \wedge \nu(x) = \mu(x) \wedge \nu(x) \\
&= (\mu \wedge \nu)(x).
\end{aligned}
$$

This implies that $\mu \wedge \nu \leq \mu \odot_k \nu \odot_k \mu$.

$(ii) \Rightarrow (iii)$ Straightforward.

$(iii) \Rightarrow (i)$ Let μ be any fuzzy k-quasi-ideal of R. Since χ_R is a fuzzy k-ideal of R, we have by hypothesis

$$\mu = \mu \wedge \chi_R \leq \mu \odot_k \chi_R \odot_k \mu.$$

But by Theorem 5.12, $\mu \odot_k \chi_R \odot_k \mu \leq \mu$. Thus $\mu \odot_k \chi_R \odot_k \mu = \mu$.

Therefore by Theorem 5.16, R is k-regular.

Theorem 5.18. *The following assertions are equivalent for a semiring R:*

(i) R is k-regular.

(ii) $\mu \wedge \nu \leq \mu \odot_k \nu$ *for every fuzzy k-bi-ideal* μ *and every fuzzy left k-ideal* ν *of R.*

(iii) $\mu \wedge \nu \leq \mu \odot_k \nu$ *for every fuzzy k-quasi-ideal* μ *and every fuzzy left k-ideal* ν *of R.*

(iv) $\mu \wedge \nu \leq \mu \odot_k \nu$ *for every fuzzy right k-ideal* μ *and every fuzzy k-bi-ideal* ν *of R.*

(v) $\mu \wedge \nu \leq \mu \odot_k \nu$ *for every fuzzy right k-ideal* μ *and every fuzzy k-quasi-ideal* ν *of R.*

(vi) $\mu \wedge \nu \wedge \omega \leq \mu \odot_k \nu \odot_k \omega$ *for every fuzzy right k-ideal* μ*, every fuzzy k-bi-ideal* ν *and every fuzzy left k-ideal* ω *of R.*

(vii) $\mu \wedge \nu \wedge \omega \leq \mu \odot_k \nu \odot_k \omega$ *for every fuzzy right k-ideal* μ*, every fuzzy k-quasi-ideal* ν *and every fuzzy left k-ideal* ω *of R.*

Proof. $(i) \Rightarrow (ii)$ Let μ and ν be any fuzzy k-bi-ideal and fuzzy left k-ideal of R, respectively. Now let $x \in R$. Since R is k-regular, there exist $a, a' \in R$ such that $x + xax = xa'x$. Then we have

$$(\mu \odot_k \nu)(x) = \bigvee_{x+\sum_{i=1}^{m} a_i b_i = \sum_{j=1}^{n} a'_j b'_j} \left[\bigwedge_{i=1}^{m} [\mu(a_i) \wedge \nu(b_i)] \wedge \bigwedge_{j=1}^{n} \left[\mu(a'_j) \wedge \nu(b'_j) \right] \right]$$
$$\geq \mu(x) \wedge \nu(ax) \wedge \nu\left(a'x\right) \geq \mu(x) \wedge \nu(x) = (\mu \wedge \nu)(x)$$

This implies that $\mu \wedge \nu \leq \mu \odot_k \nu$.

$(ii) \Rightarrow (iii)$ Straightforward.

$(iii) \Rightarrow (i)$ Let μ and ν be any fuzzy right k-ideal and fuzzy left k-ideal of R, respectively. Then μ is a fuzzy k-quasi-ideal of R. Thus by assumption we have $\mu \wedge \nu \leq \mu \odot_k \nu$. But $\mu \odot_k \nu \leq \mu \wedge \nu$ always holds. Hence $\mu \odot_k \nu = \mu \wedge \nu$. Thus by Theorem 4.6 R is k-regular.

Similarly we can prove that $(i) \Leftrightarrow (iv) \Leftrightarrow (v)$.

$(i) \Rightarrow (vi)$ Let μ, ν, ω be any fuzzy right k-ideal, any fuzzy k-bi-ideal and any fuzzy left k-ideal of R, respectively. Now let $x \in R$. Since R is k-regular, there exist $a, a' \in R$ such that $x + xax = xa'x$. Thus we have

$$(\mu \odot_k \nu \odot_k \omega)(x) = \bigvee_{x+\sum_{i=1}^{m} a_i b_i = \sum_{j=1}^{n} a'_j b'_j} \left[\begin{array}{c} \bigwedge_{i=1}^{m} [(\mu \odot_k \nu)(a_i) \wedge \omega(b_i)] \\ \wedge \bigwedge_{j=1}^{n} \left[(\mu \odot_k \nu)(a'_j) \wedge \omega(b'_j) \right] \end{array} \right]$$
$$\geq (\mu \odot_k \nu)(x) \wedge \omega(ax) \wedge \omega(a'x)$$
$$= \bigvee_{x+\sum_{i=1}^{m} a_i b_i = \sum_{j=1}^{n} a'_j b'_j} \left[\bigwedge_{i=1}^{m} [\mu(a_i) \wedge \nu(b_i)] \wedge \bigwedge_{j=1}^{n} \left[\mu(a'_j) \wedge \nu(b'_j) \right] \right] \wedge \omega(ax) \wedge \omega(a'x)$$
$$\geq \mu(xa) \wedge \mu\left(xa'\right) \wedge \nu(x) \wedge \omega(ax) \wedge \omega(a'x)$$
$$\geq \mu(x) \wedge \nu(x) \wedge \omega(x) = (\mu \wedge \nu \wedge \omega)(x)$$

So, $\mu \wedge \nu \wedge \omega \leq \mu \odot_k \nu \odot_k \omega$.

$(vi) \Rightarrow (vii)$ Straightforward.

$(vii) \Rightarrow (i)$ Let λ and μ be any fuzzy right k-ideal and fuzzy left k-ideal of R, respectively. Since χ_R is a fuzzy k-quasi-ideal of R, so by assumption we have

$$\lambda \wedge \chi_R \wedge \mu \leq \lambda \odot \chi_R \odot \mu$$
$$\lambda \wedge \mu \leq \lambda \odot \mu.$$

But $\lambda \odot \mu \leq \lambda \wedge \mu$ is always true. Thus $\lambda \odot \mu = \lambda \wedge \mu$. Hence by Theorem 4.6 R is k-regular.

Lemma 5.2. *A semiring R is k-regular if and only if each right k-ideal and each left k-ideal of R is idempotent and for any right k-ideal A and left k-ideal B of R, $\overline{AB}$ is a k-quasi-ideal of R.*

Theorem 5.19. *A semiring R is k-regular if and only if the fuzzy right and fuzzy left k-ideals of R are idempotent and for any fuzzy right k-ideal μ and fuzzy left k-ideal ν of R, $\mu \odot_k \nu$ is a fuzzy k-quasi-ideal of R.*

Proof. Let R be k-regular. Let μ be a fuzzy right k-ideal of R. Then $\mu \odot_k \mu \leq \mu \odot_k \chi_R \leq \mu$.

Let $x \in R$. Since R is k-regular, there exist $a, a' \in R$ such that $x + xax = xa'x$. Then we have

$$
\begin{aligned}
(\mu \odot_k \mu)(x) &= \bigvee_{x+\sum_{i=1}^{m} a_i b_i = \sum_{j=1}^{n} a'_j b'_j} \left[\bigwedge_{i=1}^{m} [\mu(a_i) \wedge \mu(b_i)] \wedge \bigwedge_{j=1}^{n} \left[\mu(a'_j) \wedge \mu(b'_j) \right] \right] \\
&\geq \mu(xa) \wedge \mu(x) \wedge \mu\left(xa'\right) \\
&\geq \mu(x).
\end{aligned}
$$

This implies that $(\mu \odot_k \mu) \geq \mu$.

Hence $\mu = \mu \odot_k \mu$, so μ is idempotent. Similarly we can prove that every fuzzy left k-ideal of R is idempotent. Now let μ and ν be any fuzzy right k-ideal and fuzzy left k-ideal of R, respectively. By Theorem 4.6 , we have $\mu \odot_k \nu = \mu \wedge \nu$ and it follows from Corollary 5.6 , that $\mu \odot_k \nu$ is a fuzzy k-quasi-ideal of R.

Conversely, let A be a right k-ideal of R. Then χ_A, the characteristic function of A, is a fuzzy right k-ideal of R. And $\chi_A = \chi_A \odot_k \chi_A = \chi_{\overline{A^2}}$ implies that $A = \overline{A^2}$, that is, A is idempotent. Similarly we can show that left k-ideals of R are idempotent. Now let A be a right k-ideal and B be a left k-ideal of R. Then $\chi_{\overline{AB}} = \chi_A \odot_k \chi_B$ is a fuzzy k-quasi-ideal of R, that is, $\overline{AB}$ is a k-quasi-ideal of R. Therefore by above Lemma, R is k-regular.

5.6 k-Intra-regular Semirings

Definition 5.6. A semiring R is said to be *k-intra-regular* if for each $x \in R$, there exists $a_i, a'_i, b_j, b'_j \in R$ such that $x + \sum_{i=1}^{m} a_i x^2 b_i = \sum_{j=1}^{n} a'_j x^2 b'_j$.

In the case of rings the k-intra-regularity coincides with the intra-regularity of rings.

Example 5.5. Let R be a semiring defined by the following Cayley's tables:

$+$	0	x	1
0	0	x	1
x	x	x	x
1	1	x	1

$\cdot$	0	x	1
0	0	0	0
x	0	x	x
1	0	x	1

Then R is k-intra-regular semiring.

Lemma 5.3. *Let R be a semiring. Then the following conditions are equivalent.*

(i) R is k-intra-regular.
(ii) $A \cap B \subseteq \overline{AB}$ for every left k-ideal A and every right k-ideal B of R.

Lemma 5.4. *Let R be a semiring. Then the following conditions are equivalent.*

(i) R is k-intra-regular.
(ii) $\mu \wedge \nu \leq \mu \odot_k \nu$ for every fuzzy left k-ideal μ and every fuzzy right k-ideal ν of R.

Proof. $(i) \Rightarrow (ii)$ Let R be k-intra-regular. Let μ and ν be any fuzzy left k-ideal and fuzzy right k-ideal of R, respectively. Now let $x \in R$. Since R is k-intra-regular, there exist $a_i, a_i^{'}, b_j, b_j^{'} \in R$ such that

$$x + \sum_{i=1}^{m} a_i x^2 b_i = \sum_{j=1}^{n} a_j^{'} x^2 b_j^{'}$$

that is

$$x + \sum_{i=1}^{m} (a_i x)(x b_i) = \sum_{j=1}^{n} \left(a_j^{'} x\right)\left(x b_j^{'}\right)$$

Thus we have

$$\begin{aligned}(\mu \odot_k \nu)(x) &= \bigvee_{x + \sum_{i=1}^{m} a_i b_i = \sum_{j=1}^{n} a_j^{'} b_j^{'}} \left[\bigwedge_{i=1}^{m} [\mu(a_i) \wedge \nu(b_i)] \wedge \bigwedge_{j=1}^{n} \left[\mu(a_j^{'}) \wedge \nu(b_j^{'})\right]\right] \\ &\geq \bigwedge_{i=1}^{m} [\mu(a_i x) \wedge \nu(x b_i)] \wedge \bigwedge_{j=1}^{n} \left[\mu\left(a_j^{'} x\right) \wedge \nu\left(x b_j^{'}\right)\right] \\ &\geq \mu(x) \wedge \nu(x) = (\mu \wedge \nu)(x)\end{aligned}$$

This implies that $\mu \wedge \nu \leq \mu \odot_k \nu$.

$(ii) \Rightarrow (i)$ Let A, B be left and right k-ideals of R, respectively. Then χ_A and χ_B, the characteristic functions of A and B are fuzzy left and fuzzy right k-ideals of R, respectively. Now, by the hypothesis,

$$\chi_{A \cap B} = \chi_A \wedge \chi_B \leq \chi_A \odot_k \chi_B = \chi_{\overline{AB}}.$$

Thus $A \cap B \subseteq \overline{AB}$. So R is k-intra-regular.

Lemma 5.5. *The following assertions are equivalent for a semiring R:*

(i) R is both k-regular and k-intra-regular.
(ii) $B = \overline{B^2}$ for every k-bi-ideal B of R.
(iii) $Q = \overline{Q^2}$ for every k-quasi-ideal Q of R.

Now we give the fuzzy version of this Lemma.

Theorem 5.20. *The following assertions are equivalent for a semiring R.*

(i) R is both k-regular and k-intra-regular.
(ii) $\mu \odot_k \mu = \mu$ *for each fuzzy k-bi-ideal* μ *of R.*
(iii) $\mu \odot_k \mu = \mu$ *for each fuzzy k-quasi-ideal* μ *of R.*

Proof. The proof is similar to the proof of Theorem 5.10.

Theorem 5.21. *The following assertions are equivalent for a semiring R.*

(i) R is both k-regular and k-intra-regular.
(ii) $\mu \wedge \nu \leq \mu \odot_k \nu$ *for all fuzzy k-bi-ideal* μ, ν *of R.*
(iii) $\mu \wedge \nu \leq \mu \odot_k \nu$ *for fuzzy k-bi-ideal* μ *and k-quasi-ideal* ν *of R.*
(iv) $\mu \wedge \nu \leq \mu \odot_k \nu$ *for fuzzy k-quasi-ideal* μ *and k-bi-ideal* ν *of R.*
(v) $\mu \wedge \nu \leq \mu \odot_k \nu$ *for all fuzzy k-quasi-ideal* μ, ν *of R.*

Proof. The proof is similar to the proof of Theorem 5.11.

Chapter 6
$(\in, \in \vee q)$-Fuzzy Ideals in Semirings

Following Bhakat and Das [21], Dudek et. al. [47], Ma and Zhan [106] defined $(\in, \in \vee q)$-fuzzy ideals in semirings. In this chapter we have studied properties of these fuzzy ideals. In Section 1, we look at (α, β)-fuzzy ideals, and show that a fuzzy left (right) ideal of a semiring is an $(\in, \in)$-fuzzy left (right) ideal. Section 2 provides a characterization of $(\in, \in \vee q)$-fuzzy left (right) ideals, quasi and bi-ideals of semirings. Section 3 presents characterizations of regular semirings involving these ideals and Section 4 contains characterizations of regular and intra regular semirings by these ideals. Section 5 presents a study of $(\in, \in \vee q)$-fuzzy k-ideals, k-quasi-ideals and k-bi-ideals of semirings. A study of k-regular and k-intra regular semirings in this context is separately made in Sections 6 and 7.

Throughout this chapter R is a semiring with zero element.

6.1 (α, β)-Fuzzy Ideals

Let X be a nonempty set and $x \in X$. A fuzzy subset λ of X of the form

$$\lambda(y) = \begin{cases} t \in (0,1] & \text{if } y = x \\ 0 & \text{if } y \neq x \end{cases}$$

is said to be a *fuzzy point* with support x and value t and is denoted by x_t. For a fuzzy point x_t and a fuzzy set μ in a set X, Pu and Liu [125] gave meaning to the symbol $x_t \alpha \mu$, where $\alpha \in \{\in, q, \in \vee q, \in \wedge q\}$. A fuzzy point x_t is said to belongs to (*resp. quasi-coincident*) with a fuzzy set μ written $x_t \in \mu$ (resp.$x_t q \mu$) if $\mu(x) \geq t$ (resp.$\mu(x) + t > 1$), and in this case, $x_t \in \vee q \mu$ (resp. $x_t \in \wedge q \mu$) means that $x_t \in \mu$ or $x_t q \mu$ (resp. $x_t \in \mu$ and $x_t q \mu$). To say that $x_t \overline{\alpha} \mu$ means that $x_t \alpha \mu$ does not hold.

Definition 6.1. A fuzzy subset λ of a semiring R is called an (α, β)-*fuzzy subsemiring* of R if

(1) $x_{t_1} \alpha \lambda$ and $y_{t_2} \alpha \lambda$ implies $(x+y)_{t_1 \wedge t_2} \beta \lambda$,
(2) $x_{t_1} \alpha \lambda$ and $y_{t_2} \alpha \lambda$ implies $(xy)_{t_1 \wedge t_2} \beta \lambda$

for all $x, y \in R$ and $t_1, t_2 \in (0,1]$,
where α is any one of $\in, q, \in \vee q$ and β is any one of $\in, q, \in \vee q$ or $\in \wedge q$.

J. Ahsan et al.: Fuzzy Semirings with Applications, STUDFUZZ 278, pp. 105–122.
springerlink.com

The case $\alpha = \in \wedge q$ is omitted since for a fuzzy subset λ of R such that $\lambda(x) \leq 0.5$ for all $x \in R$. In this case for $x_t \in \wedge q\lambda$ we have $\lambda(x) \geq t$ and $\lambda(x) + t > 1$. Thus $1 < \lambda(x) + t \leq \lambda(x) + \lambda(x) = 2\lambda(x)$, which implies $\lambda(x) > 0.5$. This means that $\{x_t : x_t \in \wedge q\lambda\} = \emptyset$.

Definition 6.2. A fuzzy subset λ of a semiring R is called an (α, β)*-fuzzy left (right) ideal* of R if

(1) $x_{t_1} \alpha\lambda$ and $y_{t_2} \alpha\lambda$ implies $(x+y)_{t_1 \wedge t_2} \beta\lambda$
(2) $x_t \alpha\lambda$ and $y \in R$ implies $(yx)_t \beta\lambda$ $((xy)_t \beta\lambda)$

for all $x, y \in R$ and $t \in (0,1]$.

A fuzzy subset λ of a semiring R is called an (α, β)*-fuzzy ideal* of R if it is both an (α, β)-fuzzy left and (α, β)-fuzzy right ideal of R.

Theorem 6.1. *The support of any non-zero (α, β)-fuzzy left (right) ideal of R is a left (right) ideal of R.*

Proof. Let λ be a non-zero (α, β)-fuzzy left ideal of R. Let $x, y \in \lambda_0 = \{x \in R : \lambda(x) > 0\}$. Then $\lambda(x) > 0$ and $\lambda(y) > 0$. Suppose that $\lambda(x+y) = 0$. If $\alpha \in \{\in, \in \vee q\}$, then $x_{\lambda(x)} \alpha\lambda$ and $y_{\lambda(y)} \alpha\lambda$ but $(x+y)_{\lambda(x) \wedge \lambda(y)} \overline{\beta}\lambda$ for every $\beta \in \{\in, q, \in \vee q, \in \wedge q\}$, a contradiction. Also $x_1 q\lambda$ and $y_1 q\lambda$ but $(x+y)_1 \overline{\beta}\lambda$ for every $\beta \in \{\in, q, \in \vee q, \in \wedge q\}$, a contradiction. Hence $\lambda(x+y) > 0$.

Let $x \in \lambda_0$ and $y \in R$. Suppose that $\lambda(yx) = 0$. If $\alpha \in \{\in, \in \vee q\}$, then $x_{\lambda(x)} \alpha\lambda$ but $(yx)_{\lambda(x)} \overline{\beta}\lambda$ for every $\beta \in \{\in, q, \in \vee q, \in \wedge q\}$, a contradiction. Also $x_1 q\lambda$ but $(yx)_1 \overline{\beta}\lambda$ for every $\beta \in \{\in, q, \in \vee q, \in \wedge q\}$, a contradiction. Therefore $\lambda(yx) > 0$. Hence λ_0 is a left ideal of R.

Theorem 6.2. *If I is a left (right) ideal of R, then a fuzzy subset λ of R defined by $\lambda(x) \geq 0.5$ for all $x \in I$ and $\lambda(x) = 0$ otherwise is an $(\alpha, \in \vee q)$-fuzzy left (right) ideal of R.*

Proof. (a) In this part we show that λ is an $(\in, \in \vee q)$-fuzzy left ideal of R. Let $x, y \in R$ and $t_1, t_2 \in (0,1]$ be such that $x_{t_1}, y_{t_2} \in \lambda$. Then $\lambda(x) \geq t_1$ and $\lambda(y) \geq t_2$. Thus $x, y \in I$, so $x + y \in I$. Thus $\lambda(x+y) \geq 0.5$. If $\min\{t_1, t_2\} \leq 0.5$, then $\lambda(x+y) \geq 0.5 \geq \min\{t_1, t_2\}$. Hence $(x+y)_{\min\{t_1,t_2\}} \in \lambda$. If $\min\{t_1, t_2\} > 0.5$, then $\lambda(x+y) + \min\{t_1, t_2\} > 0.5 + 0.5 = 1$, so $(x+y)_{\min\{t_1,t_2\}} q\lambda$. Thus $(x+y)_{\min\{t_1,t_2\}} \in \vee q\lambda$. Let $x, y \in R$ and $t \in (0,1]$ be such that $x_t \in \lambda$. Then $\lambda(x) \geq t$. Thus $x \in I$, so $yx \in I$. Thus $\lambda(yx) \geq 0.5$. If $t \leq 0.5$, then $\lambda(yx) \geq 0.5 \geq t$. Hence $(yx)_t \in \lambda$. If $t > 0.5$, then $\lambda(yx) + t > 0.5 + 0.5 = 1$, so $(yx)_t q\lambda$. Thus $(yx)_t \in \vee q\lambda$. Hence λ is an $(\in, \in \vee q)$-fuzzy left ideal of R.

(b) In this part we show that λ is a $(q, \in \vee q)$-fuzzy left ideal of R. Let $x, y \in R$ and $t_1, t_2 \in (0,1]$ be such that $x_{t_1}, y_{t_2} q\lambda$. Then $\lambda(x) + t_1 > 1$ and $\lambda(y) + t_2 > 1$. Thus $x, y \in I$, so $x + y \in I$. Thus $\lambda(x+y) \geq 0.5$. If $\min\{t_1, t_2\} \leq 0.5$, then $\lambda(x+y) \geq 0.5 \geq \min\{t_1, t_2\}$. Hence $(x+y)_{\min\{t_1,t_2\}} \in \lambda$. If $\min\{t_1, t_2\} > 0.5$, then $\lambda(x+y) + \min\{t_1, t_2\} > 0.5 + 0.5 = 1$, so $(x+y)_{\min\{t_1,t_2\}} q\lambda$. Thus $(x+y)_{\min\{t_1,t_2\}} \in \vee q\lambda$. Let $x, y \in R$ and $t \in (0,1]$ be such that $x_t q\lambda$. Then $\lambda(x) + t > 1$. Thus $x \in I$, so $yx \in I$. Thus $\lambda(yx) \geq 0.5$. If $t \leq 0.5$, then $\lambda(yx) \geq 0.5 \geq t$. Hence $(yx)_t \in \lambda$. If $t > 0.5$, then

$\lambda(yx) + t > 0.5 + 0.5 = 1$, so $(yx)_t q\lambda$. Thus $(yx)_t \in \vee q\lambda$. Hence λ is a $(q, \in \vee q)$-fuzzy left ideal of R.

(c) In this part we show that λ is an $(\in \vee q, \in \vee q)$-fuzzy left ideal of R. Let $x, y \in R$ and $t_1, t_2 \in (0,1]$ be such that $x_{t_1} \in \lambda$ and $y_{t_2} q\lambda$. Then $\lambda(x) \geq t_1$ and $\lambda(y) + t_2 > 1$. Thus $x, y \in I$, so $x + y \in I$. Thus $\lambda(x+y) \geq 0.5$. If $\min\{t_1, t_2\} \leq 0.5$, then $\lambda(x+y) \geq 0.5 \geq \min\{t_1, t_2\}$. Hence $(x+y)_{\min\{t_1,t_2\}} \in \lambda$. If $\min\{t_1, t_2\} > 0.5$, then $\lambda(x+y) + \min\{t_1, t_2\} > 0.5 + 0.5 = 1$, so $(x+y)_{\min\{t_1,t_2\}} q\lambda$. Thus $(x+y)_{\min\{t_1,t_2\}} \in \vee q\lambda$. The rest of the proof is a consequence of (a) and (b).

Theorem 6.3. *A fuzzy subset λ of a semiring R is an $(\in, \in)$-fuzzy left (right) ideal of R if and only if λ is a fuzzy left (right) ideal of R.*

Proof. Let λ be an $(\in, \in)$-fuzzy left ideal of R and $x, y \in R$. If $\lambda(x) = 0$ or $\lambda(y) = 0$, then $\lambda(x+y) \geq 0 = \min\{\lambda(x), \lambda(y)\}$. If $\lambda(x) \neq 0$ and $\lambda(y) \neq 0$, then $x_{\lambda(x)}, y_{\lambda(y)} \in \lambda$. Thus by assumption $(x+y)_{\min\{\lambda(x),\lambda(y)\}} \in \lambda$, that is $\lambda(x+y) \geq \min\{\lambda(x), \lambda(y)\}$.

Let $x, y \in R$. If $\lambda(x) = 0$, then $\lambda(yx) \geq 0 = \lambda(x)$. If $\lambda(x) \neq 0$, then $x_{\lambda(x)} \in \lambda$. Thus by assumption $(yx)_{\lambda(x)} \in \lambda$, that is $\lambda(yx) \geq \lambda(x)$. This shows that λ is a fuzzy left ideal of R.

Conversely, assume that λ is a fuzzy left ideal of R. Let $x, y \in R$ and $t_1, t_2 \in (0,1]$ be such that $x_{t_1}, y_{t_2} \in \lambda$. Then $\lambda(x) \geq t_1$ and $\lambda(y) \geq t_2$. Since $\lambda(x+y) \geq \lambda(x) \wedge \lambda(y) \geq t_1 \wedge t_2$, so $(x+y)_{t_1 \wedge t_2} \in \lambda$. Also if $x_t \in \lambda$. Then $\lambda(x) \geq t$. Since $\lambda(yx) \geq \lambda(x) \geq t$, so $(yx)_t \in \lambda$. This shows that λ is an $(\in, \subset)$ fuzzy left ideal of R.

Theorem 6.4. *Every $(\in \vee q, \in \vee q)$-fuzzy left (right) ideal of a semiring R is an $(\in, \in \vee q)$-fuzzy left (right) ideal of R.*

Proof. The proof follows from the fact that if $x_t \in \lambda$, then $x_t \in \vee q\lambda$.

Theorem 6.5. *Every $(q, \in \vee q)$-fuzzy left (right) ideal of a semiring R is an $(\in, \in \vee q)$-fuzzy left (right) ideal of R.*

Proof. It is easy to see that each (α, β)-fuzzy left (right) ideal of R is an $(\alpha, \in \vee q)$-fuzzy left (right) ideal of R.

The above theorems shows that every $(\alpha, \in \vee q)$-fuzzy left (right) ideal of R is an $(\in, \in \vee q)$-fuzzy left (right) ideal of R. Thus $(\in, \in \vee q)$-fuzzy left (right) ideal of R plays central role in the study of (α, β)-fuzzy left (right) ideal of R.

6.2 $(\in, \in \vee q)$-Fuzzy Ideals

We start this section with the following Theorem.

Theorem 6.6. *For any fuzzy subset λ of R and for all $x, y \in R$ and $t, r \in (0,1]$, $(1a)$ is equivalent to $(1b)$, $(2a)$ is equivalent to $(2b)$, $(3a)$ is equivalent to $(3b)$ and $(4a)$ is equivalent to $(4b)$ where*

$(1a)$ $x_t, y_r \in \lambda \Rightarrow (x+y)_{\min\{t,r\}} \in \vee q\lambda$.

$(1b)$ $\lambda(x+y) \geq \min\{\lambda(x), \lambda(y), 0.5\}$.

$(2a)$ $x_t, y_r \in \lambda \Rightarrow (xy)_{\min\{t,r\}} \in \vee q\lambda$.
$(2b)$ $\lambda(xy) \geq \min\{\lambda(x), \lambda(y), 0.5\}$.
$(3a)$ $x_t \in \lambda$ *and* $y \in R \Rightarrow (xy)_t \in \vee q\lambda$.
$(3b)$ $\lambda(xy) \geq \min\{\lambda(x), 0.5\}$.
$(4a)$ $x_t \in \lambda$ *and* $y \in R \Rightarrow (yx)_t \in \vee q\lambda$.
$(4b)$ $\lambda(yx) \geq \min\{\lambda(x), 0.5\}$
Similarly, for all $a,b,x,y,z \in R$ *such that* $x+a=b$, *and for any* $t,r \in (0,1]$, $(5a)$ *is equivalent to* $(5b)$, $(6a)$ *is equivalent to* $(6b)$ *where*
$(5a)$ $a_t, b_r \in \lambda \Rightarrow (x)_{\min\{t,r\}} \in \vee q\lambda$.
$(5b)$ $\lambda(x) \geq \min\{\lambda(a), \lambda(b), 0.5\}$.
$(6a)$ $x_t, z_r \in \lambda \Rightarrow (xyz)_{\min\{t,r\}} \in \vee q\lambda$.
$(6b)$ $\lambda(xyz) \geq \min\{\lambda(x), \lambda(z), 0.5\}$.

Proof. We prove only $(1a)$ if and only if $(1b)$. The proofs of the other parts are similar to this.

$(1a) \Rightarrow (1b)$ First we consider the case when $\min\{\lambda(x), \lambda(y)\} < 0.5$. Let $x, y \in R$ be such that $\lambda(x+y) < \min\{\lambda(x), \lambda(y), 0.5\} = \min\{\lambda(x), \lambda(y)\}$. Choose $t \in (0,1]$ such that $\lambda(x+y) < t = \min\{\lambda(x), \lambda(y)\}$. Then $x_t, y_t \in \lambda$ but $(x+y)_t \overline{\in} \lambda$. Also $\lambda(x+y)+t < 1$. This implies $(x+y)_t \overline{q} \lambda$. Thus $(x+y)_t \overline{\in} \wedge \overline{q}\lambda$, which is a contradiction. Hence $\lambda(x+y) \geq \min\{\lambda(x), \lambda(y), 0.5\}$.

Now we consider the second case when $\min\{\lambda(x), \lambda(y)\} \geq 0.5$. Let $x, y \in R$ be such that $\lambda(x+y) < \min\{\lambda(x), \lambda(y), 0.5\} = 0.5$. Then $x_{0.5}, y_{0.5} \in \lambda$ but $(x+y)_{0.5} \overline{\in} \wedge \overline{q}\lambda$, which is a contradiction. Hence $\lambda(x+y) \geq \min\{\lambda(x), \lambda(y), 0.5\}$.

$(1b) \Rightarrow (1a)$ Let $x_{t_1}, y_{t_2} \in \lambda$. Then by hypothesis $\lambda(x+y) \geq \min\{\lambda(x), \lambda(y), 0.5\} \geq \min\{t_1, t_2, 0.5\}$. If $\min\{t_1, t_2\} > 0.5$, then $\lambda(x+y) \geq 0.5$. This implies $\lambda(x+y) + \min\{t_1, t_2\} > 1$. Thus $(x+y)_{\min\{t_1,t_2\}} q\lambda$. If $\min\{t_1, t_2\} \leq 0.5$, then $\lambda(x+y) \geq \min\{t_1, t_2\}$. Thus $(x+y)_{\min\{t_1,t_2\}} \in \lambda$. Hence $(x+y)_{\min\{t_1,t_2\}} \in \vee q\lambda$.

Definition 6.3. A fuzzy subset λ of R is called an $(\in, \in \vee q)$-*fuzzy left (resp. right) ideal* of R if it satisfies $(1b)$ and $(4b)$ (resp. $(1b)$ and $(3b)$).

Definition 6.4. A fuzzy subset λ of R is called an $(\in, \in \vee q)$-*fuzzy bi-ideal* of R if it satisfies $(1a)$, $(2a)$ and $(6a)$ (resp. $(1b)$, $(2b)$ and $(6b)$).

Definition 6.5. A fuzzy subset λ of R is called an $(\in, \in \vee q)$-*fuzzy quasi-ideal* of R if it satisfies $(1b)$ and $(7b)$, where

$(7b)$ $\lambda(x) \geq \min\{(\lambda \odot_k \mathscr{R})(x), (\mathscr{R} \odot_k \lambda))(x), 0.5\}$ for all $x \in R$,

where $\mathscr{R}$ is the fuzzy subset of R mapping every element of R on 1.

Theorem 6.7. *For any fuzzy subset* λ *of* R *and for all* $x, y \in R$ *and* $t, r \in (0,1]$, $(1aa)$ *implies* $(1b)$, $(2aa)$ *implies* $(2b)$, $(3aa)$ *implies* $(3b)$ *and* $(4aa)$ *implies* $(4b)$ *where*
$(1aa)$ $x_t, y_r q\lambda \Rightarrow (x+y)_{\min\{t,r\}} \in \vee q\lambda$.
$(2aa)$ $x_t, y_r q\lambda \Rightarrow (xy)_{\min\{t,r\}} \in \vee q\lambda$.
$(3aa)$ $x_t q\lambda$ *and* $y \in R \Rightarrow (xy)_t \in \vee q\lambda$.
$(4aa)$ $x_t q\lambda$ *and* $y \in R \Rightarrow (yx)_t \in \vee q\lambda$.
Similarly for all $a,b,x,y,z \in R$ *such that* $x+a=b$, *and for any* $t,r \in (0,1]$, $(5aa)$ *implies* $(5b)$, $(6aa)$ *implies* $(6b)$ *where:*

(5*aa*) $a_t, b_r q\lambda \Rightarrow (x)_{\min\{t,r\}} \in \vee q\lambda$.
(6*aa*) $x_t, z_r q\lambda$ *implies* $(xyz)_{\min\{t,r\}} \in \vee q\lambda$.

Proof. (1*aa*) $\Rightarrow$ (1*b*) Suppose λ is a fuzzy subset of R which satisfies (1*aa*). Let $x, y \in R$ be such that $\lambda(x+y) < \min\{\lambda(x), \lambda(y), 0.5\}$. Then $1 - \lambda(x+y) > 1 - \min\{\lambda(x), \lambda(y), 0.5\}$. This implies that $1 - \lambda(x+y) > \max\{1 - \lambda(x), 1 - \lambda(y), 0.5\}$. Take $t \in (0,1]$ such that $1 - \lambda(x+y) > t > \max\{1 - \lambda(x), 1 - \lambda(y), 0.5\}$. Then $t > 1 - \lambda(x)$ and $t > 1 - \lambda(y)$, that is $\lambda(x) + t > 1$ and $\lambda(y) + t > 1$. This implies $x_t, y_t q\lambda$. But $\lambda(x+y) + t < 1$ and $\lambda(x+y) < 0.5 < t$, that is $(x+y)_t \overline{q}\lambda$ and $(x+y)_t \overline{\in}\lambda$. This contradicts our hypothesis. Hence $\lambda(x+y) \geq \min\{\lambda(x), \lambda(y), 0.5\}$.

Similarly we can prove the other parts.

From the Theorem 6.6 and 6.7 we deduce that every $(q, \in \vee q)$-fuzzy left (right, bi-) ideal of R is an $(\in, \in \vee q)$-fuzzy left (right, bi-) ideal of R.

The proofs of the following results are straightforward and hence omitted.

Lemma 6.1. *A nonempty subset A of R is a left (right) ideal, of R if and only if* χ_A *is an* $(\in, \in \vee q)$*-fuzzy left (right) ideal of R.*

Similarly a nonempty subset A of R is a quasi-ideal (bi-ideal) of R if and only if χ_A *is an* $(\in, \in \vee q)$*-fuzzy quasi-ideal (bi-ideal) of R.*

Theorem 6.8. *Let* λ *be an* $(\in, \in \vee q)$*-fuzzy left (right) ideal of R. Then* $\lambda \wedge 0.5$ *is an* $(\in, \in \vee q)$*-fuzzy left (right) ideal of R, where* $(\lambda \wedge 0.5)(x) = \lambda(x) \wedge 0.5$ *for all* $x \in R$.

Proof. Let λ be an $(\in, \in \vee q)$-fuzzy left ideal of R and $x, y \in R$. Then

$$\begin{aligned}
(\lambda \wedge 0.5)(x+y) &= \lambda(x+y) \wedge 0.5 \\
&\geq (\min\{\lambda(x), \lambda(y), 0.5\}) \wedge 0.5 \\
&= \min\{\lambda(x) \wedge 0.5, \lambda(y) \wedge 0.5\} \wedge 0.5 \\
&= \min\{(\lambda \wedge 0.5)(x), (\lambda \wedge 0.5)(y), 0.5\} \\
&\Rightarrow (\lambda \wedge 0.5)(x+y) \geq \min\{(\lambda \wedge 0.5)(x), (\lambda \wedge 0.5)(y), 0.5\}.
\end{aligned}$$

Similarly we can show that

$$(\lambda \wedge 0.5)(xy) \geq \min\{(\lambda \wedge 0.5)(y), 0.5\}.$$

This shows that $(\lambda \wedge 0.5)$ is an $(\in, \in \vee q)$-fuzzy left ideal of R.

Similarly we can show that

Theorem 6.9. *If* λ *is an* $(\in, \in \vee q)$*-fuzzy bi-ideal of R, then* $(\lambda \wedge 0.5)$ *is an* $(\in, \in \vee q)$*-fuzzy bi-ideal of R.*

Definition 6.6. Let λ, μ be fuzzy subsets of R. Then the fuzzy subsets $\lambda \wedge_{0.5} \mu$, $\lambda \circ_{0.5} \mu$ and $\lambda +_{0.5} \mu$ of R are defined as following:

$$
\begin{aligned}
(\lambda \wedge_{0.5} \mu)(x) &= \min\{\lambda(x), \mu(x), 0.5\} \\
(\lambda \circ_{0.5} \mu)(x) &= (\lambda \circ \mu)(x) \wedge 0.5 \\
\lambda +_{0.5} \mu &= (\lambda + \mu) \wedge 0.5
\end{aligned}
$$

for all $x \in R$.

Lemma 6.2. *Let A, B be nonempty subsets of R. Then $\chi_A +_{0.5} \chi_B = \chi_{A+B} \wedge 0.5$.*

Proof. Let A, B be subsets of R and $x \in R$. If $x \in A + B$, then there exist $a \in A$ and $b \in B$ such that $x = a + b$. Thus

$$
\begin{aligned}
(\chi_A +_{0.5} \chi_B)(x) &= \sup_{x=c+d} \{\chi_A(c) \wedge \chi_B(d)\} \wedge 0.5 \\
&= 1 \wedge 0.5 = \chi_{A+B}(x) \wedge 0.5
\end{aligned}
$$

If $x \notin A + B$ then there do not exist $a \in A$ and $b \in B$ such that $x = a + b$. Thus

$$(\chi_A +_{0.5} \chi_B)(x) = 0 \wedge 0.5 = \chi_{A+B}(x) \wedge 0.5 .$$

Hence $\chi_A +_{0.5} \chi_B = \chi_{A+B}(x) \wedge 0.5$.

Theorem 6.10. *A fuzzy subset λ of a semiring R satisfies condition $(1b)$ if and only if it satisfies condition $(8b)$, where*
$(8b)$ $\lambda +_{0.5} \lambda \leq \lambda \wedge 0.5$

Proof. Suppose λ satisfies condition $(1b)$ and $x \in R$. Then

$$
\begin{aligned}
(\lambda +_{0.5} \lambda)(x) &= \left(\sup_{x=a+b} \{\min\{\lambda(a), \lambda(b)\}\} \right) \wedge 0.5 \\
&= \left(\sup_{x=a+b} \{\min\{\lambda(a), \lambda(b), 0.5\} \right) \wedge 0.5 \\
&\leq \left(\sup_{x=a+b} \lambda(a+b) \right) \wedge 0.5 \\
&\leq \lambda(x) \wedge 0.5 .
\end{aligned}
$$

Thus $\lambda +_{0.5} \lambda \leq \lambda \wedge 0.5$.

Conversely, assume that $\lambda +_{0.5} \lambda \leq \lambda \wedge 0.5$ and $x, y \in R$. Then

$$
\begin{aligned}
\lambda(x+y) \geq \lambda(x+y) \wedge 0.5 \geq (\lambda +_{0.5} \lambda)(x+y) & \\
&= \left(\sup_{x+y=a+b} \{\min\{\lambda(a), \lambda(b)\}\} \right) \wedge 0.5 \\
&\geq \min\{\lambda(x), \lambda(y)\} \wedge 0.5 \\
&= \min\{\lambda(x), \lambda(y), 0.5\}
\end{aligned}
$$

Thus λ satisfies condition $(1b)$.

Theorem 6.11. *A fuzzy subset λ of a semiring R satisfies condition $(4b)$ (resp. $(3b)$) if and only if it satisfies condition*
$(9b)$ $\mathscr{R} \circ_{0.5} \lambda \leq \lambda \wedge 0.5$ *(resp.* $\lambda \circ_{0.5} \mathscr{R} \leq \lambda \wedge 0.5$*).*

Proof. Suppose λ satisfies condition $(4b)$. We show that λ satisfies condition $(9b)$. Let $x \in R$. If $(\mathscr{R} \circ_{0.5} \lambda)(x) = 0$, then $(\mathscr{R} \circ_{0.5} \lambda)(x) \leq \lambda(x) \wedge 0.5$. Otherwise, there exist elements $a_i, b_i \in R$ such that $x = \sum_{i=1}^{m} a_i b_i$. Then

$$\begin{aligned}
(\mathscr{R} \circ_{0.5} \lambda)(x) &= \sup_{x=\Sigma_{i=1}^{m} a_i b_i} \left\{ \bigwedge_{i=1}^{m} \{\mathscr{R}(a_i), \lambda(b_i)\} \right\} \wedge 0.5 \\
&= \sup_{x=\Sigma_{i=1}^{m} a_i b_i} \left\{ \bigwedge_{i=1}^{m} \lambda(b_i) \right\} \wedge 0.5 \\
&= \sup_{x=\Sigma_{i=1}^{m} a_i b_i} \left\{ \bigwedge_{i=1}^{m} \{\lambda(b_i) \wedge 0.5\} \right\} \wedge 0.5 \\
&\leq \sup_{x=\Sigma_{i=1}^{m} a_i b_i} \left\{ \bigwedge_{i=1}^{m} \lambda(a_i b_i) \right\} \wedge 0.5 \\
&= \sup_{x=\Sigma_{i=1}^{m} a_i b_i} \left\{ \bigwedge_{i=1}^{m} \{\lambda(a_i b_i) \wedge 0.5\} \right\} \wedge 0.5 \\
&\leq \sup_{x=\Sigma_{i=1}^{m} a_i b_i} \lambda\left(\sum_{i=1}^{m} a_i b_i \right) \wedge 0.5 \\
&\leq \lambda(x) \wedge 0.5 .
\end{aligned}$$

This implies $\mathscr{R} \circ_{0.5} \lambda \leq \lambda \wedge 0.5$.

Conversely, assume that λ satisfies condition $(9b)$. We show that λ satisfies condition $(4b)$. Let $x, y \in R$. Then

$$\begin{aligned}
\lambda(xy) \geq \lambda(xy) \wedge 0.5 &\geq (\mathscr{R} \circ_{0.5} \lambda)(xy) \\
&= \sup_{xy=\Sigma_{i=1}^{m} a_i b_i} \left\{ \bigwedge_{i=1}^{m} \{\mathscr{R}(a_i) \wedge \lambda(b_i)\} \right\} \wedge 0.5 \\
&= \sup_{xy=\Sigma_{i=1}^{m} a_i b_i} \left\{ \bigwedge_{i=1}^{m} \lambda(b_i) \right\} \wedge 0.5 \\
&\geq \lambda(y) \wedge 0.5.
\end{aligned}$$

This shows that λ satisfies condition $(4b)$.

Similarly we can prove that

Theorem 6.12. *A fuzzy subset λ of a semiring R satisfies condition $(6b)$ if and only if it satisfies condition $(10b)$, where*
$(10b)$ $\lambda \circ_{0.5} \mathscr{R} \circ_{0.5} \lambda \leq \lambda \wedge 0.5$ *. By Theorem 6.10 and Theorem 6.11, the following result holds.*

Theorem 6.13. *A fuzzy subset λ of R is an $(\in, \in \vee q)$-fuzzy left (resp. right) ideal of R if and only if λ satisfies conditions $(8b)$ and $(9b)$.*

By using Theorem 6.10, we can prove the following:

Theorem 6.14. *A fuzzy subset λ of R is an $(\in, \in \vee q)$-fuzzy quasi-ideal of R if and only if λ satisfies conditions $(7b)$ and $(8b)$.*

Theorem 6.15. *A fuzzy subset λ of R is an $(\in, \in \vee q)$-fuzzy bi-ideal of R if and only if λ satisfies conditions $(2b)$, $(8b)$ and $(10b)$.*

From Theorem 6.13 and Theorem 6.14 we deduce that

Theorem 6.16. *Every $(\in, \in \vee q)$-fuzzy left (right) ideal of R is an $(\in, \in \vee q)$-fuzzy quasi-ideal of R.*

From Theorem 6.14 and Theorem 6.15 we deduce that

Theorem 6.17. *Every $(\in, \in \vee q)$-fuzzy quasi-ideal of R is an $(\in, \in \vee q)$-fuzzy bi-ideal of R.*

Remark 6.1. The converse of the Theorem 6.16 and Theorem 6.17 are not true in general.

Example 6.1. Let $\mathbb{Z}_0$ be the set of all non negative integers and $R = \left\{ \begin{pmatrix} p & q \\ r & s \end{pmatrix} : p, q, r, s \in \mathbb{Z}_0 \right\}$ and $A = \left\{ \begin{pmatrix} p & 0 \\ 0 & 0 \end{pmatrix} : p \in \mathbb{Z}_0 \right\}$. Then R is a semiring under the usual addition and multiplication of matrices, and A is a quasi-ideal of R but A is not a left (right) ideal of R. Thus by Lemma 6.1, χ_A is an $(\in, \in \vee q)$-fuzzy quasi-ideal of R but χ_A is not an $(\in, \in \vee q)$-fuzzy left (right) ideal of R.

Example 6.2. Let $\mathbb{Z}^+$ and $\mathbb{R}^+$ be the sets of all positive integers and positive real numbers, respectively. And

$$R = \left\{ \begin{pmatrix} 0 & 0 \\ 0 & 0 \end{pmatrix} \right\} \cup \left\{ \begin{pmatrix} p & 0 \\ q & r \end{pmatrix} : p, q \in \mathbb{R}^+, r \in \mathbb{Z}^+ \right\}$$
$$A = \left\{ \begin{pmatrix} 0 & 0 \\ 0 & 0 \end{pmatrix} \right\} \cup \left\{ \begin{pmatrix} p & 0 \\ q & r \end{pmatrix} : p, q \in \mathbb{R}^+, r \in \mathbb{Z}^+, p < q \right\}$$
$$B = \left\{ \begin{pmatrix} 0 & 0 \\ 0 & 0 \end{pmatrix} \right\} \cup \left\{ \begin{pmatrix} p & 0 \\ q & r \end{pmatrix} : p, q \in \mathbb{R}^+, r \in \mathbb{Z}^+, q > 3 \right\}$$

Then R is a semiring under the usual addition and multiplication of matrices, and A is a right ideal, B is a left ideal, AB is a bi-ideal of R and it is not a quasi-ideal of R. Thus by Lemma 6.1, χ_{AB} is an $(\in, \in \vee q)$-fuzzy bi-ideal of R but it is not an $(\in, \in \vee q)$-fuzzy quasi-ideal of R.

Lemma 6.3. *If λ and μ are $(\in, \in \vee q)$-fuzzy right and left ideals of R respectively, then $\lambda \circ_{0.5} \mu \leq \lambda \wedge_{0.5} \mu$.*

Proof. Let $x \in R$. If $(\lambda \circ_{0.5} \mu)(x) = 0$, then $(\lambda \circ_{0.5} \mu)(x) \leq (\lambda \wedge_{0.5} \mu)(x)$. Otherwise, there exist elements $a_i, b_i \in R$ such that $x = \sum_{i=1}^{m} a_i b_i$. Then

$$\begin{aligned}
(\lambda \circ_{0.5} \mu)(x) &= (\lambda \circ \mu)(x) \wedge 0.5 \\
&= \sup_{x=\sum_{i=1}^{m} a_i b_i} \left\{ \bigwedge_{i=1}^{m} \{\lambda(a_i) \wedge \mu(b_i)\} \right\} \wedge 0.5 \\
&= \sup_{x=\sum_{i=1}^{m} a_i b_i} \left\{ \bigwedge_{i=1}^{m} \{\lambda(a_i) \wedge 0.5 \wedge \mu(b_i) \wedge 0.5\} \right\} \wedge 0.5 \\
&\leq \sup_{x=\sum_{i=1}^{m} a_i b_i} \left\{ \bigwedge_{i=1}^{m} \{\lambda(a_i b_i) \wedge \mu(a_i b_i)\} \right\} \wedge 0.5 \\
&= \sup_{x=\sum_{i=1}^{m} a_i b_i} \left\{ \bigwedge_{i=1}^{m} \{\lambda(a_i b_i) \wedge 0.5\} \wedge \bigwedge_{i=1}^{m} \{\mu(a_i b_i) \wedge 0.5\} \right\} \wedge 0.5 \\
&\leq \sup_{x=\sum_{i=1}^{m} a_i b_i} \left\{ \lambda\left(\sum_{i=1}^{m} a_i b_i\right) \wedge \mu\left(\sum_{i=1}^{m} a_i b_i\right) \right\} \wedge 0.5 \\
&\leq (\lambda \wedge_{0.5} \mu)(x).
\end{aligned}$$

Theorem 6.18. *A fuzzy subset λ of a semiring R is an $(\in, \in \vee q)$-fuzzy left (right) ideal of R if and only if each nonempty $U(\lambda;t)$ is a left (right) ideal of R, for all $t \in (0, 0.5]$.*

Proof. Suppose λ be an $(\in, \in \vee q)$-fuzzy left ideal of R and $U(\lambda;t) \neq \emptyset$ for $t \in (0, 0.5]$. Let $x, y \in U(\lambda;t)$. Then $\lambda(x) \geq t$ and $\lambda(y) \geq t$. Since $\lambda(x+y) \geq \min(\lambda(x), \lambda(y), 0.5)$, we have $\lambda(x+y) \geq t$. This implies $x+y \in U(\lambda;t)$. Let $r \in R$. Since $\lambda(rx) \geq \min\{\lambda(x), 0.5\}$, we have $\lambda(rx) \geq t$. This implies $rx \in U(\lambda;t)$. Hence $U(\lambda;t)$ is a left ideal of R.

Conversely, assume that $U(\lambda;t)$ is a left ideal of R for all $t \in (0, 0.5]$. Suppose $x, y \in R$ such that $\lambda(x+y) < \min(\lambda(x), \lambda(y), 0.5)$. Select $t \in (0, 0.5]$ such that $\lambda(x+y) < t \leq \min(\lambda(x), \lambda(y), 0.5)$. Then $x, y \in U(\lambda;t)$ but $x+y \notin U(\lambda;t)$, a contradiction. Hence $\lambda(x+y) \geq \min(\lambda(x), \lambda(y), 0.5)$.

Similarly, we can prove $\lambda(rx) \geq \min(\lambda(x), 0.5)$.

6.3 Regular Semirings

In this section we characterize regular semirings by the properties of their $(\in, \in \vee q)$-fuzzy ideals, $(\in, \in \vee q)$-fuzzy *bi*-ideals and $(\in, \in \vee q)$-fuzzy *quasi*-ideals.

Theorem 6.19. *For a semiring R the following conditions are equivalent.*

(i) R is regular.

(ii) $(\lambda \wedge_{0.5} \mu) = (\lambda \circ_{0.5} \mu)$ for every $(\in, \in \vee q)$-fuzzy right ideal λ and every $(\in, \in \vee q)$-fuzzy left ideal μ of R.

Proof. (*i*) $\Rightarrow$ (*ii*) Let λ be an $(\in, \in \vee q)$-fuzzy right ideal and μ be an $(\in, \in \vee q)$-fuzzy left ideal of R and $x \in R$. Then there exists $a \in R$ such that $x = xax$. Now

$$\begin{aligned}(\lambda \circ_{0.5} \mu)(x) &= \sup_{x=\Sigma_{i=1}^{m} a_i b_i} \left\{ \bigwedge_{i=1}^{m} \{\lambda(a_i) \wedge \mu(b_i)\} \right\} \wedge 0.5 \\ &\geq \{\lambda(xa) \wedge \mu(x)\} \wedge 0.5 \\ &\geq \{\lambda(x) \wedge \mu(x)\} \wedge 0.5 \\ &= (\lambda \wedge_{0.5} \mu)(x).\end{aligned}$$

Thus $\lambda \circ_{0.5} \mu \geq \lambda \wedge_{0.5} \mu$. By Lemma 6.3, $\lambda \circ_{0.5} \mu \leq \lambda \wedge_{0.5} \mu$. Hence $\lambda \wedge_{0.5} \mu = \lambda \circ_{0.5} \mu$.

(*ii*) $\Rightarrow$ (*i*) Let A and B be right and left ideals of R, respectively. Then by Lemma 6.1, χ_A is an $(\in, \in \vee q)$-fuzzy right and χ_B is an $(\in, \in \vee q)$-fuzzy left ideal of R. By hypothesis $\chi_A \circ_{0.5} \chi_B = \chi_A \wedge_{0.5} \chi_B \Rightarrow \chi_{AB} \wedge 0.5 = \chi_{A \cap B} \wedge 0.5$
$\Rightarrow A \cap B = AB \Rightarrow R$ is regular.

Theorem 6.20. *For a semiring R, the following conditions are equivalent.*
(i) R is regular.
(ii) $\lambda \wedge 0.5 \leq (\lambda \circ_{0.5} \mathscr{R} \circ_{0.5} \lambda)$ for every $(\in, \in \vee q)$-fuzzy bi-ideal λ of R.
(iii) $\lambda \wedge 0.5 \leq (\lambda \circ_{0.5} \mathscr{R} \circ_{0.5} \lambda)$ for every $(\in, \in \vee q)$-fuzzy quasi-ideal λ of R.

Proof. (*i*) $\Rightarrow$ (*ii*) Let λ be an $(\in, \in \vee q)$-fuzzy bi-ideal of R and $x \in R$. Then there exists $a \in R$ such that $x = xax$. Now

$$\begin{aligned}(\lambda \circ_{0.5} \mathscr{R} \circ_{0.5} \lambda)(x) &= \sup_{x=\Sigma_{i=1}^{m} a_i b_i} \left\{ \bigwedge_{i=1}^{m} \{(\lambda \circ_{0.5} \mathscr{R})(a_i) \wedge \lambda(b_i)\} \right\} \wedge 0.5 \\ &\geq \{(\lambda \circ_{0.5} \mathscr{R})(xa) \wedge \lambda(x)\} \wedge 0.5 \\ &= \left\{ \left\{ \sup_{xa=\Sigma_{i=1}^{m} c_i d_i} \left\{ \bigwedge_{i=1}^{m} \{\lambda(c_i) \wedge \mathscr{R}(d_i)\} \right\} \wedge 0.5 \right\} \wedge \lambda(x) \right\} \wedge 0.5 \\ &\geq \{(\lambda(xax) \wedge 0.5) \wedge \lambda(x)\} \wedge 0.5 \qquad \text{(because } xa = xaxa\text{)} \\ &\geq \lambda(x) \wedge 0.5.\end{aligned}$$

(*ii*) $\Rightarrow$ (*iii*) This is straightforward.

(*iii*) $\Rightarrow$ (*i*) Let Q be any quasi-ideal of R. Then by Lemma 6.1, χ_Q is an $(\in, \in \vee q)$-fuzzy quasi-ideal of R. Now by hypothesis $\chi_Q \wedge 0.5 \leq (\chi_Q \circ_{0.5} \mathscr{R} \circ_{0.5} \chi_Q) = \chi_{QRQ} \wedge 0.5$.
$\Rightarrow Q \subseteq QRQ$. Also $QRQ \subseteq RQ \cap QR = Q$. Thus $Q = QRQ$.
Hence R is regular.

Theorem 6.21. *For a semiring R, the following conditions are equivalent.*
(i) R is regular.
(ii) $(\lambda \wedge_{0.5} \mu) \leq (\lambda \circ_{0.5} \mu \circ_{0.5} \lambda)$ for every $(\in, \in \vee q)$-fuzzy bi-ideal λ and every $(\in, \in \vee q)$-fuzzy ideal μ of R.
(iii) $(\lambda \wedge_{0.5} \mu) \leq (\lambda \circ_{0.5} \mu \circ_{0.5} \lambda)$ for every $(\in, \in \vee q)$-fuzzy quasi-ideal λ and every $(\in, \in \vee q)$-fuzzy ideal μ of R.

Proof. (*i*) $\Rightarrow$ (*ii*) Let λ be an $(\in, \in \vee q)$-fuzzy bi-ideal and μ be an $(\in, \in \vee q)$-fuzzy ideal *of* R. Then for $x \in R$, there exists $a \in R$ such that $x = xax$. Now

$$(\lambda \circ_{0.5} \mu \circ_{0.5} \lambda)(x)$$
$$= \sup_{x=\Sigma_{i=1}^{m} a_i b_i} \left\{ \bigwedge_{i=1}^{m} \{(\lambda \circ_{0.5} \mu)(a_i) \wedge \lambda(b_i)\} \right\} \wedge 0.5$$
$$\geq \{(\lambda \circ_{0.5} \mu)(xa) \wedge \lambda(x)\} \wedge 0.5$$
$$= \left\{ \left\{ \sup_{xa=\Sigma_{i=1}^{m} c_i d_i} \left\{ \bigwedge_{i=1}^{m} \{\lambda(c_i) \wedge \mu(d_i)\} \right\} \wedge 0.5 \right\} \wedge \lambda(x) \right\} \wedge 0.5$$
$$\geq \{(\{\lambda(x) \wedge \mu(axa) \wedge \lambda(x)\} \wedge 0.5)\} \wedge 0.5 \ \text{(because } xa = xaxa)$$
$$\geq (\lambda \wedge \mu)(x) \wedge 0.5 = (\lambda \wedge_{0.5} \mu)(x).$$

(*ii*) $\Rightarrow$ (*iii*) This is straightforward.

(*iii*) $\Rightarrow$ (*i*) Let λ be any $(\in, \in \vee q)$-fuzzy quasi ideal of R. Since $\mathscr{R}$ is an $(\in, \in \vee q)$-fuzzy ideal *of* R, so by hypothesis $(\lambda \wedge_{0.5} \mathscr{R}) \leq (\lambda \circ_{0.5} \mathscr{R} \circ_{0.5} \lambda)$ $\Rightarrow \lambda \wedge 0.5 \leq (\lambda \circ_{0.5} \mathscr{R} \circ_{0.5} \lambda)$. Therefore by Theorem 6.20, R is regular.

Theorem 6.22. *For a semiring R, the following conditions are equivalent.*

(*i*) *R is regular.*

(*ii*) $(\lambda \wedge_{0.5} \mu) \leq (\lambda \circ_{0.5} \mu)$ *for every $(\in, \in \vee q)$-fuzzy bi-ideal λ and every $(\in, \in \vee q)$-fuzzy left ideal μ of R.*

(*iii*) $(\lambda \wedge_{0.5} \mu) \leq (\lambda \circ_{0.5} \mu)$ *for every $(\in, \in \vee q)$-fuzzy quasi-ideal λ and every $(\in, \in \vee q)$-fuzzy left ideal μ of R.*

(*iv*) $(\lambda \wedge_{0.5} \mu) \leq (\lambda \circ_{0.5} \mu)$ *for every $(\in, \in \vee q)$-fuzzy right ideal λ and every $(\in, \in \vee q)$-fuzzy bi-ideal μ of R.*

(*v*) $(\lambda \wedge_{0.5} \mu) \leq (\lambda \circ_{0.5} \mu)$ *for every $(\in, \in \vee q)$-fuzzy right ideal λ and every $(\in, \in \vee q)$-fuzzy quasi-ideal μ of R.*

(*vi*) $(\lambda \wedge_{0.5} \mu \wedge_{0.5} \nu) \leq (\lambda \circ_{0.5} \mu \circ_{0.5} \nu)$ *for every $(\in, \in \vee q)$-fuzzy right ideal λ, every $(\in, \in \vee q)$-fuzzy bi-ideal μ and every $(\in, \in \vee q)$-fuzzy left ideal ν of R.*

(*vii*) $(\lambda \wedge_{0.5} \mu \wedge_{0.5} \nu) \leq (\lambda \circ_{0.5} \mu \circ_{0.5} \nu)$ *for every $(\in, \in \vee q)$-fuzzy right ideal λ, every $(\in, \in \vee q)$-fuzzy quasi-ideal μ and every $(\in, \in \vee q)$-fuzzy left ideal ν of R.*

Proof. (*i*) $\Rightarrow$ (*ii*) Let λ be any $(\in, \in \vee q)$-fuzzy bi-ideal and μ any $(\in, \in \vee q)$-fuzzy left ideal of R. Now for any $a \in R$ there exists $x \in R$ such that $a = axa$. Thus we have

$$(\lambda \circ_{0.5} \mu)(a) = \sup_{a=\Sigma_{i=1}^{m} a_i b_i} \left\{ \bigwedge_{i=1}^{m} \{\lambda(a_i) \wedge \mu(b_i)\} \right\} \wedge 0.5$$
$$\geq \{\lambda(a) \wedge \mu(xa)\} \wedge 0.5$$
$$\geq \{\lambda(a) \wedge \mu(a) \wedge 0.5\} \wedge 0.5 = (\lambda \wedge_{0.5} \mu)(a).$$

Thus $(\lambda \circ_{0.5} \mu) \geq (\lambda \wedge_{0.5} \mu)$.

(*ii*) $\Rightarrow$ (*iii*) This is obvious because every $(\in, \in \vee q)$-fuzzy quasi-ideal is an $(\in, \in \vee q)$-fuzzy bi-ideal.

(*iii*) $\Rightarrow$ (*i*) Let λ be an $(\in, \in \vee q)$-fuzzy right ideal and μ be an $(\in, \in \vee q)$-fuzzy left ideal of R. Since every $(\in, \in \vee q)$-fuzzy right ideal is an $(\in, \in \vee q)$-fuzzy quasi-ideal, so by hypothesis we have $(\lambda \circ_{0.5} \mu) \geq (\lambda \wedge_{0.5} \mu)$. But by Lemma 6.3, $(\lambda \circ_{0.5} \mu) \leq (\lambda \wedge_{0.5} \mu)$. Hence $(\lambda \circ_{0.5} \mu) = (\lambda \wedge_{0.5} \mu)$. Thus by Theorem 6.19, R is regular.

Similarly we can show that $(i) \Leftrightarrow (iv) \Leftrightarrow (v)$.

$(i) \Rightarrow (vi)$ Let λ be an $(\in, \in \vee q)$-fuzzy right ideal, μ be an $(\in, \in \vee q)$-fuzzy bi-ideal and ν be an $(\in, \in \vee q)$-fuzzy left ideal of R. Now for any $a \in R$ there exists $x \in R$ such that $a = axa$. Thus we have

$$
\begin{aligned}
&(\lambda \circ_{0.5} \mu \circ_{0.5} \nu)(a) \\
&= \sup_{a=\Sigma_{i=1}^{m} a_i b_i} \left\{ \bigwedge_{i=1}^{m} \{(\lambda \circ_{0.5} \mu)(a_i) \wedge \nu(b_i)\} \right\} \wedge 0.5 \\
&\geq \{(\lambda \circ_{0.5} \mu)(a) \wedge \nu(xa)\} \wedge 0.5 \\
&\geq \{(\lambda \circ_{0.5} \mu)(a) \wedge \nu(a)\} \wedge 0.5 \\
&\geq \left\{ \left\{ \sup_{a=\Sigma_{i=1}^{m} a_i b_i} \left\{ \bigwedge_{i=1}^{m} \{\lambda(a_i) \wedge \mu(b_i)\} \right\} \wedge 0.5 \right\} \wedge \nu(a) \right\} \wedge 0.5 \\
&\geq \{\lambda(ax) \wedge \mu(a) \wedge \nu(a)\} \wedge 0.5 \\
&\geq \{\lambda(a) \wedge \mu(a) \wedge \nu(a)\} \wedge 0.5 = (\lambda \wedge_{0.5} \mu \wedge_{0.5} \nu)(a).
\end{aligned}
$$

Hence $(\lambda \wedge_{0.5} \mu \wedge_{0.5} \nu) \leq (\lambda \circ_{0.5} \mu \circ_{0.5} \nu)$.

$(vi) \Rightarrow (vii)$ Obvious.

$(vii) \Rightarrow (i)$ Let λ be an $(\in, \in \vee q)$-fuzzy right ideal, and ν be an $(\in, \in \vee q)$-fuzzy left ideal of R. Then

$$(\lambda \wedge_{0.5} \nu) = (\lambda \wedge_{0.5} \mathscr{R} \wedge_{0.5} \nu) \leq (\lambda \circ_{0.5} \mathscr{R} \circ_{0.5} \nu) \leq (\lambda \circ_{0.5} \nu).$$

But by Lemma 6.3 $(\lambda \circ_{0.5} \nu) \leq (\lambda \wedge_{0.5} \nu)$. Hence $(\lambda \circ_{0.5} \nu) = (\lambda \wedge_{0.5} \nu)$ for every $(\in, \in \vee q)$-fuzzy right ideal λ and for every $(\in, \in \vee q)$-fuzzy left ideal ν of R. Thus by Theorem 6.19, R is regular.

6.4 Intra-regular Semirings

In this section we characterize regular and intra-regular semirings by the properties of their $(\in, \in \vee q)$-fuzzy ideals, $(\in, \in \vee q)$-fuzzy *bi*-ideals and $(\in, \in \vee q)$-fuzzy quasi-ideals.

Lemma 6.4. *A semiring R is intra-regular if and only if $\lambda \wedge_{0.5} \mu \leq \lambda \circ_{0.5} \mu$ for every $(\in, \in \vee q)$-fuzzy left ideal λ and for every $(\in, \in \vee q)$-fuzzy right ideal μ of R.*

Proof. Let R be an intra-regular semiring and λ be an $(\in, \in \vee q)$-fuzzy left ideal and μ be an $(\in, \in \vee q)$-fuzzy right ideal of R. Now for every $x \in R$, there exist $a_i, a_i' \in R$ such that $x = \Sigma_{i=1}^{m} a_i x^2 a_i'$. Thus we have

$$
\begin{aligned}
&(\lambda \circ_{0.5} \mu)(x) \\
&= \sup_{a=\Sigma_{i=1}^{m} a_i b_i} \left\{ \bigwedge_{i=1}^{m} \{\lambda(a_i) \wedge \mu(b_i)\} \right\} \wedge 0.5 \\
&\geq \bigwedge_{i=1}^{m} \left\{ \lambda(a_i x) \wedge \mu\left(x a_i'\right) \right\} \wedge 0.5 \quad (\text{because } x = \Sigma_{i=1}^{m} (a_i x)(x a_i')) \\
&\geq (\lambda \wedge_{0.5} \mu)(x).
\end{aligned}
$$

Conversely, assume that A and B are left and right ideals of R, respectivly. Then by Lemma 6.1, χ_A and χ_B are $(\in, \in \vee q)$-fuzzy left ideal and $(\in, \in \vee q)$-fuzzy right ideal of R, respectively. Thus by hypothesis $\chi_A \wedge_{0.5} \chi_B \leq \chi_A \circ_{0.5} \chi_B \Rightarrow \chi_{A \cap B} \wedge 0.5 \leq \chi_{AB} \wedge 0.5 \Rightarrow A \cap B \subseteq AB$. Hence by Theorem 5.4, R is intra-regular.

Theorem 6.23. *The following conditions are equivalent for a semiring R*

(i) R is both regular and intra-regular.

(ii) $\lambda \wedge 0.5 = \lambda \circ_{0.5} \lambda$ *for every* $(\in, \in \vee q)$*-fuzzy bi-ideal* λ *of R.*

(iii) $\lambda \wedge 0.5 = \lambda \circ_{0.5} \lambda$ *for every* $(\in, \in \vee q)$*-fuzzy quasi-ideal* λ *of R.*

Proof. $(i) \Rightarrow (ii)$ Let λ be an $(\in, \in \vee q)$-fuzzy bi-ideal of R and $x \in R$. Since R is both regular and intra-regular, there exist elements $a, p_i, p_i' \in R$ such that

$x = xax$ and $x = \sum_{i=1}^{m}(p_i x x p_i')$.

Thus $x = xax = xaxax = xa\left(\sum_{i=1}^{m}(p_i x x p_i')\right)ax = \sum_{i=1}^{m}(xap_i x)(xp_i' ax)$. Now

$$(\lambda \circ_{0.5} \lambda)(x)$$
$$= \sup_{x=\Sigma_{i=1}^{m} a_i b_i} \left\{ \bigwedge_{i=1}^{m} \{\lambda(a_i) \wedge \lambda(b_i)\} \right\} \wedge 0.5$$
$$\geq \bigwedge_{i=1}^{m} \{\lambda(xap_i x) \wedge \lambda(xp_i' ax)\} \wedge 0.5$$
$$\geq \{\lambda(x) \wedge 0.5.$$

This implies that $\lambda \circ_{0.5} \lambda \geq \lambda \wedge 0.5$.

On the other hand $\lambda \circ_{0.5} \lambda \leq \lambda \wedge 0.5$.

Consequently $\lambda \circ_{0.5} \lambda = \lambda \wedge 0.5$.

$(ii) \Rightarrow (iii)$ This is straightforward because every $(\in, \in \vee q)$-fuzzy quasi-ideal of R is an $(\in, \in \vee q)$-fuzzy bi-ideal of R.

$(iii) \Rightarrow (i)$ Let Q be a quasi-ideal of R. Then χ_Q is an $(\in, \in \vee q)$-fuzzy quasi-ideal of R. Thus by hypothesis

$\chi_Q \wedge 0.5 = \chi_Q \circ_{0.5} \chi_Q = \chi_{Q^2} \wedge 0.5 \Rightarrow Q = Q^2$. Hence by Theorem 5.9, R is both regular and intra-regular.

Theorem 6.24. *The following conditions are equivalent for a semiring R.*

(i) R is both regular and intra-regular.

(ii) $\lambda \wedge_{0.5} \mu \leq \lambda \circ_{0.5} \mu$ *for all* $(\in, \in \vee q)$*-fuzzy bi-ideals* λ *and* μ *of R.*

(iii) $\lambda \wedge_{0.5} \mu \leq \lambda \circ_{0.5} \mu$ *for every* $(\in, \in \vee q)$*-fuzzy bi-ideal* λ *and every* $(\in, \in \vee q)$*-fuzzy quasi-ideal* μ *of R.*

(iv) $\lambda \wedge_{0.5} \mu \leq \lambda \circ_{0.5} \mu$ *for every* $(\in, \in \vee q)$*-fuzzy quasi-ideal* λ *and every* $(\in, \in \vee q)$*-fuzzy bi-ideal* μ *of R.*

(v) $\lambda \wedge_{0.5} \mu \leq \lambda \circ_{0.5} \mu$ *for all* $(\in, \in \vee q)$*-fuzzy quasi-ideals* λ *and* μ *of R.*

Proof. $(i) \Rightarrow (ii)$ Let λ and μ be $(\in, \in \vee q)$-fuzzy bi-ideals of R and $x \in R$. Since R is both regular and intra-regular, there exist elements $a, p_i, p_i' \in R$ such that

$x = xax$ and $x = \sum_{i=1}^{m}(p_i x x p_i')$.

Thus $x = xax = xaxax = xa\left(\sum_{i=1}^{m}(p_i x x p_i')\right)ax = \sum_{i=1}^{m}(xap_i x)(xp_i' ax)$.

$$(\lambda \circ_{0.5} \mu)(x)$$
$$= \sup_{x=\Sigma_{i=1}^{m} a_i b_i} \left\{ \bigwedge_{i=1}^{m} \{\lambda(a_i) \wedge \mu(b_i)\} \right\} \wedge 0.5$$
$$\geq \bigwedge_{i=1}^{m} \{\{\lambda(xap_i x) \wedge \mu(xp_i' ax)\}\} \wedge 0.5$$
$$\geq \{\lambda(x) \wedge \mu(x) \wedge 0.5\} \wedge 0.5 = \lambda(x) \wedge_{0.5} \mu(x).$$

This implies that $\lambda \circ_{0.5} \mu \geq \lambda \wedge_{0.5} \mu$.

$(ii) \Rightarrow (iii) \Rightarrow (v)$ and $(ii) \Rightarrow (iv) \Rightarrow (v)$ are clear.

$(v) \Rightarrow (i)$ Let λ be an $(\in,\in\vee q)$-fuzzy left ideals of R and μ be an $(\in,\in\vee q)$-fuzzy right ideal of R. Then λ and μ are $(\in,\in\vee q)$-fuzzy bi-ideals of R. So by hypothesis $\lambda \wedge_{0.5} \mu \leq \lambda \circ_{0.5} \mu$ but $\lambda \wedge_{0.5} \mu \geq \lambda \circ_{0.5} \mu$ by Lemma 6.3. Thus $\lambda \wedge_{0.5} \mu = \lambda \circ_{0.5} \mu$. Hence by Theorem 6.19, R is regular. On the other hand by hypothesis we also have $\lambda \wedge_{0.5} \mu \leq \mu \circ_{0.5} \lambda$. By Lemma 6.4, R is intra-regular.

6.5 $(\in,\in\vee q)$-Fuzzy k-Ideals

Definition 6.7. A fuzzy subset λ of R is called an $(\in,\in\vee q)$-*fuzzy left (resp. right) k-ideal* of R if it satisfies $(1b)$, $(4b)$ and $(5b)$ (resp. $(1b)$, $(3b)$ and $(5b)$).

Definition 6.8. A fuzzy subset λ of R is called an $(\in,\in\vee q)$-*fuzzy k-bi-ideal* of R if it satisfies $(1b)$, $(2b)$, $(5b)$ and $(6b)$.

Definition 6.9. A fuzzy subset λ of R is called an $(\in,\in\vee q)$-*fuzzy k-quasi-ideal* of R if it satisfies $(1b)$, $(5b)$ and $(11b)$, where

$(11b)$ $\lambda(x) \geq \min\{(\lambda \odot_k \mathscr{R})(x), (\mathscr{R} \odot_k \lambda))(x), 0.5\}$ for all $x \in R$.

The proofs of the following results are straightforward and hence omitted.

Lemma 6.5. *A nonempty subset A of R is a left k-ideal (right k-ideal, k-ideal, k-bi-ideal, k-quasi-ideal) of R if and only if* χ_A *is an* $(\in,\in\vee q)$*-fuzzy left k-ideal (right k-ideal, k-ideal, k-bi-ideal, k-quasi-ideal) of R.*

Theorem 6.25. *Let* λ *be an* $(\in,\in\vee q)$*-fuzzy left (right) k-ideal of R. Then* $\lambda \wedge 0.5$ *is an* $(\in,\in\vee q)$*-fuzzy left (right) k-ideal of R, where* $(\lambda \wedge 0.5)(x) = \lambda(x) \wedge 0.5$ *for all* $x \in R$.

Proof. The proof is similar to the proof of Theorem 6.8.

Theorem 6.26. *If* λ *is an* $(\in,\in\vee q)$*-fuzzy k-bi-ideal of R, then* $(\lambda \wedge 0.5)$ *is an* $(\in,\in\vee q)$*-fuzzy k-bi-ideal of R.*

Definition 6.10. Let λ,μ be fuzzy subsets of R. Then the fuzzy subsets $\lambda \odot_{0.5} \mu$ and $\lambda \oplus_{0.5} \mu$ of R are defined as following:

$(\lambda \odot_{0.5} \mu)(x) = (\lambda \odot_k \mu)(x) \wedge 0.5$

$\lambda \oplus_{0.5} \mu = (\lambda +_k \mu) \wedge 0.5$ for all $x \in R$.

Lemma 6.6. *Let A, B be nonempty subsets of R. Then* $\chi_A \oplus_{0.5} \chi_B = \chi_{\overline{A+B}} \wedge 0.5$.

Proof. Let A,B be subsets of R and $x \in R$. If $x \in \overline{A+B}$ then there exist $a_1,a_2 \in A$ and $b_1,b_2 \in B$ such that $x+(a_1+b_1) = (a_2+b_2)$. Thus

$$(\chi_A +_{0.5} \chi_B)(x)$$
$$= \sup_{x+(a_1'+b_1')=(a_2'+b_2')} (\min\{\chi_A(a_1'),\chi_A(a_2'),\chi_B(b_1'),\chi_B(b_2')\} \wedge 0.5)$$
$$= 1 \wedge 0.5 = \chi_{\overline{A+B}}(x) \wedge 0.5.$$

If $x \notin \overline{A+B}$ then there do not exist $a_1, a_2 \in A$ and $b_1, b_2 \in B$ such that $x + (a_1 + b_1) = (a_2 + b_2)$. Thus
$(\chi_A +_{0.5} \chi_B)(x) = 0 \wedge 0.5 = \chi_{\overline{A+B}}(x) \wedge 0.5$.
Hence $\chi_A +_{0.5} \chi_B = \chi_{\overline{A+B}} \wedge 0.5$.

Lemma 6.7. *A fuzzy subset λ of a semiring R satisfies conditions $(1b)$ and $(5b)$ if and only if it satisfies condition $(12b)$, where*
$(12b)$ $\lambda \oplus_{0.5} \lambda \leq \lambda \wedge 0.5$.

Proof. Suppose λ satisfies conditions $(1b)$ and $(5b)$. Let $x \in R$. Then $(\lambda \oplus_{0.5} \lambda)(x)$

$$= \left(\sup_{x+(a_1+b_1)=(a_2+b_2)} \{\min\{\lambda(a_1), \lambda(a_2), \lambda(b_1), \lambda(b_2)\}\} \right) \wedge 0.5$$

$$= \left(\sup_{x+(a_1+b_1)=(a_2+b_2)} \{\min\{\min\{\lambda(a_1), \lambda(b_1), 0.5\}, \min\{\lambda(a_2), \lambda(b_2), 0.5\}\} \right) \wedge 0.5$$

$$\leq \left(\sup_{x+(a_1+b_1)=(a_2+b_2)} \{\min\{\lambda(a_1+b_1), \lambda(a_2+b_2)\}\} \right) \wedge 0.5 \quad \text{(by condition (1b))}$$

$$\leq \lambda(x) \wedge 0.5 \qquad \text{(by condition (5b))}.$$

Thus $\lambda \oplus_{0.5} \lambda \leq \lambda \wedge 0.5$.
Conversely, assume that $\lambda \oplus_{0.5} \lambda \leq \lambda \wedge 0.5$ and $x, y \in R$. Then
$\lambda(x+y) \geq \lambda(x+y) \wedge 0.5 \geq (\lambda \oplus_{0.5} \lambda)(x+y)$

$$= \left(\sup_{(x+y)+(a_1+b_1)=(a_2+b_2)} \{\min\{\lambda(a_1), \lambda(a_2), \lambda(b_1), \lambda(b_2)\}\} \right) \wedge 0.5$$

$\geq \min\{\lambda(0), \lambda(x), \lambda(0), \lambda(y)\} \wedge 0.5$ because $(x+y) + (0+0) = (x+y)$
$= \min\{\lambda(x), \lambda(y), 0.5\}$
Thus λ satisfies condition $(1b)$.
Let $a, b, x \in R$ such that $x + a = b$. Then
$\lambda(x) \geq \lambda(x) \wedge 0.5 \geq (\lambda \oplus_{0.5} \lambda)(x)$

$$= \left(\sup_{x+(a_1+b_1)=(a_2+b_2)} \{\min\{\lambda(a_1), \lambda(a_2), \lambda(b_1), \lambda(b_2)\}\} \right) \wedge 0.5$$

$\geq \min\{\lambda(a), \lambda(0), \lambda(b)\} \wedge 0.5$ because $x + (a+0) = (b+0)$
$= \min\{\lambda(a), \lambda(b), 0.5\}$.
This shows that λ satisfies condition $(5b)$.

Theorem 6.27. *A fuzzy subset λ of a semiring R is an $(\in, \in \vee q)$-fuzzy left (resp. right) k-ideal of R if and only if λ satisfies conditions*
$(12b)$ $\lambda \oplus_{0.5} \lambda \leq \lambda \wedge 0.5$
$(13b)$ $\mathscr{R} \odot_{0.5} \lambda \leq \lambda \wedge 0.5$ (*resp.* $\lambda \odot_{0.5} \mathscr{R} \leq f \wedge 0.5$).

Proof. Suppose λ is an $(\in, \in \vee q)$-fuzzy left k-ideal of R. Then by Lemma 6.7, λ satisfies condition $(12b)$. Now we show that λ satisfies condition $(13b)$. Let $x \in R$. If $(\mathscr{R} \odot_{0.5} \lambda)(x) = 0$, then $(\mathscr{R} \odot_{0.5} \lambda)(x) \leq \lambda(x) \wedge 0.5$. Otherwise, there exist elements $a_i, b_i, c_j, d_j \in R$ such that $x + \sum_{i=1}^{m} a_i b_i = \sum_{j=1}^{n} c_j d_j$. Then we have

$$\begin{aligned}
&(\mathscr{R}\odot_{0.5}\lambda)(x)\\
&=\sup_{x+\Sigma_{i=1}^{m}a_ib_i=\Sigma_{j=1}^{n}a'_jb'_j}\left\{\min\left\{\mathscr{R}(a_i),\lambda(b_i),\mathscr{R}\left(a'_j\right),\lambda\left(b'_j\right)\right\}\right\}\wedge 0.5\\
&=\sup_{x+\Sigma_{i=1}^{m}a_ib_i=\Sigma_{j=1}^{n}a'_jb'_j}\left\{\min\left\{\lambda(b_i),\lambda\left(b'_j\right)\right\}\right\}\wedge 0.5\\
&=\sup_{x+\Sigma_{i=1}^{m}a_ib_i=\Sigma_{j=1}^{n}a'_jb'_j}\left\{\min\left\{\lambda(b_i)\wedge 0.5,\lambda\left(b'_j\right)\wedge 0.5\right\}\right\}\wedge 0.5\\
&\le\sup_{x+\Sigma_{i=1}^{m}a_ib_i=\Sigma_{j=1}^{n}a'_jb'_j}\left\{\min\left\{\lambda(a_ib_i),\lambda\left(a'_jb'_j\right)\right\}\right\}\wedge 0.5\\
&=\sup_{x+\Sigma_{i=1}^{m}a_ib_i=\Sigma_{j=1}^{n}a'_jb'_j}\left\{\min\left\{\lambda(a_ib_i)\wedge 0.5,\lambda\left(a'_jb'_j\right)\wedge 0.5\right\}\right\}\wedge 0.5\\
&\le\sup_{x+\Sigma_{i=1}^{m}a_ib_i=\Sigma_{j=1}^{n}a'_jb'_j}\left(\min\left\{\lambda\left(\Sigma_{i=1}^{m}a_ib_i\right),\lambda\left(\Sigma_{j=1}^{n}a'_jb'_j\right)\right\}\right)\wedge 0.5\\
&\le\lambda(x)\wedge 0.5.\\
&\Rightarrow\mathscr{R}\odot_{0.5}\lambda\le\lambda\wedge 0.5.
\end{aligned}$$

Conversely, assume that λ satisfies conditions $(12b)$ and $(13b)$. Then by Lemma 6.7, λ satisfies conditions $(1b)$ and $(5b)$. We show that λ satisfies condition $(4b)$. Let $x,y\in R$. Then we have

$$\begin{aligned}
\lambda(xy)\wedge 0.5&\ge(\mathscr{R}\odot_{0.5}\lambda)(xy)\\
&=\sup_{xy+\Sigma_{i=1}^{m}a_ib_i=\Sigma_{j=1}^{n}a'_jb'_j}\left\{\min\left\{\mathscr{R}(a_i),\lambda(b_i),\mathscr{R}\left(a'_j\right),\lambda\left(b'_j\right)\right\}\right\}\wedge 0.5\\
&=\sup_{xy+\Sigma_{i=1}^{m}a_ib_i=\Sigma_{j=1}^{n}a'_jb'_j}\left\{\min\left\{\lambda(b_i),\lambda\left(b'_j\right)\right\}\right\}\wedge 0.5\\
&\ge\lambda(y)\wedge 0.5\qquad\text{because } xy+0y=xy.
\end{aligned}$$

This shows that λ satisfies condition $(4b)$. So λ is an $(\in,\in\vee q)$-fuzzy left k-ideal of R.

By using Lemma 6.7, we can prove the following:

Theorem 6.28. *A fuzzy subset λ of R is an $(\in,\in\vee q)$-fuzzy k-quasi-ideal of R if and only if λ satisfies conditions $(11b)$ and $(12b)$.*

Theorem 6.29. *Every $(\in,\in\vee q)$-fuzzy left (right) k-ideal of R is an $(\in,\in\vee q)$-fuzzy k-quasi-ideal of R.*

Theorem 6.30. *Every $(\in,\in\vee q)$-fuzzy k-quasi-ideal of R is an $(\in,\in\vee q)$-fuzzy k-bi-ideal of R.*

Lemma 6.8. *If λ and μ are $(\in,\in\vee q)$-fuzzy right and left k-ideals of R respectively, then $\lambda\odot_{0.5}\mu\le\lambda\wedge_{0.5}\mu$.*

Proof. The proof is similar to the proof of Lemma 6.3.

6.6 *k*-Regular Semirings

In this section we characterize *k*-regular semirings by the properties of their $(\in, \in \vee q)$-fuzzy *k*-ideals,$(\in, \in \vee q)$-fuzzy *k–bi*-ideals and $(\in, \in \vee q)$-fuzzy *k–quasi*-ideals.

Theorem 6.31. *For a semiring R the following conditions are equivalent.*
(*i*) *R is k-regular.*
(*ii*) $(\lambda \wedge_{0.5} \mu) = (\lambda \odot_{0.5} \mu)$ *for every* $(\in, \in \vee q)$*-fuzzy right k-ideal* λ *and every* $(\in, \in \vee q)$*-fuzzy left k-ideal* μ *of R.*

Proof. The proof is similar to the proof of Theorem 6.19.

Theorem 6.32. *For a semiring R, the following conditions are equivalent.*
(*i*) *R is k-regular.*
(*ii*) $\lambda \wedge 0.5 \leq (\lambda \odot_{0.5} \mathscr{R} \odot_{0.5} \lambda)$ *for every* $(\in, \in \vee q)$*-fuzzy k-bi-ideal* λ *of R.*
(*iii*) $\lambda \wedge 0.5 \leq (\lambda \odot_{0.5} \mathscr{R} \odot_{0.5} \lambda)$ *for every* $(\in, \in \vee q)$*-fuzzy k-quasi-ideal* λ *of R.*

Proof. The proof is similar to the proof of Theorem 6.20.

Theorem 6.33. *For a semiring R, the following conditions are equivalent.*
(*i*) *R is k-regular.*
(*ii*) $(\lambda \wedge_{0.5} \mu) \leq (\lambda \odot_{0.5} \mu \odot_{0.5} \lambda)$ *for every* $(\in, \in \vee q)$*-fuzzy k-bi-ideal* λ *and every* $(\in, \in \vee q)$*-fuzzy k-ideal* μ *of R.*
(*iii*) $(\lambda \wedge_{0.5} \mu) \leq (\lambda \odot_{0.5} \mu \odot_{0.5} \lambda)$ *for every* $(\in, \in \vee q)$*-fuzzy k-quasi-ideal* λ *and every* $(\in, \in \vee q)$*-fuzzy k-ideal* μ *of R.*

Proof. The proof is similar to the proof of Theorem 6.21.

Theorem 6.34. *For a semiring R, the following conditions are equivalent.*
(*i*) *R is k-regular.*
(*ii*) $(\lambda \wedge_{0.5} \mu) \leq (\lambda \odot_{0.5} \mu)$ *for every* $(\in, \in \vee q)$*-fuzzy k-bi-ideal* λ *and every* $(\in, \in \vee q)$*-fuzzy left k-ideal* μ *of R.*
(*iii*) $(\lambda \wedge_{0.5} \mu) \leq (\lambda \odot_{0.5} \mu)$ *for every* $(\in, \in \vee q)$*-fuzzy k-quasi-ideal* λ *and every* $(\in, \in \vee q)$*-fuzzy left k-ideal* μ *of R.*
(*iv*) $(\lambda \wedge_{0.5} \mu) \leq (\lambda \odot_{0.5} \mu)$ *for every* $(\in, \in \vee q)$*-fuzzy right k-ideal* λ *and every* $(\in, \in \vee q)$*-fuzzy k-bi-ideal* μ *of R.*
(*v*) $(\lambda \wedge_{0.5} \mu) \leq (\lambda \odot_{0.5} \mu)$ *for every* $(\in, \in \vee q)$*-fuzzy right k-ideal* λ *and every* $(\in, \in \vee q)$*-fuzzy k-quasi-ideal* μ *of R.*
(*vi*) $(\lambda \wedge_{0.5} \mu \wedge_{0.5} \nu) \leq (\lambda \odot_{0.5} \mu \odot_{0.5} \nu)$ *for every* $(\in, \in \vee q)$*-fuzzy right k-ideal* λ*,every* $(\in, \in \vee q)$*-fuzzy k-bi-ideal* μ *and every* $(\in, \in \vee q)$*-fuzzy left k-ideal* ν *of R.*
(*vii*) $(\lambda \wedge_{0.5} \mu \wedge_{0.5} \nu) \leq (\lambda \odot_{0.5} \mu \odot_{0.5} \nu)$ *for every* $(\in, \in \vee q)$*-fuzzy right k-ideal* λ*,every* $(\in, \in \vee q)$*-fuzzy k-quasi-ideal* μ *and every* $(\in, \in \vee q)$*-fuzzy left k-ideal* ν *of R.*

Proof. The proof is similar to the proof of Theorem 6.22.

6.7 *k*-Intra-regular Semirings

In this section we characterize *k*-regular and *k*-intra-regular semirings by the properties of their $(\in, \in \vee q)$-fuzzy *k*-ideals,$(\in, \in \vee q)$-fuzzy *k*-bi-ideals and $(\in, \in \vee q)$-fuzzy *k*-quasi-ideals.

Lemma 6.9. *A semiring R is k-intra-regular if and only if* $\lambda \wedge_{0.5} \mu \leq \lambda \odot_{0.5} \mu$ *for every* $(\in, \in \vee q)$*-fuzzy left k-ideal* λ *and for every* $(\in, \in \vee q)$*-fuzzy right k-ideal* μ *of R.*

Proof. The proof is similar to the proof of Lemma 6.4.

Theorem 6.35. *The following conditions are equivalent for a semiring R*
(*i*) *R is both k-regular and k-intra-regular.*
(*ii*) $\lambda \wedge 0.5 = \lambda \odot_{0.5} \lambda$ *for every* $(\in, \in \vee q)$*-fuzzy k-bi-ideal* λ *of R.*
(*iii*) $\lambda \wedge 0.5 = \lambda \odot_{0.5} \lambda$ *for every* $(\in, \in \vee q)$*-fuzzy k-quasi-ideal* λ *of R.*

Proof. The proof is similar to the proof of Theorem 6.23.

Theorem 6.36. *The following conditions are equivalent for a semiring R*
(*i*) *R is both k-regular and k-intra-regular.*
(*ii*) $\lambda \wedge_{0.5} \mu \leq \lambda \odot_{0.5} \mu$ *for all* $(\in, \in \vee q)$*-fuzzy k-bi-ideals* λ *and* μ *of R.*
(*iii*) $\lambda \wedge_{0.5} \mu \leq \lambda \odot_{0.5} \mu$ *for every* $(\in, \in \vee q)$*-fuzzy k-bi-ideal* λ *and every* $(\in, \in \vee q)$*-fuzzy k-quasi-ideals* μ *of R.*
(*iv*) $\lambda \wedge_{0.5} \mu \leq \lambda \odot_{0.5} \mu$ *for every* $(\in, \in \vee q)$*-fuzzy k-quasi-ideal* λ *and every* $(\in, \in \vee q)$*-fuzzy k-bi-ideals* μ *of R.*
(*v*) $\lambda \wedge_{0.5} \mu \leq \lambda \odot_{0.5} \mu$ *for all* $(\in, \in \vee q)$*-fuzzy k-quasi-ideals* λ *and* μ *of R.*

Proof. The proof is similar to the proof of Theorem 6.24.

Chapter 7
$(\overline{\in},\overline{\in}\vee\overline{q})$-Fuzzy Ideals in Semirings

This chapter is devoted to a study of $(\overline{\in},\overline{\in}\vee\overline{q})$-fuzzy ideals, fuzzy quasi-ideals and bi-ideals of semirings on the pattern of Chapter 6.

7.1 $(\overline{\in},\overline{\in}\vee\overline{q})$-Fuzzy Ideals

We start this section with the following Theorem.

Theorem 7.1. *For any fuzzy subset λ of a semiring R and for all $x,y\in R$ and $t,r\in(0,1]$, $[1a]$ is equivalent to $[1b]$, $[2a]$ is equivalent to $[2b]$, $[3a]$ is equivalent to $[3b]$ and $[4a]$ is equivalent to $[4b]$ where:*

$[1a]$ $(x+y)_{\min\{t,r\}}\overline{\in}\lambda$ *implies* $x_t\overline{\in}\vee\overline{q}\lambda$ *or* $y_r\overline{\in}\vee\overline{q}\lambda$

$[1b]$ $\max\{\lambda(x+y),0.5\}\geq\min\{\lambda(x),\lambda(y)\}$

$[2a]$ $(xy)_{\min\{t,r\}}\overline{\in}\lambda$ *implies* $x_t\overline{\in}\vee\overline{q}\lambda$ *or* $y_r\overline{\in}\vee\overline{q}\lambda$

$[2b]$ $\max\{\lambda(xy),0.5\}\geq\min\{\lambda(x),\lambda(y)\}$

$[3a]$ $(xy)_t\overline{\in}\lambda$ *implies* $x_t\overline{\in}\vee\overline{q}\lambda$

$[3b]$ $\max\{\lambda(xy),0.5\}\geq\lambda(x)$

$[4a]$ $(xy)_t\overline{\in}\lambda$ *implies* $y_t\overline{\in}\vee\overline{q}\lambda$

$[4b]$ $\max\{\lambda(xy),0.5\}\geq\lambda(y)$

Similarly for all $a,b,x,y,z\in R$ such that $x+a=b$, and for all $t,r\in(0,1]$, $[5a]$ is equivalent to $[5b]$, $[6a]$ is equivalent to $[6b]$, $[7a]$ is equivalent to $[7b]$ and $[8a]$ is equivalent to $[8b]$ where

$[5a]$ $x_{\min\{t,r\}}\overline{\in}\lambda$ *implies* $a_t\overline{\in}\vee\overline{q}\lambda$ *or* $b_r\overline{\in}\vee\overline{q}\lambda$

$[5b]$ $\max\{\lambda(x),0.5\}\geq\min\{\lambda(a),\lambda(b)\}$

$[6a]$ $(xyz)_{\min\{t,r\}}\overline{\in}\lambda$ *implies* $x_t\overline{\in}\vee\overline{q}\lambda$ *or* $z_r\overline{\in}\vee\overline{q}\lambda$

$[6b]$ $\max\{\lambda(xyz),0.5\}\geq\min\{\lambda(x),\lambda(z)\}$

$[7a]$ $x_t\overline{\in}\lambda$ *implies* $x_t\overline{\in}\vee\overline{q}((\lambda\circ\mathscr{R})\wedge(\mathscr{R}\circ\lambda))$

$[7b]$ $\max\{\lambda(x),0.5\}\geq\min\{(\lambda\circ\mathscr{R})(x),(\mathscr{R}\circ\lambda)(x)\}$

$[8a]$ $x_t\overline{\in}\lambda$ *implies* $x_t\overline{\in}\vee\overline{q}((\lambda\odot_k\mathscr{R})\wedge(\mathscr{R}\odot_k\lambda))$

$[8b]$ $\max\{\lambda(x),0.5\}\geq\min\{(\lambda\odot_k\mathscr{R})(x),(\mathscr{R}\odot_k\lambda)(x)\}$

J. Ahsan et al.: Fuzzy Semirings with Applications, STUDFUZZ 278, pp. 123–139.
springerlink.com

Proof. We prove only [1a] if and only if [1b]. The proofs of the other parts are similar to this.

[1a] $\Rightarrow$ [1b] Suppose there exist $x,y \in R$ such that $\max\{\lambda(x+y),0.5\} < \min\{\lambda(x),\lambda(y)\}$. Choose $t \in (0,1]$ such that $\max\{\lambda(x+y),0.5\} < t = \min\{\lambda(x),\lambda(y)\}$. Then $(x+y)_t\overline{\in}\lambda$. So by hypothesis $x_t\overline{\in}\vee\overline{q}\lambda$ or $y_r\overline{\in}\vee\overline{q}\lambda$. But $x_t, y_t \in \lambda$. Thus $x_t\overline{q}\lambda$ or $y_t\overline{q}\lambda$, that is either $\lambda(x)+t \leq 1$ or $\lambda(y)+t \leq 1$. But this is possible only when $t \leq 0.5$. This is a contradiction. Hence $\max\{\lambda(x+y),0.5\} \geq \min\{\lambda(x),\lambda(y)\}$.

[1b] $\Rightarrow$ [1a] Let $(x+y)_{\min\{t,r\}}\overline{\in}\lambda$. Then $\lambda(x+y) < \min\{t,r\}$. If $\max\{\lambda(x+y),0.5\} = \lambda(x+y)$, then $\min\{t,r\} > \lambda(x+y) \geq \min\{\lambda(x),\lambda(y)\}$. Thus $\lambda(x) < t$ or $\lambda(y) < r$, that is $x_t\overline{\in}\lambda$ or $y_r\overline{\in}\lambda$. Hence $x_t\overline{\in}\vee\overline{q}\lambda$ or $y_r\overline{\in}\vee\overline{q}\lambda$. If $\max\{\lambda(x+y),0.5\} = 0.5$, then by hypothesis $0.5 \geq \min\{\lambda(x),\lambda(y)\}$. If $x_t \in \lambda$ or $y_r \in \lambda$, then $t \leq \lambda(x) \leq 0.5$ or $r \leq \lambda(y) \leq 0.5$. Thus $\lambda(x)+t \leq 1$ or $\lambda(y)+r \leq 1$, that is $x_t\overline{q}\lambda$ or $y_r\overline{q}\lambda$. Hence $x_t\overline{\in}\vee\overline{q}\lambda$ or $y_r\overline{\in}\vee\overline{q}\lambda$.

Definition 7.1. A fuzzy subset λ of a semiring R is said to be an *$(\overline{\in},\overline{\in}\vee\overline{q})$-fuzzy left (resp. right) ideal* of R if it satisfies [1a] and [4a] (resp. [3a]).

Definition 7.2. A fuzzy subset λ of a semiring R is said to be an *$(\overline{\in},\overline{\in}\vee\overline{q})$-fuzzy bi-ideal* of R if it satisfies [1a], [2a] and [6a].

Definition 7.3. A fuzzy subset λ of a semiring R is said to be an *$(\overline{\in},\overline{\in}\vee\overline{q})$-fuzzy quasi-ideal* of R if it satisfies [1a] and [7a].

The proof of the following Lemma is straightforward and hence omitted.

Lemma 7.1. *A nonempty subset A of a semiring R is a left (right, bi-, quasi-) ideal of R if and only if χ_A is an $(\overline{\in},\overline{\in}\vee\overline{q})$-fuzzy left (right, bi-, quasi-) ideal of R.*

Definition 7.4. Let λ and μ be fuzzy subsets of a semiring R. Then the fuzzy subsets $\lambda\vee 0.5$, $\lambda\wedge^{0.5}\mu$, $\lambda\odot^{0.5}\mu$, $\lambda\circ^{0.5}\mu$, $\lambda+^{0.5}\mu$ and $\lambda\oplus^{0.5}\mu$ of R are defined as

$$\begin{aligned}(\lambda\vee 0.5)(x) &= \lambda(x)\vee 0.5\\ (\lambda\wedge^{0.5}\mu)(x) &= (\lambda\wedge\mu)(x)\vee 0.5\\ (\lambda\circ^{0.5}\mu)(x) &= (\lambda\circ\mu)(x)\vee 0.5\\ (\lambda\odot^{0.5}\mu)(x) &= (\lambda\odot_k\mu)(x)\vee 0.5\\ (\lambda+^{0.5}\mu)(x) &= (\lambda+\mu)(x)\vee 0.5\\ (\lambda\oplus^{0.5}\mu)(x) &= (\lambda+_k\mu)(x)\vee 0.5 \quad \text{for all } x\in R.\end{aligned}$$

Theorem 7.2. *If λ is an $(\overline{\in},\overline{\in}\vee\overline{q})$-fuzzy left (right) ideal of a semiring R, then $(\lambda\vee 0.5)$ is an $(\overline{\in},\overline{\in}\vee\overline{q})$-fuzzy left (right) ideal of R.*

Proof. Let λ be an $(\overline{\in},\overline{\in}\vee\overline{q})$-fuzzy left ideal of R and $x,y\in R$. Then

$$\begin{aligned}\max\{(\lambda\vee 0.5)(x+y),0.5\} &= \lambda(x+y)\vee 0.5\\ &\geq \min\{\lambda(x),\lambda(y)\}\vee 0.5\\ &= \min\{\lambda(x)\vee 0.5,\lambda(y)\vee 0.5\}\\ &= \min\{(\lambda\vee 0.5)(x),(\lambda\vee 0.5)(y)\}.\end{aligned}$$

Similarly we can show that

$$\max\{(\lambda\vee 0.5)(xy),0.5\}\geq(\lambda\vee 0.5)(y).$$

This shows that $(\lambda\vee 0.5)$ is an $(\overline{\in},\overline{\in}\vee\overline{q})$-fuzzy left ideal of R.

Similarly we can prove that:

Theorem 7.3. *If λ is an $(\overline{\in},\overline{\in}\vee\overline{q})$-fuzzy bi-ideal of R, then $\lambda\vee 0.5$ is an $(\overline{\in},\overline{\in}\vee\overline{q})$-fuzzy bi-ideal of R.*

Lemma 7.2. *Let A,B be subsets of R. Then*

$$C_A+^{0.5}C_B=C_{A+B}\vee 0.5.$$

Proof. Let A,B be subsets of R and $x\in R$. If $x\in A+B$, then there exist $a\in A$ and $b\in B$ such that $x=a+b$. Thus

$$\begin{aligned}(\chi_A+^{0.5}\chi_B)(x)&=\sup_{x=c+d}\{\chi_A(c)\wedge\chi_B(d)\}\vee 0.5\\&=1\vee 0.5=\chi_{A+B}(x)\vee 0.5.\end{aligned}$$

If $x\notin A+B$, then there do not exist $a\in A$ and $b\in B$ such that $x=a+b$. Thus

$$(\chi_A+^{0.5}\chi_B)(x)=0\vee 0.5=\chi_{A+B}(x)\vee 0.5$$

Hence $\chi_A+^{0.5}\chi_B=\chi_{A+B}(x)\vee 0.5$.

Lemma 7.3. *A fuzzy subset λ of a semiring R satisfies condition $[1b]$ if and only if it satisfies condition*
$[9b]$ $\lambda+^{0.5}\lambda\leq\lambda\vee 0.5$.

Proof. Suppose λ satisfies condition $[1b]$ and $x\in R$. Then

$$\begin{aligned}(\lambda+^{0.5}\lambda)(x)&=\left(\sup_{x=a+b}\{\min\{\lambda(a),\lambda(b)\}\}\right)\vee 0.5\\&\leq\left(\sup_{x=a+b}\{\max(\lambda(a+b),0.5)\}\right)\vee 0.5\quad\text{by condition }[1b]\\&=\left(\sup_{x=a+b}\{\max\{\lambda(x),0.5\}\right)\vee 0.5\\&\leq\lambda(x)\vee 0.5\end{aligned}$$

Thus $\lambda+^{0.5}\lambda\leq\lambda\vee 0.5$.

Conversely, assume that $\lambda +^{0.5} \lambda \leq \lambda \vee 0.5$ and $x,y \in R$. Then

$$\begin{aligned}\lambda(x+y)\vee 0.5 &\geq (\lambda +^{0.5} \lambda)(x+y)\\ &= \left(\sup_{(x+y)=a+b} \{\min\{\lambda(a),\lambda(b)\}\right)\vee 0.5\\ &\geq \min\{\lambda(x),\lambda(y)\}\vee 0.5\\ &\geq \min\{\lambda(x),\lambda(y)\}.\end{aligned}$$

Thus λ satisfies condition $[1b]$.

Theorem 7.4. *A fuzzy subset λ of a semiring R is an $(\overline{\in},\overline{\in}\vee\overline{q})$-fuzzy left (resp. right) ideal of R if and only if λ satisfies the conditions*
$[9b]$ $\lambda +^{0.5} \lambda \leq \lambda \vee 0.5$.
$[10b]$ $\mathscr{R} \circ^{0.5} \lambda \leq \lambda \vee 0.5$ $(resp.\ \lambda \circ^{0.5} \mathscr{R} \leq \lambda \vee 0.5)$.

Proof. Suppose λ is an $(\overline{\in},\overline{\in}\vee\overline{q})$-fuzzy left ideal of R. Then λ satisfies condition $[1b]$ and $[4b]$. Thus by Lemma 7.3, λ satisfies $[9b]$. Let $x \in R$. If $\left(\mathscr{R} \circ^{0.5} \lambda\right)(x) = (\mathscr{R} \circ \lambda)(x) \vee 0.5 = 0 \vee 0.5 = 0.5$, then $\left(\mathscr{R} \circ^{0.5} \lambda\right)(x) \leq \lambda(x) \vee 0.5$. Otherwise, there exist elements $a_i, b_i \in R$ such that $x = \sum_{i=1}^{m} a_i b_i$. Then

$$\begin{aligned}&(\mathscr{R} \circ^{0.5} \lambda)(x)\\ &= \left(\sup_{x=\Sigma_{i=1}^{m} a_i b_i} \left\{\bigwedge_{i=1}^{m} \{\mathscr{R}(a_i) \wedge \lambda(b_i)\}\right\}\right) \vee 0.5\\ &= \sup_{x=\Sigma_{i=1}^{m} a_i b_i} \left\{\bigwedge_{i=1}^{m} \lambda(b_i)\right\} \vee 0.5\\ &\leq \sup_{x=\Sigma_{i=1}^{m} a_i b_i} \left\{\bigwedge_{i=1}^{m} (\lambda(a_i b_i) \vee 0.5)\right\} \vee 0.5\\ &= \sup_{x=\Sigma_{i=1}^{m} a_i b_i} \left\{\bigwedge_{i=1}^{m} \lambda(a_i b_i)\right\} \vee 0.5\\ &\leq \sup_{x=\Sigma_{i=1}^{m} a_i b_i} (\lambda(\Sigma_{i=1}^{m} a_i b_i) \vee 0.5) \vee 0.5 \qquad \text{by } [1b]\\ &= \lambda(x) \vee 0.5.\\ &\Rightarrow \mathscr{R} \circ^{0.5} \lambda \leq \lambda \vee 0.5.\end{aligned}$$

Conversely, assume that λ satisfies condition $[9b]$ and $[10b]$. Then by Lemma 7.3, λ satisfies condition $[1b]$. We show that λ satisfies condition $[4b]$. Let $x,y \in R$. Then

$$\begin{aligned}\lambda(xy) \vee 0.5 &\geq \left(\mathscr{R} \circ^{0.5} \lambda\right)(xy)\\ &= \sup_{xy=\Sigma_{i=1}^{m} a_i b_i} \left\{\bigwedge_{i=1}^{m} \{\mathscr{R}(a_i) \wedge \lambda(b_i)\}\right\} \vee 0.5\\ &= \sup_{xy=\Sigma_{i=1}^{m} a_i b_i} \left\{\bigwedge_{i=1}^{m} \lambda(b_i)\right\} \vee 0.5\\ &\geq \lambda(y) \vee 0.5 \geq \lambda(y).\end{aligned}$$

This shows that λ satisfies condition $[4b]$. Hence λ is an $(\overline{\in}, \overline{\in} \vee \overline{q})$-fuzzy left ideal of R.

Lemma 7.4. *A fuzzy subset λ of a semiring R satisfies conditions $[1b]$ and $[2b]$ if and only if it satisfies conditions $[9b]$ and*
$[11b]$ $\lambda \circ^{0.5} \lambda \leq \lambda \vee 0.5$.

Proof. Suppose λ satisfies conditions $[1b]$ and $[2b]$. Then by Lemma 7.3, λ satisfies condition $[9b]$. Let $x \in R$. Then

$$(\lambda \circ^{0.5} \lambda)(x) = \left(\sup_{x=\Sigma_{i=1}^m a_i b_i} \left\{ \bigwedge_{i=1}^{m} \{\lambda(a_i) \wedge \lambda(b_i)\} \right\} \right) \vee 0.5$$
$$= \left(\sup_{x=\Sigma_{i=1}^m a_i b_i} \left\{ \bigwedge_{i=1}^{m} (\{\max(\lambda(a_i b_i), 0.5)\}) \right\} \right) \vee 0.5 \text{ by condition } [2b]$$
$$= \left(\sup_{x=\Sigma_{i=1}^m a_i b_i} \left\{ \bigwedge_{i=1}^{m} \lambda(a_i b_i) \right\} \right) \vee 0.5$$
$$= \left(\sup_{x=\Sigma_{i=1}^m a_i b_i} \left\{ \lambda(\Sigma_{i=1}^m a_i b_i) \vee 0.5 \right\} \right) \vee 0.5 \quad \text{by condition } [1b]$$
$$\leq \lambda(x) \vee 0.5$$

Thus $\lambda \circ^{0.5} \lambda \leq \lambda \vee 0.5$, that is λ satisfies condition $[11b]$.

Conversely, assume that λ satisfies conditions $[9b]$ and $[11b]$. Then by Lemma 7.3, λ satisfies $[1b]$. Let $x, y \in R$. Then

$$\lambda(xy) \vee 0.5 \geq (\lambda \circ^{0.5} \lambda)(xy)$$
$$= \left(\sup_{xy=\Sigma_{i=1}^m a_i b_i} \left\{ \bigwedge_{i=1}^{m} \{\lambda(a_i) \wedge \lambda(b_i)\} \right\} \right) \vee 0.5$$
$$\geq \min\{\lambda(x), \lambda(y)\} \vee 0.5$$
$$\geq \min\{\lambda(x), \lambda(y)\}.$$

Thus λ satisfies condition $[2b]$.

Lemma 7.5. *A fuzzy subset λ of a semiring R satisfies conditions $[1b]$ and $[6b]$ if and only if it satisfies conditions $[9b]$ and*
$[12b]$ $\lambda \circ^{0.5} \mathscr{R} \circ^{0.5} \lambda \leq \lambda \vee 0.5$.

Proof. Suppose λ satisfies conditions $[1b]$ and $[6b]$. Then by Lemma 7.3, λ satisfies condition $[9b]$. Let $x \in R$. Then

$$(\lambda \circ^{0.5} \mathscr{R} \circ^{0.5} \lambda)(x) = \left(\sup_{x=\Sigma_{i=1}^m a_i b_i} \left\{ \bigwedge_{i=1}^{m} \{(\lambda \circ^{0.5} \mathscr{R})(a_i) \wedge \lambda(b_i)\} \right\} \right) \vee 0.5$$
$$= \left(\sup_{x=\Sigma_{i=1}^m a_i b_i} \left\{ \bigwedge_{i=1}^{m} \left\{ \left(\left(\sup_{a_i=\Sigma_{j=1}^n c_j d_j} \left\{ \bigwedge_{j=1}^{n} \{\lambda(c_j) \wedge \mathscr{R}(d_j)\} \right\} \right) \vee 0.5 \right) \wedge \lambda(b_i) \right\} \right\} \right) \vee 0.5$$
$$= \left(\sup_{x=\Sigma_{i=1}^m a_i b_i} \left\{ \bigwedge_{i=1}^{m} \left\{ \left(\sup_{a_l=\Sigma_{j=1}^n c_j d_j} \left\{ \bigwedge_{j=1}^{n} \lambda(c_j) \right\} \right) \wedge \lambda(b_i) \right\} \right\} \right) \vee 0.5$$

$$= \left(\sup_{x=\Sigma_{i=1}^{m}a_ib_i}\left\{\bigwedge_{i=1}^{m}\left\{\sup_{a_i=\Sigma_{j=1}^{n}c_jd_j}\bigwedge_{j=1}^{n}\lambda(c_j)\wedge\lambda(b_i)\right\}\right\}\right)\vee 0.5$$

$$\leq \left(\sup_{x=\Sigma_{i=1}^{m}a_ib_i}\left\{\bigwedge_{i=1}^{m}\left\{\sup_{a_i=\Sigma_{j=1}^{n}c_jd_j}\bigwedge_{j=1}^{n}(\lambda(c_jd_jb_i)\vee 0.5)\right\}\right\}\right)\vee 0.5$$

$$= \left(\sup_{x=\Sigma_{i=1}^{m}a_ib_i}\left\{\bigwedge_{i=1}^{m}\left\{\sup_{a_i=\Sigma_{j=1}^{n}c_jd_j}\bigwedge_{j=1}^{n}\lambda(c_jd_jb_i)\right\}\right\}\right)\vee 0.5$$

$$\leq \left(\lambda\left(\sum_{i=1}^{m}\sum_{j=1}^{n}c_jd_jb_i\right)\vee 0.5\right)\vee 0.5$$

$$\leq \lambda(x)\vee 0.5$$

Thus $\lambda\circ^{0.5}\mathscr{R}\circ^{0.5}\lambda\leq\lambda\vee 0.5$, that is λ satisfies condition $[12b]$.

Conversely, assume that λ satisfies conditions $[9b]$ and $[12b]$. Then by Lemma 7.3, λ satisfies $[1b]$. Let $x,y,z\in R$. Then

$$\lambda(xyz)\vee 0.5\geq(\lambda\circ^{0.5}\mathscr{R}\circ^{0.5}\lambda)(xyz)$$

$$= \left(\sup_{xyz=\Sigma_{i=1}^{m}a_ib_i}\left\{\bigwedge_{i=1}^{m}\left\{\lambda\circ^{0.5}\mathscr{R}(a_i)\wedge\lambda(b_i)\right\}\right\}\right)\vee 0.5$$

$$\geq \min\left\{\left(\lambda\circ^{0.5}\mathscr{R}\right)(xy),\lambda(z)\right\}\vee 0.5$$

$$\geq \min\left\{\left(\sup_{xy=\Sigma_{j=1}^{n}c_jd_j}\left\{\bigwedge_{j=1}^{n}\left\{\lambda(c_j)\wedge\mathscr{R}(d_j)\right\}\right\}\right)\vee 0.5,\lambda(z)\right\}\vee 0.5$$

$$\geq \min\left\{\left(\sup_{xy=\Sigma_{j=1}^{n}c_jd_j}\left\{\bigwedge_{j=1}^{n}\lambda(c_j)\right\}\right),\lambda(z)\right\}\vee 0.5$$

$$\geq \min\{\lambda(x),\lambda(z)\}\vee 0.5$$

$$\geq \min\{\lambda(x),\lambda(z)\}.$$

Thus λ satisfies condition $[6b]$.

Theorem 7.5. *A fuzzy subset λ of a semiring R is an $(\overline{\in},\overline{\in}\vee\overline{q})$-fuzzy bi-ideal of R if and only if λ satisfies conditions $[9b], [11b]$ and $[12b]$.*

Proof. The proof follows from Lemmas 7.4 and 7.5.

Theorem 7.6. *A fuzzy subset λ of a semiring R is an $(\overline{\in},\overline{\in}\vee\overline{q})$-fuzzy quasi-ideal of R if and only if λ satisfies conditions $[7b]$ and $[9b]$.*

Proof. The proof follows from Lemma 7.3.

From Theorem 7.4, Theorem7.5 and Theorem 7.6 we deduce the following results.

Theorem 7.7. *Every $(\overline{\in},\overline{\in}\vee\overline{q})$-fuzzy left (right) ideal of a semiring R is an $(\overline{\in},\overline{\in}\vee\overline{q})$-fuzzy quasi-ideal of R.*

Theorem 7.8. *Every $(\overline{\in},\overline{\in}\vee\overline{q})$-fuzzy quasi-ideal of R is an $(\overline{\in},\overline{\in}\vee\overline{q})$-fuzzy bi-ideal of R.*

Lemma 7.6. *If λ and μ are $(\overline{\in},\overline{\in}\vee\overline{q})$-fuzzy right and left ideals of R, respectively, then $\lambda \circ^{0.5} \mu \leq \lambda \wedge^{0.5} \mu$.*

Proof. Let λ and μ be $(\overline{\in},\overline{\in}\vee\overline{q})$-fuzzy right and left ideals of R, respectively. Then

$$\begin{aligned}
&(\lambda \circ^{0.5} \mu)(x)\\
&= \left(\sup_{x=\sum_{i=1}^m a_i b_i} \left\{ \bigwedge_{i=1}^{m} \{\lambda(a_i) \wedge \mu(b_i)\} \right\} \right) \vee 0.5\\
&\leq \left(\sup_{x=\sum_{i=1}^m a_i b_i} \left\{ \bigwedge_{i=1}^{m} \{(\lambda(a_ib_i) \vee 0.5) \wedge (\mu(a_ib_i) \vee 0.5)\} \right\} \right) \vee 0.5\\
&\leq \left(\sup_{x=\sum_{i=1}^m a_i b_i} \left\{ \bigwedge_{i=1}^{m} \lambda(a_ib_i) \wedge \mu(a_ib_i) \right\} \right) \vee 0.5\\
&\leq \left(\sup_{x=\sum_{i=1}^m a_i b_i} \{\{(\lambda(\textstyle\sum_{i=1}^m a_ib_i) \vee 0.5) \wedge (\mu(\sum_{i=1}^m a_ib_i) \vee 0.5)\}\} \right) \vee 0.5\\
&\leq \left(\sup_{x=\sum_{i=1}^m a_i b_i} \{\{\lambda(\textstyle\sum_{i=1}^m a_ib_i) \wedge \mu(\sum_{i=1}^m a_ib_i)\}\} \right) \vee 0.5\\
&\leq \{\lambda(x) \wedge \mu(x)\} \vee 0.5 = (\lambda \wedge^{0.5} \mu)(x).
\end{aligned}$$

Thus $\lambda \circ^{0.5} \mu \leq \lambda \wedge^{0.5} \mu$.

7.2 Regular Semirings

In this section we characterize regular semirings by the properties of their $(\overline{\in},\overline{\in}\vee\overline{q})$-fuzzy left (right) ideals, $(\overline{\in},\overline{\in}\vee\overline{q})$-fuzzy bi-ideals and $(\overline{\in},\overline{\in}\vee\overline{q})$-fuzzy quasi-ideals.

Theorem 7.9. *For a semiring R, the following conditions are equivalent.*
(i) R is regular.
(ii) $(\lambda \wedge^{0.5} \mu) = (\lambda \circ^{0.5} \mu)$ for every $(\overline{\in},\overline{\in}\vee\overline{q})$-fuzzy right ideal λ and every $(\overline{\in},\overline{\in}\vee\overline{q})$-fuzzy left ideal μ of R.

Proof. *(i)* $\Rightarrow$ *(ii)* Let λ be an $(\overline{\in},\overline{\in}\vee\overline{q})$-fuzzy right ideal and μ be an $(\overline{\in},\overline{\in}\vee\overline{q})$-fuzzy left ideal of R. Then by Lemma 7.6, $\lambda \circ^{0.5} \mu \leq \lambda \wedge^{0.5} \mu$. Let $a \in R$. Then there exists $x \in R$ such that $a = axa$. Thus

$$\begin{aligned}
\left(\lambda \circ^{0.5} \mu\right)(a) &= \left(\sup_{x=\sum_{i=1}^m a_i b_i} \left\{ \bigwedge_{i=1}^{m} \{\lambda(a_i) \wedge \mu(b_i)\} \right\} \right) \vee 0.5\\
&\geq \{(\lambda(a) \wedge \mu(xa)\} \vee 0.5\\
&= \{(\lambda(a) \wedge (\mu(xa) \vee 0.5)\} \vee 0.5\\
&\geq \{\lambda(a) \wedge \mu(a)\} \vee 0.5 = \left(\lambda \wedge^{0.5} \mu\right)(a).
\end{aligned}$$

Thus $(\lambda \circ^{0.5} \mu) \geq (\lambda \wedge^{0.5} \mu)$. Hence $(\lambda \circ^{0.5} \mu) = (\lambda \wedge^{0.5} \mu)$.

$(ii) \Rightarrow (i)$ Let A and B be right and left ideals of R, respectively. Then χ_A and χ_B are $(\overline{\in},\overline{\in}\vee\overline{q})$-fuzzy right ideal and $(\overline{\in},\overline{\in}\vee\overline{q})$-fuzzy left ideal of R. Hence by hypothesis $\chi_A \circ^{0.5} \chi_B = \chi_A \wedge^{0.5} \chi_B$. Thus $(\chi_A \circ \chi_B) \vee 0.5 = (\chi_A \wedge \chi_B) \vee 0.5$. This implies $\chi_{AB} \vee 0.5 = \chi_{A\cap B} \vee 0.5$. Hence $AB = A \cap B$. This shows that R is regular.

Theorem 7.10. *For a semiring R the following conditions are equivalent.*
(i) R is regular.
(ii) $\lambda \vee 0.5 \leq (\lambda \circ^{0.5} \mathscr{R} \circ^{0.5} \lambda)$ for every $(\overline{\in},\overline{\in}\vee\overline{q})$-fuzzy bi-ideal λ of R.
(iii) $\lambda \vee 0.5 \leq (\lambda \circ^{0.5} \mathscr{R} \circ^{0.5} \lambda)$ for every $(\overline{\in},\overline{\in}\vee\overline{q})$-fuzzy quasi-ideal λ of R.

Proof. $(i) \Rightarrow (ii)$ Let λ be an $(\overline{\in},\overline{\in}\vee\overline{q})$-fuzzy bi-ideal of R and $a \in R$. Then there exists $x \in R$ such that $a = axa$. Thus

$$
\begin{aligned}
&(\lambda \circ^{0.5} \mathscr{R} \circ^{0.5} \lambda)(a) \\
&= \left(\sup_{a=\sum_{i=1}^{m} a_i b_i} \left\{ \bigwedge_{i=1}^{m} \{ (\lambda \circ^{0.5} \mathscr{R})(a_i) \wedge \lambda(b_i)) \right\} \right) \vee 0.5 \\
&\geq \{ (\lambda \circ^{0.5} \mathscr{R})(ax) \wedge \lambda(a) \} \vee 0.5 \\
&= \left(\left(\left(\sup_{ax=\sum_{i=1}^{m} c_i d_i} \left\{ \bigwedge_{i=1}^{m} \{ \lambda(c_i) \wedge \mathscr{R}(d_i) \} \right\} \right) \vee 0.5 \right) \wedge \lambda(a) \right) \vee 0.5 \\
&\geq \{ ((\lambda(axa) \wedge \mathscr{R}(x)) \vee 0.5) \wedge \lambda(a) \} \vee 0.5 \quad \text{(because } ax = axax\text{)} \\
&\geq \lambda(a) \vee 0.5.
\end{aligned}
$$

$(ii) \Rightarrow (iii)$ This is straightforward.

$(iii) \Rightarrow (i)$ Let Q be a quasi-ideal of R. Then χ_Q is an $(\overline{\in},\overline{\in}\vee\overline{q})$-fuzzy quasi-ideal of R. Thus by hypothesis $\chi_Q \vee 0.5 \leq (\chi_Q \circ^{0.5} \mathscr{R} \circ^{0.5} \chi_Q) = \chi_{QRQ} \vee 0.5$. Hence $Q \subseteq QRQ$, but $QRQ \subseteq Q$ always holds. Therefore $Q = QRQ$. Thus R is regular.

Theorem 7.11. *For a semiring R, the following conditions are equivalent.*
(i) R is regular.
(ii) $(\lambda \wedge^{0.5} \mu) \leq (\lambda \circ^{0.5} \mu \circ^{0.5} \lambda)$ for every $(\overline{\in},\overline{\in}\vee\overline{q})$-fuzzy bi-ideal λ and every $(\overline{\in},\overline{\in}\vee\overline{q})$-fuzzy ideal μ of R.
(iii) $(\lambda \wedge^{0.5} \mu) \leq (\lambda \circ^{0.5} \mu \circ^{0.5} \lambda)$ for every $(\overline{\in},\overline{\in}\vee\overline{q})$-fuzzy quasi-ideal λ and every $(\overline{\in},\overline{\in}\vee\overline{q})$-fuzzy ideal μ of R.

Proof. $(i) \Rightarrow (ii)$ Let λ be an $(\overline{\in},\overline{\in}\vee\overline{q})$-fuzzy bi-ideal and μ be an $(\overline{\in},\overline{\in}\vee\overline{q})$-fuzzy ideal of R. Then for $x \in R$, there exists $a \in R$ such that $x = xax$. Thus

$$
\begin{aligned}
&(\lambda \circ^{0.5} \mu \circ^{0.5} \lambda)(x) \\
&= \left(\sup_{x=\sum_{i=1}^{m} a_i b_i} \left\{ \bigwedge_{i=1}^{m} \{ (\lambda \circ^{0.5} \mu)(a_i) \wedge \lambda(b_i) \} \right\} \right) \vee 0.5 \\
&\geq \{ (\lambda \circ^{0.5} \mu)(xa) \wedge \lambda(x) \} \vee 0.5 \\
&= \left(\left(\sup_{xa=\sum_{i=1}^{m} c_i d_i} \left\{ \bigwedge_{i=1}^{m} \{ \lambda(c_i) \wedge \mu(d_i) \} \right\} \right) \vee 0.5 \wedge \lambda(x) \right) \vee 0.5 \\
&\geq \{ ((\lambda(x) \wedge \mu(axa)) \vee 0.5) \wedge \lambda(x) \} \vee 0.5 \quad \text{(because } xa = xaxa\text{)} \\
&\geq (\lambda(x) \wedge \mu(x)) \vee 0.5 \\
&\geq (\lambda \wedge \mu)(x) \vee 0.5.
\end{aligned}
$$

$(ii) \Rightarrow (iii)$ Straightforward.

$(iii) \Rightarrow (i)$ Let λ be an $(\overline{\in},\overline{\in}\vee\overline{q})$-fuzzy quasi-ideal of R. Since $\mathscr{R}$ is an $(\overline{\in},\overline{\in}\vee\overline{q})$-fuzzy ideal of R, so by hypothesis $\left(\lambda \wedge^{0.5} \mathscr{R}\right) \leq \left(\lambda \circ^{0.5} \mathscr{R} \circ^{0.5} \lambda\right)$. This implies $\lambda \vee 0.5 \leq \left(\lambda \circ^{0.5} \mathscr{R} \circ^{0.5} \lambda\right)$. Hence by Theorem 7.10, R is regular.

Theorem 7.12. *For a semiring R, the following conditions are equivalent.*

(i) R is regular.

(ii) $\left(\lambda \wedge^{0.5} \mu\right) \leq \left(\lambda \circ^{0.5} \mu\right)$ *for every* $(\overline{\in},\overline{\in}\vee\overline{q})$*-fuzzy bi-ideal* λ *and every* $(\overline{\in},\overline{\in}\vee\overline{q})$*-fuzzy left ideal* μ *of R.*

(iii) $\left(\lambda \wedge^{0.5} \mu\right) \leq \left(\lambda \circ^{0.5} \mu\right)$ *for every* $(\overline{\in},\overline{\in}\vee\overline{q})$*-fuzzy quasi-ideal* λ *and every* $(\overline{\in},\overline{\in}\vee\overline{q})$*-fuzzy left ideal* μ *of R.*

(iv) $\left(\lambda \wedge^{0.5} \mu\right) \leq \left(\lambda \circ^{0.5} \mu\right)$ *for every* $(\overline{\in},\overline{\in}\vee\overline{q})$*-fuzzy right ideal* λ *and every* $(\overline{\in},\overline{\in}\vee\overline{q})$*-fuzzy bi-ideal* μ *of R.*

(v) $\left(\lambda \wedge^{0.5} \mu\right) \leq \left(\lambda \circ^{0.5} \mu\right)$ *for every* $(\overline{\in},\overline{\in}\vee\overline{q})$*-fuzzy right ideal* λ *and every* $(\overline{\in},\overline{\in}\vee\overline{q})$*-fuzzy quasi-ideal* μ *of R.*

(vi) $\left(\lambda \wedge^{0.5} \mu \wedge^{0.5} \nu\right) \leq \left(\lambda \circ^{0.5} \mu \circ^{0.5} \nu\right)$ *for every* $(\overline{\in},\overline{\in}\vee\overline{q})$*-fuzzy right ideal* λ*, every* $(\overline{\in},\overline{\in}\vee\overline{q})$*-fuzzy bi-ideal* μ *and every* $(\overline{\in},\overline{\in}\vee\overline{q})$*-fuzzy left ideal* ν *of R.*

(vii) $\left(\lambda \wedge^{0.5} \mu \wedge^{0.5} \nu\right) \leq \left(\lambda \circ^{0.5} \mu \circ^{0.5} \nu\right)$ *for every* $(\overline{\in},\overline{\in}\vee\overline{q})$*-fuzzy right ideal* λ*, every* $(\overline{\in},\overline{\in}\vee\overline{q})$*-fuzzy quasi-ideal* μ *and every* $(\overline{\in},\overline{\in}\vee\overline{q})$*-fuzzy right ideal* ν *of R.*

Proof. $(i) \Rightarrow (ii)$ Let λ be an $(\overline{\in},\overline{\in}\vee\overline{q})$-fuzzy bi-ideal and μ an $(\overline{\in},\overline{\in}\vee\overline{q})$-fuzzy left ideal of R. Now for any $a \in R$ there exists $x \in R$ such that $a = axa$. Thus

$$
\begin{aligned}
&\left(\lambda \circ^{0.5} \mu\right)(a) \\
&= \sup_{a=\Sigma_{i=1}^{m} a_i b_i} \left\{ \bigwedge_{i=1}^{m} \{\lambda(a_i) \wedge \mu(b_i)\} \right\} \vee 0.5 \\
&\geq \{\lambda(a) \wedge \mu(xa)\} \vee 0.5 \text{ because } a = axa \\
&\geq \{\lambda(a) \wedge (\mu(xa) \vee 0.5)\} \vee 0.5 \\
&\geq \{\lambda(a) \wedge \mu(a)\} \vee 0.5 = \{(\lambda \wedge \mu)(a) \vee 0.5\} = \left(\lambda \wedge^{0.5} \mu\right)(a).
\end{aligned}
$$

So $\left(\lambda \circ^{0.5} \mu\right) \geq \left(\lambda \wedge^{0.5} \mu\right)$.

$(ii) \Rightarrow (iii)$ This is obvious because every $(\overline{\in},\overline{\in}\vee\overline{q})$-fuzzy quasi-ideal is an $(\overline{\in},\overline{\in}\vee\overline{q})$-fuzzy bi-ideal.

$(iii) \Rightarrow (i)$ Let λ be an $(\overline{\in},\overline{\in}\vee\overline{q})$-fuzzy right ideal and μ be an $(\overline{\in},\overline{\in}\vee\overline{q})$-fuzzy left ideal of R. Since every $(\overline{\in},\overline{\in}\vee\overline{q})$-fuzzy right ideal is an $(\overline{\in},\overline{\in}\vee\overline{q})$-fuzzy quasi-ideal, we have by (iii) that $\left(\lambda \circ^{0.5} \mu\right) \geq \left(\lambda \wedge^{0.5} \mu\right)$. But by Lemma 7.6, $\left(\lambda \circ^{0.5} \mu\right) \leq \left(\lambda \wedge^{0.5} \mu\right)$. Hence $\left(\lambda \circ^{0.5} \mu\right) = \left(\lambda \wedge^{0.5} \mu\right)$ for every $(\overline{\in},\overline{\in}\vee\overline{q})$-fuzzy right ideal λ and every $(\overline{\in},\overline{\in}\vee\overline{q})$-fuzzy left ideal μ of R. Thus by Theorem 7.9, R is regular.

Similarly we can show that $(i) \Leftrightarrow (iv) \Leftrightarrow (v)$.

$(i) \Rightarrow (vi)$ Let λ be an $(\overline{\in},\overline{\in}\vee\overline{q})$-fuzzy right ideal, μ be an $(\overline{\in},\overline{\in}\vee\overline{q})$-fuzzy bi-ideal and ν be an $(\overline{\in},\overline{\in}\vee\overline{q})$-fuzzy left ideal of R. Now for any $a \in R$ there exists $x \in R$ such that $a = axa$. Hence

$$\begin{aligned}
&\left(\left(\lambda \circ^{0.5} \mu \circ^{0.5} \nu\right)\right)(a) \\
&= \sup_{a=\Sigma_{i=1}^{m} a_i b_i} \left\{ \bigwedge_{i=1}^{m} \left\{ \left(\lambda \circ^{0.5} \mu\right)(a_i) \wedge \nu(b_i) \right\} \right\} \vee 0.5 \\
&\geq \left\{ \left(\lambda \circ^{0.5} \mu\right)(a) \wedge \nu(xa) \right\} \vee 0.5 \\
&\geq \left\{ \left(\lambda \circ^{0.5} \mu\right)(a) \wedge (\nu(xa) \vee 0.5) \right\} \vee 0.5 \\
&\geq \left\{ \left(\lambda \circ^{0.5} \mu\right)(a) \wedge \nu(a) \right\} \vee 0.5 = \\
&\geq \left[\left(\sup_{a=\Sigma_{i=1}^{m} a_i b_i} \left\{ \bigwedge_{i=1}^{m} \{\lambda(a_i) \wedge \mu(b_i)\} \right\} \vee 0.5 \right) \wedge \nu(a) \right] \vee 0.5 \\
&\geq [\lambda(ax) \wedge \mu(a) \wedge \nu(a)] \vee 0.5 \\
&\geq [\lambda(a) \wedge \mu(a) \wedge \nu(a)] \vee 0.5 \\
&\geq [(\lambda \wedge \mu \wedge \nu)(a)] \vee 0.5 \\
&\geq \left(\lambda \wedge^{0.5} \mu \wedge^{0.5} \nu\right)(a).
\end{aligned}$$

Thus $\left(\lambda \wedge^{0.5} \mu \wedge^{0.5} \nu\right) \leq \left(\lambda \circ^{0.5} \mu \circ^{0.5} \nu\right)$.

$(vi) \Rightarrow (vii)$ Obvious.

$(vii) \Rightarrow (i)$ Let λ be an $(\overline{\in},\overline{\in}\vee\overline{q})$-fuzzy right ideal and ν be an $(\in,\in\vee q)$-fuzzy left ideal of R. Then

$\left(\lambda \wedge^{0.5} \nu\right) = \left(\lambda \wedge^{0.5} \mathscr{R} \wedge^{0.5} \nu\right) \leq \left(\lambda \circ^{0.5} \mathscr{R} \circ^{0.5} \nu\right) \leq \left(\lambda \circ^{0.5} \nu\right)$.

But by Lemma 7.6, $\left(\lambda \circ^{0.5} \nu\right) \leq \left(\lambda \wedge^{0.5} \nu\right)$. Hence $\left(\lambda \circ^{0.5} \nu\right) = \left(\lambda \wedge^{0.5} \nu\right)$ for every $(\overline{\in},\overline{\in}\vee\overline{q})$-fuzzy right ideal λ and for every $(\overline{\in},\overline{\in}\vee\overline{q})$-fuzzy left ideal ν of R. Thus by Theorem 7.9, R is regular.

7.3 Intra-regular Semirings

In this section we characterize intra-regular and regular and intra-regular semirings by the properties of their $(\overline{\in},\overline{\in}\vee\overline{q})$-fuzzy left (right) ideals, $(\overline{\in},\overline{\in}\vee\overline{q})$-fuzzy bi-ideals and $(\overline{\in},\overline{\in}\vee\overline{q})$-fuzzy quasi-ideals.

Lemma 7.7. *A semiring R is intra-regular if and only if $\lambda \wedge^{0.5} \mu \leq \lambda \circ^{0.5} \mu$ for every $(\overline{\in},\overline{\in}\vee\overline{q})$-fuzzy left ideal λ and for every $(\overline{\in},\overline{\in}\vee\overline{q})$-fuzzy right ideal μ of R.*

Proof. Let λ and μ be $(\overline{\in},\overline{\in}\vee\overline{q})$-fuzzy left ideal and $(\overline{\in},\overline{\in}\vee\overline{q})$-fuzzy right ideal of R, respectively. Let $a \in R$. Then there exist $x_i, x_i' \in R$ such that $a = \Sigma_{i=1}^{m} x_i a^2 x_i'$. Thus we have

$$
\begin{aligned}
&\left(\lambda \circ^{0.5} \mu\right)(a) \\
&= \sup_{a=\Sigma_{i=1}^{m} a_i b_i} \left\{ \bigwedge_{i=1}^{m} \{\lambda(a_i) \wedge \mu(b_i)\} \right\} \vee 0.5 \\
&\geq \bigwedge_{i=1}^{m} \{\lambda(x_i a) \wedge \mu(a x_i')\} \vee 0.5 \\
&\geq \bigwedge_{i=1}^{m} \{(\lambda(x_i a) \vee 0.5) \wedge (\mu(a x_i') \vee 0.5)\} \vee 0.5 \\
&\geq \{\lambda(a) \wedge \mu(a)\} \vee 0.5 = \left(\lambda \wedge^{0.5} \mu\right)(a).
\end{aligned}
$$

So $\lambda \circ^{0.5} \mu \geq \lambda \wedge^{0.5} \mu$.

Conversely, assume that A and B be left and right ideals of R, respectively. Then χ_A and χ_B are $(\overline{\in}, \overline{\in} \vee \overline{q})$-fuzzy left ideal and $(\overline{\in}, \overline{\in} \vee \overline{q})$-fuzzy right ideal of R, respectively. Thus by hypothesis $\chi_A \wedge^{0.5} \chi_B \leq \chi_A \circ^{0.5} \chi_B \Rightarrow \chi_{A \cap B} \vee 0.5 \leq \chi_{AB} \vee 0.5 \Rightarrow A \cap B \subseteq AB$. Thus by Theorem 5.7, R is intra-regular.

Theorem 7.13. *The following conditions are equivalent for a semiring R.*
(i) R is both regular and intra-regular.
(ii) $\lambda \vee 0.5 = \lambda \circ^{0.5} \lambda$ for every $(\overline{\in}, \overline{\in} \vee \overline{q})$-fuzzy bi-ideal λ of R.
(iii) $\lambda \vee 0.5 = \lambda \circ^{0.5} \lambda$ for every $(\overline{\in}, \overline{\in} \vee \overline{q})$-fuzzy quasi-ideal λ of R.

Proof. $(i) \Rightarrow (ii)$ Let λ be an $(\overline{\in}, \overline{\in} \vee \overline{q})$-fuzzy bi-ideal of R and $x \in R$. Since R is both regular and intra-regular, there exist elements $a, p_i, p_i' \in R$ such that

$x = xax$ and $x = \Sigma_{i=1}^{m} p_i x x p_i'$.

Thus $x = xax = xaxax = xa\left(\Sigma_{i=1}^{m} p_i x x p_i'\right) ax = \Sigma_{i=1}^{m} (xap_i x)(xp_i' ax)$.

Now

$$
\begin{aligned}
&(\lambda \circ^{0.5} \lambda)(x) \\
&= \sup_{x=\Sigma_{i=1}^{m} a_i b_i} \left\{ \bigwedge_{i=1}^{m} \{\lambda(a_i) \wedge \lambda(b_i)\} \right\} \vee 0.5 \\
&\geq \bigwedge_{i=1}^{m} \{\lambda(xap_i x) \wedge \lambda(xp_i' ax)\} \vee 0.5 \\
&\geq \lambda(x) \vee 0.5.
\end{aligned}
$$

Thus $\lambda \circ^{0.5} \lambda \geq \lambda \vee 0.5$.

On the other hand $\lambda \circ^{0.5} \lambda \leq \lambda \vee 0.5$.

Hence $\lambda \vee 0.5 = \lambda \circ^{0.5} \lambda$.

$(ii) \Rightarrow (iii)$ Obvious.

$(iii) \Rightarrow (i)$ Let Q be a quasi-ideal of R. Then χ_Q is an $(\overline{\in}, \overline{\in} \vee \overline{q})$-fuzzy quasi-ideal of R. Thus by hypothesis

$\chi_Q \vee 0.5 = \chi_Q \circ^{0.5} \chi_Q = \chi_Q \circ \chi_Q \vee 0.5 = \chi_{Q^2} \vee 0.5$.

$\Rightarrow Q = Q^2$. Hence by Theorem 5.9, R is both regular and intra-regular.

Theorem 7.14. *The following conditions are equivalent for a semiring R.*
(i) R is both regular and intra-regular.
(ii) $\lambda \wedge^{0.5} \mu \leq \lambda \circ^{0.5} \mu$ for all $(\overline{\in}, \overline{\in} \vee \overline{q})$-fuzzy bi-ideals λ and μ of R.

(iii) $\lambda\wedge^{0.5}\mu\leq\lambda\circ^{0.5}\mu$ *for every* $(\overline{\in},\overline{\in}\vee\overline{q})$*-fuzzy bi-ideal* λ *and every* $(\overline{\in},\overline{\in}\vee\overline{q})$*-fuzzy quasi-ideal* μ *of* R.

(iv) $\lambda\wedge^{0.5}\mu\leq\lambda\circ^{0.5}\mu$ *for every* $(\overline{\in},\overline{\in}\vee\overline{q})$*-fuzzy quasi-ideal* λ *and every* $(\overline{\in},\overline{\in}\vee\overline{q})$*-fuzzy bi-ideal* μ *of* R.

(v) $\lambda\wedge^{0.5}\mu\leq\lambda\circ^{0.5}\mu$ *for all* $(\overline{\in},\overline{\in}\vee\overline{q})$*-fuzzy quasi-ideals* λ *and* μ *of* R.

Proof. $(i)\Rightarrow(ii)$ Let λ and μ be $(\overline{\in},\overline{\in}\vee\overline{q})$-fuzzy bi-ideals of R and $x\in R$. Since R is both regular and intra-regular, there exist elements $a,p_i,p_i'\in R$ such that

$x=xax$ and $x=\sum_{i=1}^{m}p_ixxp_i'$.

Thus $x=xax=xaxax=xa\left(\sum_{i=1}^{m}p_ixxp_i'\right)ax=\sum_{i=1}^{m}(xap_ix)(xp_i'ax)$.

Now

$$
\begin{aligned}
&(\lambda\circ^{0.5}\mu)(x)\\
&=\sup_{x=\sum_{i=1}^{m}a_ib_i}\left\{\bigwedge_{i=1}^{m}\{\lambda(a_i)\wedge\mu(b_i)\}\right\}\vee 0.5\\
&\geq\bigwedge_{i=1}^{m}\{\lambda(xap_ix)\wedge\mu(xp_i'ax)\}\vee 0.5\\
&\geq\{\lambda(x)\wedge\mu(x)\}\vee 0.5\\
&\geq(\lambda\wedge^{0.5}\mu)(x).
\end{aligned}
$$

Thus $\lambda\circ^{0.5}\mu\geq(\lambda\wedge^{0.5}\mu)$.

$(ii)\Rightarrow(iii)\Rightarrow(v)$ and $(ii)\Rightarrow(iv)\Rightarrow(v)$ are clear.

$(v)\Rightarrow(i)$ Let λ be an $(\overline{\in},\overline{\in}\vee\overline{q})$-fuzzy left ideal and μ be an $(\overline{\in},\overline{\in}\vee\overline{q})$-fuzzy right ideal of R. Then λ and μ are $(\overline{\in},\overline{\in}\vee\overline{q})$-fuzzy bi-ideals of R. So by hypothesis $\lambda\wedge^{0.5}\mu\leq\lambda\circ^{0.5}\mu$. But by Lemma 7.6, $\lambda\wedge^{0.5}\mu\geq\lambda\circ^{0.5}\mu$ Thus $\lambda\wedge^{0.5}\mu=\lambda\circ^{0.5}\mu$. Hence by Theorem 7.9, R is regular. On the other hand by hypothesis we also have $\lambda\wedge^{0.5}\mu\leq\mu\circ^{0.5}\lambda$. By Lemma 7.7, R is intra-regular.

7.4 $(\overline{\in},\overline{\in}\vee\overline{q})$-Fuzzy k-Ideals

Definition 7.5. A fuzzy subset λ of a semiring R is said to be an $(\overline{\in},\overline{\in}\vee\overline{q})$*-fuzzy left (resp. right) k-ideal* of R if it satisfies $[1a]$, $[4a]$ and $[5a]$ (resp. $[1a]$, $[3a]$ and $[5a]$).

Definition 7.6. A fuzzy subset λ of a semiring R is said to be an $(\overline{\in},\overline{\in}\vee\overline{q})$*-fuzzy k-bi-ideal* of R if it satisfies $[1a]$, $[2a]$, $[5a]$ and $[6a]$.

Definition 7.7. A fuzzy subset λ of a semiring R is said to be an $(\overline{\in},\overline{\in}\vee\overline{q})$*-fuzzy k-quasi-ideal* of R if it satisfies $[1a]$, $[5a]$ and $[8a]$.

The proof of the following Lemma is straightforward and hence omitted.

Lemma 7.8. *A nonempty subset* A *of* R *is a left k-ideal (right k-ideal, k-ideal, k-bi-ideal, k-quasi-ideal) of* R *if and only if* χ_A *is an* $(\overline{\in},\overline{\in}\vee\overline{q})$*-fuzzy left k-ideal (right k-ideal, k-ideal, k-bi-ideal, k-quasi-ideal) of* R.

Theorem 7.15. *(i) If* λ *is an* $(\overline{\in},\overline{\in}\vee\overline{q})$*-fuzzy left (right) k-ideal of a semiring* R*, then* $(\lambda\vee 0.5)$ *is an* $(\overline{\in},\overline{\in}\vee\overline{q})$*-fuzzy left (right) k-ideal of* R.

(ii) If λ *is an* $(\overline{\in},\overline{\in}\vee\overline{q})$*-fuzzy k-bi-ideal of a semiring* R*, then* $\lambda\vee 0.5$ *is an* $(\overline{\in},\overline{\in}\vee\overline{q})$*-fuzzy k-bi-ideal of* R.

Proof. The proof is similar to the proof of Theorem 7.2.

Lemma 7.9. *Let A, B be subsets of R. Then*

$$C_A \oplus^{0.5} C_B = C_{\overline{A+B}} \vee 0.5.$$

Proof. Let A, B be subsets of R and $x \in R$. If $x \in \overline{A+B}$, then there exist $a_1, a_2 \in A$ and $b_1, b_2 \in B$ such that $x + (a_1 + b_1) = (a_2 + b_2)$. Thus
$(\chi_A \oplus_{0.5} \chi_B)(x)$
$= \sup_{x+(a_1'+b_1')=(a_2'+b_2')} (\min\{\chi_A(a_1'), \chi_A(a_2'), \chi_B(b_1'), \chi_B(b_2')\} \vee 0.5)$
$= 1 \vee 0.5 = \chi_{\overline{A+B}}(x) \vee 0.5.$
If $x \notin \overline{A+B}$, then there do not exist $a_1, a_2 \in A$ and $b_1, b_2 \in B$ such that $x + (a_1 + b_1) = (a_2 + b_2)$. Thus
$(\chi_A \oplus_{0.5} \chi_B)(x) = 0 \vee 0.5 = \chi_{\overline{A+B}}(x) \vee 0.5.$
Hence $\chi_A \oplus_{0.5} \chi_B = \chi_{\overline{A+B}} \vee 0.5.$

Lemma 7.10. *A fuzzy subset λ of a semiring R satisfies conditions $[1b]$ and $[5b]$ if and only if it satisfies condition*
$[13b]$ $\lambda \oplus^{0.5} \lambda \le \lambda \vee 0.5.$

Proof. Suppose λ satisfies conditions $[1b]$ and $[5b]$. Let $x \in R$. Then

$$(\lambda \oplus^{0.5} \lambda)(x) = \left(\sup_{x+(a_1+b_1)=(a_2+b_2)} \{\min\{\lambda(a_1), \lambda(a_2), \lambda(b_1), \lambda(b_2)\}\} \right) \vee 0.5$$

$$= \left(\sup_{x+(a_1+b_1)=(a_2+b_2)} \min\{\min\{\lambda(a_1), \lambda(b_1)\}, \min\{\lambda(a_2), \lambda(b_2)\}\} \right) \vee 0.5$$

$$\le \left(\sup_{x+(a_1+b_1)=(a_2+b_2)} \min\{\{\max(\lambda(a_1+b_1), 0.5), \max(\lambda(a_2+b_2), 0.5)\}\} \right) \vee 0.5$$

$$\le \left(\sup_{x+(a_1+b_1)=(a_2+b_2)} \{\min\{\lambda(a_1+b_1), \lambda(a_2+b_2)\} \vee 0.5\} \right) \vee 0.5$$

$\le \lambda(x) \vee 0.5.$
Thus $\lambda \oplus^{0.5} \lambda \le \lambda \vee 0.5.$
Conversely, assume that $\lambda \oplus^{0.5} \lambda \le \lambda \vee 0.5$ and $x, y \in R$. Then
$\lambda(x+y) \vee 0.5 = (\lambda \oplus^{0.5} \lambda)(x+y)$

$$= \left(\sup_{(x+y)+(a_1+b_1)=(a_2+b_2)} \{\min\{\lambda(a_1), \lambda(a_2), \lambda(b_1), \lambda(b_2)\}\} \right) \vee 0.5$$

$\ge \min\{\lambda(0), \lambda(x), \lambda(0), \lambda(y)\} \vee 0.5$ because $(x+y) + (0+0) = (x+y)$
$\ge \{(\lambda(x) \wedge \lambda(y))\}.$
Thus λ satisfies condition $[1b]$.
Let $a, b, x \in R$ such that $x + a = b$. Then
$\lambda(x) \vee 0.5 \ge (\lambda \oplus^{0.5} \lambda)(x)$

$$= \left(\sup_{x+(a_1+b_1)=(a_2+b_2)} \{\min\{\lambda(a_1), \lambda(a_2), \lambda(b_1), \lambda(b_2)\}\} \right) \vee 0.5$$

$\geq \min\{\lambda(a),\lambda(0),\lambda(b)\}\vee 0.5$ because $x+(a+0)=(b+0)$
$=\min\{\lambda(a),\lambda(b)\}$.
This shows that λ satisfies $[5b]$.

Theorem 7.16. *A fuzzy subset λ of a semiring R is an $(\overline{\in},\overline{\in}\vee\overline{q})$-fuzzy left (resp. right) k-ideal of R if and only if λ satisfies conditions*
$[13b]$ $\lambda\oplus^{0.5}\lambda\leq\lambda\vee 0.5$
$[14b]$ $\mathscr{R}\odot^{0.5}\lambda\leq\lambda\vee 0.5$ $\left(\textit{resp. } \lambda\odot^{0.5}\mathscr{R}\leq\lambda\vee 0.5\right)$.

Proof. Suppose λ be an $(\overline{\in},\overline{\in}\vee\overline{q})$-fuzzy left k-ideal of R. Then by Lemma 7.10, λ satisfies condition $[13b]$. Now we show that λ satisfies condition $[14b]$. Let $x\in R$. If $(\mathscr{R}\odot^{0.5}\lambda)(x)=0\vee 0.5$, then $(\mathscr{R}\odot^{0.5}\lambda)(x)\leq\lambda(x)\vee 0.5$. Otherwise, there exist elements $a_i,b_i,c_j,d_j\in R$ such that $x+\sum_{i=1}^{m}a_ib_i=\sum_{j=1}^{n}c_jd_j$. Then we have

$$\begin{aligned}
&(\mathscr{R}\odot^{0.5}\lambda)(x)\\
&=\sup_{x+\Sigma_{i=1}^{m}a_ib_i=\Sigma_{j=1}^{n}a'_jb'_j}\left\{\min\left\{\mathscr{R}(a_i),\lambda(b_i),\mathscr{R}\left(a'_j\right),\lambda\left(b'_j\right)\right\}\right\}\vee 0.5\\
&=\sup_{x+\Sigma_{i=1}^{m}a_ib_i=\Sigma_{j=1}^{n}a'_jb'_j}\left\{\min\left\{\lambda(b_i),\lambda\left(b'_j\right)\right\}\right\}\vee 0.5\\
&\leq\sup_{x+\Sigma_{i=1}^{m}a_ib_i=\Sigma_{j=1}^{n}a'_jb'_j}\left\{\min\left\{\lambda(a_ib_i)\vee 0.5,\lambda\left(a'_jb'_j\right)\vee 0.5\right\}\right\}\vee 0.5\\
&\leq\sup_{x+\Sigma_{i=1}^{m}a_ib_i=\Sigma_{j=1}^{n}a'_jb'_j}\left(\min\left\{\lambda\left(\Sigma_{i=1}^{m}a_ib_i\right)\vee 0.5,\lambda\left(\Sigma_{j=1}^{n}a'_jb'_j\right)\vee 0.5\right\}\right)\wedge 0.5\\
&\leq\lambda(x)\vee 0.5.\\
&\Rightarrow\mathscr{R}\odot_{0.5}\lambda\leq\lambda\wedge 0.5.
\end{aligned}$$

Conversely, assume that λ satisfies conditions $[13b]$ and $[14b]$. Then by Lemma 7.10, λ satisfies conditions $[1b]$ and $[5b]$. We show that λ satisfies condition $[4b]$. Let $x,y\in R$. Then we have

$$\begin{aligned}
\lambda(xy)\vee 0.5&\geq(\mathscr{R}\odot_{0.5}\lambda)(xy)\\
&=\sup_{xy+\Sigma_{i=1}^{m}a_ib_i=\Sigma_{j=1}^{n}a'_jb'_j}\left\{\min\left\{\mathscr{R}(a_i),\lambda(b_i),\mathscr{R}\left(a'_j\right),\lambda\left(b'_j\right)\right\}\right\}\vee 0.5\\
&=\sup_{xy+\Sigma_{i=1}^{m}a_ib_i=\Sigma_{j=1}^{n}a'_jb'_j}\left\{\min\left\{\lambda(b_i),\lambda\left(b'_j\right)\right\}\right\}\vee 0.5\\
&\geq\lambda(y)\vee 0.5\qquad\text{because } xy+0y=xy.
\end{aligned}$$

This shows that λ satisfies condition $[4b]$. So λ is an $(\overline{\in},\overline{\in}\vee\overline{q})$-fuzzy left k-ideal of R.

Similarly we can prove that:

Lemma 7.11. *A fuzzy subset λ of a semiring R satisfies conditions $[1b],[2b]$ and $[5b]$ if and only if it satisfies $[13b]$ and*
$[14b]$ $\lambda\odot^{0.5}\lambda\leq\lambda\vee 0.5$.

Lemma 7.12. *A fuzzy subset λ of a semiring R satisfies conditions $[1b], [5b]$ and $[6b]$ if and only if it satisfies $[13b]$ and*
$[15b]$ $\lambda \odot^{0.5} \mathscr{R} \odot^{0.5} \lambda \leq \lambda \vee 0.5$.

By using the above Lemmas we can prove the following Theorems

Theorem 7.17. *A fuzzy subset λ of a semiring R is an $(\overline{\in}, \overline{\in} \vee \overline{q})$-fuzzy k-bi-ideal of R if and only if λ satisfies conditions $[13b], [14b]$ and $[15b]$.*

Theorem 7.18. *A fuzzy subset λ of a semiring R is an $(\overline{\in}, \overline{\in} \vee \overline{q})$-fuzzy k-quasi-ideal of R if and only if λ satisfies conditions $[8b]$ and $[13b]$.*

The proofs of the following Results are straight forward.

Theorem 7.19. *Every $(\overline{\in}, \overline{\in} \vee \overline{q})$-fuzzy left (right) k-ideal of a semiring R is an $(\overline{\in}, \overline{\in} \vee \overline{q})$-fuzzy k-quasi-ideal of R.*

Theorem 7.20. *Every $(\overline{\in}, \overline{\in} \vee \overline{q})$-fuzzy k-quasi-ideal of a semiring R is an $(\overline{\in}, \overline{\in} \vee \overline{q})$-fuzzy k-bi-ideal of R.*

Lemma 7.13. *If λ and μ are $(\overline{\in}, \overline{\in} \vee \overline{q})$-fuzzy right and left k-ideals of R, respectively, then $\lambda \odot^{0.5} \mu \leq \lambda \wedge^{0.5} \mu$.*

Proof. Let λ and μ be $(\overline{\in}, \overline{\in} \vee \overline{q})$-fuzzy right and left k-ideals of R, respectively. Then

$$\left(\lambda \odot^{0.5} \mu\right)(x)$$
$$= \left(\sup_{x+\sum_{i=1}^{m} a_i b_i = \sum_{j=1}^{n} c_j d_j} \{\min\{\lambda(a_i), \mu(b_i), \lambda(c_j), \mu(d_j)\}\} \right) \vee 0.5$$
$$\leq \left(\sup_{x+\sum_{i=1}^{m} a_i b_i = \sum_{j=1}^{n} c_j d_j} \left\{ \begin{array}{c} \min\{\lambda(a_i b_i) \vee 0.5, \mu(a_i b_i) \vee 0.5, \\ \lambda(c_j d_j) \vee 0.5, \mu(c_j d_j) \vee 0.5\} \end{array} \right\} \right) \vee 0.5$$
$$\leq \left(\sup_{x+\sum_{i=1}^{m} a_i b_i = \sum_{j=1}^{n} c_j d_j} \left\{ \begin{array}{c} \min\{\lambda(a_i b_i), \mu(a_i b_i), \\ \lambda(c_j d_j), \mu(c_j d_j)\} \end{array} \right\} \right) \vee 0.5$$
$$\leq \left(\sup_{x+\sum_{i=1}^{m} a_i b_i = \sum_{j=1}^{n} c_j d_j} \left\{ \min \left\{ \begin{array}{c} (\lambda\left(\sum_{i=1}^{m} a_i b_i\right) \vee 0.5), (\mu\left(\sum_{i=1}^{m} a_i b_i\right) \vee 0.5), \\ (\lambda\left(\sum_{j=1}^{n} c_j d_j\right) \vee 0.5), (\mu\left(\sum_{j=1}^{n} c_j d_j\right) \vee 0.5) \end{array} \right\} \right\} \right) \vee 0.5$$
$$\leq \min\{\lambda(x), \mu(x)\} \vee 0.5 = \left(\lambda \wedge^{0.5} \mu\right)(x).$$

Thus $\lambda \odot^{0.5} \mu \leq \lambda \wedge^{0.5} \mu$.

7.5 *k*-Regular Semirings

In this section we characterize k-regular semirings by the properties of their $(\overline{\in}, \overline{\in} \vee \overline{q})$-fuzzy left (right) k-ideals, $(\overline{\in}, \overline{\in} \vee \overline{q})$-fuzzy k-bi-ideals and $(\overline{\in}, \overline{\in} \vee \overline{q})$-fuzzy k-quasi-ideals.

Theorem 7.21. *For a semiring R the following conditions are equivalent.*
(i) R is k-regular.
(ii) $(\lambda \wedge^{0.5} \mu) = (\lambda \odot^{0.5} \mu)$ *for every* $(\overline{\in}, \overline{\in} \vee \overline{q})$*-fuzzy right k-ideal* λ *and every* $(\overline{\in}, \overline{\in} \vee \overline{q})$*-fuzzy left k-ideal* μ *of R.*

Proof. The proof is similar to the proof of Theorem 7.9.

Theorem 7.22. *For a semiring R, the following conditions are equivalent.*
(i) R is k-hemiregular.
(ii) $\lambda \vee 0.5 \leq (\lambda \odot^{0.5} \mathscr{R} \odot^{0.5} \lambda)$ *for every* $(\overline{\in}, \overline{\in} \vee \overline{q})$*-fuzzy k-bi-ideal* λ *of R.*
(iii) $\lambda \vee 0.5 \leq (\lambda \odot^{0.5} \mathscr{R} \odot^{0.5} \lambda)$ *for every* $(\overline{\in}, \overline{\in} \vee \overline{q})$*-fuzzy k-quasi-ideal* λ *of R.*

Proof. The proof is similar to the proof of Theorem 7.10.

Theorem 7.23. *For a semiring R, the following conditions are equivalent.*
(i) R is k-regular.
(ii) $(\lambda \wedge^{0.5} \mu) \leq (\lambda \odot^{0.5} \mu \odot^{0.5} \lambda)$ *for every* $(\overline{\in}, \overline{\in} \vee \overline{q})$*-fuzzy k-bi-ideal* λ *and every* $(\overline{\in}, \overline{\in} \vee \overline{q})$*-fuzzy k-ideal* μ *of R.*
(iii) $(\lambda \wedge^{0.5} \mu) \leq (\lambda \odot^{0.5} \mu \odot^{0.5} \lambda)$ *for every* $(\overline{\in}, \overline{\in} \vee \overline{q})$*-fuzzy k-quasi-ideal* λ *and every* $(\overline{\in}, \overline{\in} \vee \overline{q})$*-fuzzy k-ideal* μ *of R.*

Proof. The proof is similar to the proof of Theorem 7.11.

Theorem 7.24. *For a semiring R, the following conditions are equivalent.*
(i) R is k-regular.
(ii) $(\lambda \wedge^{0.5} \mu) \leq (\lambda \odot^{0.5} \mu)$ *for every* $(\overline{\in}, \overline{\in} \vee \overline{q})$*-fuzzy k-bi-ideal* λ *and every* $(\overline{\in}, \overline{\in} \vee \overline{q})$*-fuzzy left k-ideal* μ *of R.*
(iii) $(\lambda \wedge^{0.5} \mu) \leq (\lambda \odot^{0.5} \mu)$ *for every* $(\overline{\in}, \overline{\in} \vee \overline{q})$*-fuzzy k-quasi-ideal* λ *and every* $(\overline{\in}, \overline{\in} \vee \overline{q})$*-fuzzy left k-ideal* μ *of R.*
(iv) $(\lambda \wedge^{0.5} \mu) \leq (\lambda \odot^{0.5} \mu)$ *for every* $(\overline{\in}, \overline{\in} \vee \overline{q})$*-fuzzy right k-ideal* λ *and every* $(\overline{\in}, \overline{\in} \vee \overline{q})$*-fuzzy k-bi-ideal* μ *of R.*
(v) $(\lambda \wedge^{0.5} \mu) \leq (\lambda \odot^{0.5} \mu)$ *for every* $(\overline{\in}, \overline{\in} \vee \overline{q})$*-fuzzy right k-ideal* λ *and every* $(\overline{\in}, \overline{\in} \vee \overline{q})$*-fuzzy k-quasi-ideal* μ *of R.*
(vi) $(\lambda \wedge^{0.5} \mu \wedge^{0.5} \nu) \leq (\lambda \odot^{0.5} \mu \odot^{0.5} \nu)$ *for every* $(\overline{\in}, \overline{\in} \vee \overline{q})$*-fuzzy right k-ideal* λ*, every* $(\overline{\in}, \overline{\in} \vee \overline{q})$*-fuzzy k-bi-ideal* μ *and every* $(\overline{\in}, \overline{\in} \vee \overline{q})$*-fuzzy left k-ideal* ν *of R.*
(vii) $(\lambda \wedge^{0.5} \mu \wedge^{0.5} \nu) \leq (\lambda \odot^{0.5} \mu \odot^{0.5} \nu)$ *for every* $(\overline{\in}, \overline{\in} \vee \overline{q})$*-fuzzy right k-ideal* λ*, every* $(\overline{\in}, \overline{\in} \vee \overline{q})$*-fuzzy k-quasi-ideal* μ *and every* $(\overline{\in}, \overline{\in} \vee \overline{q})$*-fuzzy left k-ideal* ν *of R.*

Proof. The proof is similar to the proof of Theorem 7.12.

7.6 *k*-Intra-regular Semirings

In this section we characterize *k*-intra-regular and *k*-regular and *k*-intra-regular semirings by the properties of their $(\overline{\in}, \overline{\in} \vee \overline{q})$-fuzzy left (right) *k*-ideals, $(\overline{\in}, \overline{\in} \vee \overline{q})$-fuzzy *k*-bi-ideals and $(\overline{\in}, \overline{\in} \vee \overline{q})$-fuzzy *k*-quasi-ideals.

Lemma 7.14. *A semiring R is k-intra-regular if and only if* $\lambda \wedge^{0.5} \mu \leq \lambda \odot^{0.5} \mu$ *for every* $(\overline{\in}, \overline{\in} \vee \overline{q})$*-fuzzy left k-ideal* λ *and for every* $(\overline{\in}, \overline{\in} \vee \overline{q})$*-fuzzy right k-ideal* μ *of R.*

Proof. The proof is similar to the proof of Lemma 7.7.

Theorem 7.25. *The following conditions are equivalent for a semiring R*

(i) *R is both k-regular and k-intra-regular.*

(ii) $\lambda \vee 0.5 = \lambda \odot^{0.5} \lambda$ *for every* $(\overline{\in}, \overline{\in} \vee \overline{q})$*-fuzzy k-bi-ideal* λ *of R.*

(iii) $\lambda \vee 0.5 = \lambda \odot^{0.5} \lambda$ *for every* $(\overline{\in}, \overline{\in} \vee \overline{q})$*-fuzzy k-quasi-ideal* λ *of R.*

Proof. The proof is similar to the proof of Theorem 7.13.

Theorem 7.26. *The following conditions are equivalent for a semiring R*

(i) *R is both k-regular and k-intra-regular.*

(ii) $\lambda \wedge^{0.5} \mu \leq \lambda \odot^{0.5} \mu$ *for all* $(\overline{\in}, \overline{\in} \vee \overline{q})$*-fuzzy k-bi-ideals* λ *and* μ *of R.*

(iii) $\lambda \wedge^{0.5} \mu \leq \lambda \odot^{0.5} \mu$ *for every* $(\overline{\in}, \overline{\in} \vee \overline{q})$*-fuzzy k-bi-ideal* λ *and every* $(\overline{\in}, \overline{\in} \vee \overline{q})$*-fuzzy k-quasi-ideals* μ *of R.*

(iv) $\lambda \wedge^{0.5} \mu \leq \lambda \odot^{0.5} \mu$ *for every* $(\overline{\in}, \overline{\in} \vee \overline{q})$*-fuzzy k-quasi-ideal* λ *and every* $(\overline{\in}, \overline{\in} \vee \overline{q})$*-fuzzy k-bi-ideals* μ *of R.*

(v) $\lambda \wedge^{0.5} \mu \leq \lambda \odot^{0.5} \mu$ *for all* $(\overline{\in}, \overline{\in} \vee \overline{q})$*-fuzzy k-quasi-ideals* λ *and* μ *of R.*

Proof. The proof is similar to the proof of Theorem 7.14.

Chapter 8
Fuzzy Ideals with Thresholds

Modelled on the previous two chapters, we make a similar study of fuzzy ideals, fuzzy quasi-ideals and fuzzy bi-ideals with thresholds in this chapter.

8.1 Fuzzy Ideals with Threshold (α,β)

Through out this chapter $\alpha,\beta \in (0,1]$ and $\alpha < \beta$.

Definition 8.1. A fuzzy subset λ of a semiring R is called a *fuzzy subsemiring with thresholds* (α,β) of R if it satisfies the following conditions

(1) $\max\{\lambda(x+y),\alpha\} \geq \min\{\lambda(x),\lambda(y),\beta\}$

(2) $\max\{\lambda(xy),\alpha\} \geq \min\{\lambda(x),\lambda(y),\beta\}$

for all $x,y \in R$.

Definition 8.2. A fuzzy subset λ of a semiring R is called a *fuzzy left (right)ideal with thresholds* (α,β) of R if it satisfies the conditions (1) and (3), where

(3) $\max\{\lambda(xy),\alpha\} \geq \min\{\lambda(y),\beta\}$ $\quad(\max\{\lambda(xy),\alpha\} \geq \min\{\lambda(x),\beta\})$

for all $x,y \in R$.

A fuzzy subset λ of a semiring R is called a *fuzzy ideal with thresholds* (α,β) of R if it is both a fuzzy left ideal and a fuzzy right ideal with thresholds (α,β) of R.

Definition 8.3. A fuzzy subset λ of a semiring R is called a *fuzzy bi-ideal with thresholds* (α,β) of R if it satisfies the conditions (1), (2) and (4), where

(4) $\max\{\lambda(xyz),\alpha\} \geq \min\{\lambda(x),\lambda(z),\beta\}$

for all $x,y,z \in R$.

Definition 8.4. A fuzzy subset λ of a semiring R is called a *fuzzy quasi-ideal with thresholds* (α,β) of R if it satisfies the conditions (1) and (5), where

(5) $\max\{\lambda(x),\alpha\} \geq \min\{\lambda\circ\mathscr{R}(x),\mathscr{R}\circ\lambda(x),\beta\}$

for all $x \in R$, where $\mathscr{R}$ is the fuzzy subset of R mapping every element of R onto 1.

Lemma 8.1. *A nonempty subset A of a semiring R is a left (right) ideal of R if and only if the characteristic function χ_A of A is a fuzzy left (right) ideal with thresholds (α,β) of R.*

J. Ahsan et al.: Fuzzy Semirings with Applications, STUDFUZZ 278, pp. 141–159.
springerlink.com

Proof. Let A be a left ideal of R and $x,y \in R$. If $x,y \in A$, then $x+y \in A$. Thus $\max\{\chi_A(x+y),\alpha\} = 1 \geq \beta = \min\{\chi_A(x),\chi_A(y),\beta\}$. Similarly, if $y \in A$, then $xy \in A$. Thus $\max\{\chi_A(xy),\alpha\} = 1 \geq \beta = \min\{\chi_A(x),\beta\}$. If one of x,y does not belongs to A, then $\min\{\chi_A(x),\chi_A(y),\beta\} = 0 \leq \max\{\chi_A(x+y),\alpha\}$. Similarly, if $y \notin A$,then $\min\{\chi_A(x),\beta\} = 0 \leq \max\{\chi_A(xy),\alpha\}$. Thus χ_A is a fuzzy left ideal with thresholds (α,β) of R.

Similarly we can prove the following lemma

Lemma 8.2. *A nonempty subset A of a semiring R is a quasi-ideal (bi-ideal) of R if and only if the characteristic function χ_A of A is a fuzzy quasi-ideal (bi-ideal) with thresholds (α,β) of R.*

Theorem 8.1. *A fuzzy subset λ of a semiring R is a fuzzy subsemiring with thresholds (α,β) of R if and only if $U(\lambda;t) \neq \emptyset$ is a subsemiring of R for all $t \in (\alpha,\beta]$.*

Proof. Let λ be a fuzzy subsemiring with thresholds (α,β) of R and $x,y \in U(\lambda;t)$. Then $\lambda(x) \geq t$ and $\lambda(y) \geq t$. As $\max\{\lambda(x+y),\alpha\} \geq \min\{\lambda(x),\lambda(y),\beta\}$, so $\max\{\lambda(x+y),\alpha\} \geq \min\{t,t,\beta\} = t$. Thus $\lambda(x+y) \geq t$. This implies $x+y \in U(\lambda;t)$. Similarly $xy \in U(\lambda;t)$. Hence $U(\lambda;t)$ is a subsemiring of R.

Conversely, assume that each $U(\lambda;t) \neq \emptyset$ is a subsemiring of R for all $t \in (\alpha,\beta]$. Let $x,y \in R$ be such that $\max\{\lambda(x+y),\alpha\} < \min\{\lambda(x),\lambda(y),\beta\}$. Select a $t \in (\alpha,\beta]$ such that $\max\{\lambda(x+y),\alpha\} < t \leq \min\{\lambda(x),\lambda(y),\beta\}$. Then $x,y \in U(\lambda;t)$ but $x+y \notin U(\lambda;t)$, a contradiction. Hence $\max\{\lambda(x+y),\alpha\} \geq \min\{\lambda(x),\lambda(y),\beta\}$. Similarly we can show that $\max\{\lambda(xy),\alpha\} \geq \min\{\lambda(x),\lambda(y),\beta\}$. Hence λ is a fuzzy subsemiring with thresholds (α,β) of R.

Similarly we can show the following result:

Theorem 8.2. *A fuzzy subset λ of a semiring R is a fuzzy left (right, quasi-, bi-) ideal with thresholds (α,β) of R if and only if $U(\lambda;t) \neq \emptyset$ is a left (right, quasi-, bi-) ideal of R for all $t \in (\alpha,\beta]$.*

Theorem 8.3. *Let λ be a fuzzy left (right, quasi-, bi-) ideal with thresholds (α,β) of R. Then $\lambda \wedge \beta$ is a left (right, quasi-, bi-) ideal with thresholds (α,β) of R.*

Proof. Let λ be a fuzzy left ideal with thresholds (α,β) of R and $x,y \in R$. Then

$\max\{(\lambda\wedge\beta)(x+y),\alpha\} = [\lambda(x+y)\wedge\beta]\vee\alpha$
$= [\lambda(x+y)\vee\alpha]\wedge\beta \geq \min\{\lambda(x),\lambda(y),\beta\}$
$= \min\{(\lambda\wedge\beta)(x),(\lambda\wedge\beta)(y),\beta\}$.

Similarly we can show that
$\max\{(\lambda\wedge\beta)(xy),\alpha\} \geq \min\{(\lambda\wedge\beta)(y),\beta\}$.
Thus $\lambda\wedge\beta$ is a fuzzy left ideal with thresholds (α,β) of R.
Similarly we can prove the other cases.

Definition 8.5. Let λ,μ be fuzzy subsets of a semiring R. Then the fuzzy subsets $\lambda\wedge_\alpha^\beta\mu, \lambda\circ_\alpha^\beta\mu, \lambda\odot_\alpha^\beta\mu, \lambda+_\alpha^\beta\mu$ and $\lambda\oplus_\alpha^\beta\mu$ of R are defined as following:

$$\left(\lambda \wedge_{\alpha}^{\beta} \mu\right)(x) = \{((\lambda \wedge \mu)(x)) \wedge \beta\} \vee \alpha$$
$$\left(\lambda \circ_{\alpha}^{\beta} \mu\right)(x) = \{((\lambda \circ \mu)(x)) \wedge \beta\} \vee \alpha$$
$$\left(\lambda \odot_{\alpha}^{\beta} \mu\right)(x) = \{((\lambda \odot \mu)(x)) \wedge \beta\} \vee \alpha$$
$$\left(\lambda +_{\alpha}^{\beta} \mu\right)(x) = \{((\lambda + \mu)(x)) \wedge \beta\} \vee \alpha$$
$$\left(\lambda \oplus_{\alpha}^{\beta} \mu\right)(x) = \{((\lambda +_k \mu)(x)) \wedge \beta\} \vee \alpha$$

for all $x \in R$.

Lemma 8.3. *Let A, B be subsets of R. Then $\chi_A +_{\alpha}^{\beta} \chi_B = (\chi_{A+B} \wedge \beta) \vee \alpha$.*

Proof. Let A, B be subsets of R and $x \in R$. If $x \in A + B$, then there exist $a \in A$ and $b \in B$ such that $x = a + b$. Thus

$$(\chi_A +_{\alpha}^{\beta} \chi_B)(x)$$
$$= \left\{\left(\sup_{x=c+d} \{\chi_A(c) \wedge \chi_B(d)\}\right) \wedge \beta\right\} \vee \alpha$$
$$= (1 \wedge \beta) \vee \alpha = (\chi_{A+B}(x) \wedge \beta) \vee \alpha.$$

If $x \notin A + B$ then there do not exist $a \in A$ and $b \in B$ such that $x = a + b$. Thus

$$(\chi_A +_{\alpha}^{\beta} \chi_B)(x) = (0 \wedge \beta) \vee \alpha = (\chi_{A+B}(x) \wedge \beta) \vee \alpha.$$

Hence $\chi_A +_{\alpha}^{\beta} \chi_B = (\chi_{A+B} \wedge \beta) \vee \alpha$.

Theorem 8.4. *A fuzzy subset λ of a semiring R satisfies condition (1) if and only if it satisfies condition (6), where*

(6) $\lambda +_{\alpha}^{\beta} \lambda \leq (\lambda \wedge \beta) \vee \alpha$.

Proof. Suppose λ satisfies condition (1) and $x \in R$. Then

$$(\lambda +_{\alpha}^{\beta} \lambda)(x)$$
$$= \left\{\left(\sup_{x=c+d} \{\lambda(c) \wedge \lambda(d)\}\right) \wedge \beta\right\} \vee \alpha$$
$$= \left\{\left(\sup_{x=c+d} \{\min\{\lambda(c), \lambda(d), \beta\}\right) \wedge \beta\right\} \vee \alpha$$
$$\leq \left\{\left(\sup_{x=c+d} (\lambda(c+d) \vee \alpha)\right) \wedge \beta\right\} \vee \alpha$$
$$\leq (\lambda(x) \wedge \beta) \vee \alpha.$$

Thus $\lambda +_{\alpha}^{\beta} \lambda \leq (\lambda \wedge \beta) \vee \alpha$.

Conversely, assume that $\lambda +_{\alpha}^{\beta} \lambda \leq (\lambda \wedge \beta) \vee \alpha$ and $x, y \in R$. Then

$$\lambda(x+y) \vee \alpha \geq (\lambda(x+y) \wedge \beta) \vee \alpha \geq ((\lambda \wedge \beta) \vee \alpha)(x+y)$$
$$\geq (\lambda +_{\alpha}^{\beta} \lambda)(x+y)$$
$$= \left\{\left(\sup_{x+y=c+d} \{\lambda(c) \wedge \lambda(d)\}\right) \wedge \beta\right\} \vee \alpha$$
$$\geq (\{\lambda(x) \wedge \lambda(y)\} \wedge \beta) \vee \alpha$$
$$\geq (\lambda(x) \wedge \lambda(y) \wedge \beta).$$

Thus λ satisfies condition (1).

Theorem 8.5. *A fuzzy subset λ of a semiring R is a fuzzy left (right) ideal with thresholds (α,β) of R if and only if λ satisfies conditions (6) and (7), where*
(7) $\mathscr{R}\circ_{\alpha}^{\beta}\lambda \le (\lambda\wedge\beta)\vee\alpha$ *(resp.* $\lambda\circ_{\alpha}^{\beta}\mathscr{R}\le(\lambda\wedge\beta)\vee\alpha$ *).*

Proof. Suppose λ be a fuzzy left ideal with thresholds (α,β) of R. Then λ satisfies conditions (1) and (3). Thus by Theorem 8.4, λ satisfies condition (6). Let $x\in R$. If $(\mathscr{R}\circ_{\alpha}^{\beta}\lambda)(x)=(0\wedge\beta)\vee\alpha$, then $(\mathscr{R}\circ_{\alpha}^{\beta}\lambda)(x)\le(\lambda(x)\wedge\beta)\vee\alpha$. Otherwise, there exist elements $a_i,b_i\in R$ such that $x=\sum_{i=1}^{m}a_ib_i$. Then

$$
\begin{aligned}
&(\mathscr{R}\circ_{\alpha}^{\beta}\lambda)(x)\\
&=\left(\sup_{x=\Sigma_{i=1}^{m}a_ib_i}\left\{\bigwedge_{i=1}^{m}\{\mathscr{R}(a_i)\wedge\lambda(b_i)\}\right\}\wedge\beta\right)\vee\alpha\\
&=\left(\sup_{x=\Sigma_{i=1}^{m}a_ib_i}\left\{\bigwedge_{i=1}^{m}\lambda(b_i)\right\}\wedge\beta\right)\vee\alpha\\
&=\left(\sup_{x=\Sigma_{i=1}^{m}a_ib_i}\left\{\bigwedge_{i=1}^{m}(\lambda(b_i)\wedge\beta)\right\}\wedge\beta\right)\vee\alpha\\
&\le\left(\sup_{x=\Sigma_{i=1}^{m}a_ib_i}\left\{\bigwedge_{i=1}^{m}(\lambda(a_ib_i)\vee\alpha)\right\}\wedge\beta\right)\vee\alpha\\
&=\left(\sup_{x=\Sigma_{i=1}^{m}a_ib_i}\left\{\bigwedge_{i=1}^{m}\lambda(a_ib_i)\right\}\wedge\beta\right)\vee\alpha\\
&=\left(\sup_{x=\Sigma_{i=1}^{m}a_ib_i}\left\{\bigwedge_{i=1}^{m}(\lambda(a_ib_i)\wedge\beta)\right\}\wedge\beta\right)\vee\alpha\\
&\le\left(\sup_{x=\Sigma_{i=1}^{m}a_ib_i}\{\lambda(\Sigma_{i=1}^{m}a_ib_i)\vee\alpha\}\wedge\beta\right)\vee\alpha\\
&\le\left(\sup_{x=\Sigma_{i=1}^{m}a_ib_i}\lambda(\Sigma_{i=1}^{m}a_ib_i)\wedge\beta\right)\vee\alpha\\
&\le(\lambda(x)\wedge\beta)\vee\alpha.
\end{aligned}
$$

Thus $\mathscr{R}\circ_{\alpha}^{\beta}\lambda\le(\lambda\wedge\beta)\vee\alpha$.

Conversely, assume that λ satisfies conditions (6) and (7). Then by Theorem 8.4, λ satisfies condition (1). We show that λ satisfies condition (3). Let $x,y\in R$. Then

$$
\begin{aligned}
\lambda(xy)\vee\alpha\ge(\lambda(xy)\wedge\beta)\vee\alpha&\ge\left(\mathscr{R}\circ_{\alpha}^{\beta}\lambda\right)(xy)\\
&=\left(\sup_{x=\Sigma_{i=1}^{m}a_ib_i}\left\{\bigwedge_{i=1}^{m}\{\mathscr{R}(a_i)\wedge\lambda(b_i)\}\right\}\wedge\beta\right)\vee\alpha\\
&=\left(\sup_{x=\Sigma_{i=1}^{m}a_ib_i}\left\{\bigwedge_{i=1}^{m}\lambda(b_i)\right\}\wedge\beta\right)\vee\alpha\\
&\ge(\lambda(y)\wedge\beta)\vee\alpha\ge\lambda(y)\wedge\beta.
\end{aligned}
$$

This shows that λ satisfies condition (3). Hence λ is a fuzzy left ideal with thresholds (α,β) of R.

Theorem 8.6. *A fuzzy subset λ of a semiring R satisfies conditions (1) and (2) if and only if it satisfies conditions (6) and (8), where*
(8) $\lambda \circ_\alpha^\beta \lambda \leq (\lambda \wedge \beta) \vee \alpha.$

Proof. Suppose λ satisfies condition (1) and (2). Then by Theorem 8.4, λ satisfies condition (6). Let $x \in R$. Then

$$(\lambda \circ_\alpha^\beta \lambda)(x) = \left\{\left(\sup_{x=\Sigma_{i=1}^m a_i b_i} \left\{\bigwedge_{i=1}^m \{\lambda(a_i) \wedge \lambda(b_i)\}\right\}\right) \wedge \beta\right\} \vee \alpha$$
$$= \left\{\left(\sup_{x=\Sigma_{i=1}^m a_i b_i} \left\{\bigwedge_{i=1}^m \{\lambda(a_i) \wedge \lambda(b_i) \wedge \beta\}\right\}\right) \wedge \beta\right\} \vee \alpha$$
$$\leq \left\{\left(\sup_{x=\Sigma_{i=1}^m a_i b_i} \left\{\bigwedge_{i=1}^m \{\lambda(a_i b_i) \vee \alpha\}\right\}\right) \wedge \beta\right\} \vee \alpha \quad \text{by condition } (2)$$
$$= \left\{\left(\sup_{x=\Sigma_{i=1}^m a_i b_i} \left\{\bigwedge_{i=1}^m \lambda(a_i b_i)\right\}\right) \wedge \beta\right\} \vee \alpha$$
$$= \left\{\left(\sup_{x=\Sigma_{i=1}^m a_i b_i} \left\{\bigwedge_{i=1}^m (\lambda(a_i b_i) \wedge \beta)\right\}\right) \wedge \beta\right\} \vee \alpha$$
$$\leq \left\{\left(\sup_{x=\Sigma_{i=1}^m a_i b_i} \left\{\lambda(\Sigma_{i=1}^m a_i b_i) \vee \alpha\right\}\right) \wedge \beta\right\} \vee \alpha \quad \text{by condition } (1)$$
$$= \left\{\left(\sup_{x=\Sigma_{i=1}^m a_i b_i} \lambda(\Sigma_{i=1}^m a_i b_i)\right) \wedge \beta\right\} \vee \alpha$$
$$\leq (\lambda(x) \wedge \beta) \vee \alpha.$$

Thus $\lambda \circ_\alpha^\beta \lambda \leq (\lambda \wedge \beta) \vee \alpha$.

Conversely, assume that λ satisfies conditions (6) and (8). Then by Theorem 8.4, λ satisfies (1). Let $x, y \in R$. Then

$$\lambda(xy) \vee \alpha \geq (\lambda(xy) \wedge \beta) \vee \alpha \geq (\lambda \circ_\alpha^\beta \lambda)(xy)$$
$$= \left\{\left(\sup_{x=\Sigma_{i=1}^m a_i b_i} \left\{\bigwedge_{i=1}^m \{\lambda(a_i) \wedge \lambda(b_i)\}\right\}\right) \wedge \beta\right\} \vee \alpha$$
$$\geq \{\{\lambda(x) \wedge \lambda(y)\} \wedge \beta\} \vee \alpha$$
$$\geq \lambda(x) \wedge \lambda(y) \wedge \beta.$$

Thus λ satisfies condition (2).

Theorem 8.7. *A fuzzy subset λ of a semiring R satisfies condition (1) and (4) if and only if it satisfies condition (6) and (9), where*
(9) $\lambda \circ_\alpha^\beta \mathscr{R} \circ_\alpha^\beta \lambda \leq (\lambda \wedge \beta) \vee \alpha.$

Proof. Suppose λ satisfies condition (1) and (4). Then by Theorem 8.4, λ satisfies condition (6). Let $x \in R$. Then

$$(\lambda \circ_{\alpha}^{\beta} \mathscr{R} \circ_{\alpha}^{\beta} \lambda)(x)$$
$$= \left\{ \left(\sup_{x=\Sigma_{i=1}^{m} a_i b_i} \left\{ \bigwedge_{i=1}^{m} \left\{ \left(\lambda \circ_{\alpha}^{\beta} \mathscr{R} \right)(a_i) \wedge \lambda(b_i) \right\} \right\} \right) \wedge \beta \right\} \vee \alpha$$
$$= \left\{ \left(\sup_{x=\Sigma_{i=1}^{m} a_i b_i} \left\{ \bigwedge_{i=1}^{m} \left\{ \left(\left\{ \left(\sup_{x=\Sigma_{j=1}^{n} c_j d_j} \left\{ \bigwedge_{j=1}^{n} \{\lambda(c_j) \wedge \mathscr{R}(d_j)\} \right\} \right) \wedge \beta \right\} \vee \alpha \right) \wedge \lambda(b_i) \right\} \right\} \right) \wedge \beta \right\} \vee \alpha$$
$$= \left\{ \left(\sup_{x=\Sigma_{i=1}^{m} a_i b_i} \left\{ \bigwedge_{i=1}^{m} \left\{ \left(\sup_{x=\Sigma_{j=1}^{n} c_j d_j} \left\{ \bigwedge_{j=1}^{n} \lambda(c_j) \right\} \right) \wedge \lambda(b_i) \right\} \right\} \right) \wedge \beta \right\} \vee \alpha$$
$$= \left\{ \left(\sup_{x=\Sigma_{i=1}^{m} a_i b_i} \left\{ \bigwedge_{i=1}^{m} \left\{ \sup_{a_i=\Sigma_{j=1}^{n} c_j d_j} \left\{ \bigwedge_{j=1}^{n} \lambda(c_j) \wedge \lambda(b_i) \wedge \beta \right\} \right\} \right\} \right) \wedge \beta \right\} \vee \alpha$$
$$\leq \left\{ \left(\sup_{x=\Sigma_{i=1}^{m} a_i b_i} \left\{ \bigwedge_{i=1}^{m} \left\{ \sup_{a_i=\Sigma_{j=1}^{n} c_j d_j} \bigwedge_{j=1}^{n} \lambda(c_j d_j b_i) \vee \alpha \right\} \right\} \right) \wedge \beta \right\} \vee \alpha$$
$$= \left\{ \left(\sup_{x=\Sigma_{i=1}^{m} a_i b_i} \left\{ \bigwedge_{i=1}^{m} \left\{ \sup_{a_i=\Sigma_{j=1}^{n} c_j d_j} \bigwedge_{j=1}^{n} \lambda(c_j d_j b_i) \right\} \right\} \right) \wedge \beta \right\} \vee \alpha$$
$$\leq \left(\lambda \left(\sum_{i=1}^{m} \sum_{j=1}^{n} c_j d_j b_i \right) \wedge \beta \right) \vee \alpha$$
$$\leq (\lambda(x) \wedge \beta) \vee \alpha.$$

Thus $\lambda \circ_{\alpha}^{\beta} \mathscr{R} \circ_{\alpha}^{\beta} \lambda \leq (\lambda \wedge \beta) \vee \alpha$.

Conversely, assume that λ satisfies conditions (6) and (10). Then by Theorem 8.4, λ satisfies (1). Let $x, y, z \in R$. Then

$$\lambda(xyz) \vee \alpha \geq (\lambda(xyz) \wedge \beta) \vee \alpha \geq (\lambda \circ_{\alpha}^{\beta} \mathscr{R} \circ_{\alpha}^{\beta} \lambda)(xyz)$$
$$= \left\{ \left(\sup_{xyz=\Sigma_{i=1}^{m} a_i b_i} \left\{ \bigwedge_{i=1}^{m} \left\{ \left(\lambda \circ_{\alpha}^{\beta} \mathscr{R} \right)(a_i) \wedge \lambda(b_i) \right\} \right\} \right) \wedge \beta \right\} \vee \alpha$$
$$\geq \left(\left\{ \left(\lambda \circ_{\alpha}^{\beta} \mathscr{R} \right)(xy) \wedge \lambda(z) \right\} \wedge \beta \right) \vee \alpha$$
$$= \left(\left\{ \left(\left\{ \left(\sup_{xy=\Sigma_{j=1}^{n} c_j d_j} \left\{ \bigwedge_{j=1}^{n} \{\lambda(c_j) \wedge \mathscr{R}(d_j)\} \right\} \right) \wedge \beta \right\} \vee \alpha \right) \wedge \lambda(z) \right\} \wedge \beta \right) \vee$$
$$\alpha$$
$$= \left(\left\{ \left(\left\{ \left(\sup_{xy=\Sigma_{j=1}^{n} c_j d_j} \left\{ \bigwedge_{j=1}^{n} \lambda(c_j) \right\} \right) \wedge \beta \right\} \vee \alpha \right) \wedge \lambda(z) \right\} \wedge \beta \right) \vee \alpha$$
$$\geq (\{(\{\lambda(x) \wedge \beta\} \vee \alpha) \wedge \lambda(z)\} \wedge \beta) \vee \alpha$$
$$= (\{\lambda(x) \wedge \lambda(z)\} \wedge \beta) \vee \alpha$$
$$\geq \lambda(x) \wedge \lambda(z) \wedge \beta.$$

Thus λ satisfies condition (4).

Theorem 8.8. *A fuzzy subset λ of a semiring R is a fuzzy bi-ideal with thresholds (α, β) of R if and only if λ satisfies conditions* (6), (8) *and* (9).

Proof. The proof follows from Theorem 8.6 and 8.7.

Theorem 8.9. *A fuzzy subset λ of a semiring R is a fuzzy quasi-ideal with thresholds (α, β) of R if and only if λ satisfies conditions* (5) *and* (6).

Proof. The proof follows from Theorem 8.4.

From Theorem 8.5, Theorem8.8 and Theorem 8.9 we deduce the following results.

Theorem 8.10. *Every fuzzy left (right) ideal with thresholds (α,β) of a semiring R is a fuzzy quasi-ideal with thresholds (α,β) of R.*

Theorem 8.11. *Every fuzzy quasi-ideal with thresholds (α,β) of R is a fuzzy bi-ideal with thresholds (α,β) of R.*

Lemma 8.4. *If λ and μ are fuzzy right and left ideals with thresholds (α,β) of R, respectively, then $\lambda \circ_\alpha^\beta \mu \le \lambda \wedge_\alpha^\beta \mu$.*

Proof. Let λ and μ be fuzzy right and left ideals with thresholds (α,β) of R, respectively. Then

$$\begin{aligned}
&\left(\lambda \circ_\alpha^\beta \mu\right)(x)\\
&=\left\{\left(\sup_{x=\Sigma_{i=1}^m a_ib_i}\left\{\bigwedge_{i=1}^m\{\lambda(a_i)\wedge\mu(b_i)\}\right\}\right)\wedge\beta\right\}\vee\alpha\\
&=\left\{\left(\sup_{x=\Sigma_{i=1}^m a_ib_i}\left\{\bigwedge_{i=1}^m\{(\lambda(a_i)\wedge\beta)\wedge(\mu(b_i)\wedge\beta)\}\right\}\right)\wedge\beta\right\}\vee\alpha\\
&\le\left\{\left(\sup_{x=\Sigma_{i=1}^m a_ib_i}\left\{\bigwedge_{i=1}^m\{(\lambda(a_ib_i)\vee\alpha)\wedge(\mu(a_ib_i)\vee\alpha)\}\right\}\right)\wedge\beta\right\}\vee\alpha\\
&=\left\{\left(\sup_{x=\Sigma_{i=1}^m a_ib_i}\left\{\bigwedge_{i=1}^m\{\lambda(a_ib_i)\wedge\mu(a_ib_i)\}\right\}\right)\wedge\beta\right\}\vee\alpha\\
&=\left\{\left(\sup_{x=\Sigma_{i=1}^m a_ib_i}\left\{\left(\bigwedge_{i=1}^m\lambda(a_ib_i)\right)\wedge\left(\bigwedge_{i=1}^m\mu(a_ib_i)\right)\right\}\right)\wedge\beta\right\}\vee\alpha\\
&=\left\{\left(\sup_{x=\Sigma_{i=1}^m a_ib_i}\left\{\left(\bigwedge_{i=1}^m\lambda(a_ib_i)\wedge\beta\right)\wedge\left(\bigwedge_{i=1}^m\mu(a_ib_i)\wedge\beta\right)\right\}\right)\wedge\beta\right\}\vee\alpha\\
&\le\left\{\left(\sup_{x=\Sigma_{i=1}^m a_ib_i}\{(\lambda(\Sigma_{i=1}^m a_ib_i)\vee\alpha)\wedge(\mu(\Sigma_{i=1}^m a_ib_i)\vee\alpha)\}\right)\wedge\beta\right\}\vee\alpha\\
&\le\left\{\left(\sup_{x=\Sigma_{i=1}^m a_ib_i}\{\lambda(\Sigma_{i=1}^m a_ib_i)\wedge\mu(\Sigma_{i=1}^m a_ib_i)\}\right)\wedge\beta\right\}\vee\alpha\\
&\le\{\lambda(x)\wedge\mu(x)\wedge\beta\}\vee\alpha.
\end{aligned}$$

Thus $\lambda \circ_\alpha^\beta \mu \le \lambda \wedge_\alpha^\beta \mu$.

8.2 Regular Semirings

In this section we characterize regular semirings by the properties of their fuzzy left (right) ideals with thresholds (α,β), fuzzy bi-ideals with thresholds (α,β) and fuzzy quasi-ideals with thresholds (α,β).

Theorem 8.12. *For a semiring R the following conditions are equivalent.*

(i) R is regular.

(ii) $\lambda \circ_{\alpha}^{\beta} \mu = \lambda \wedge_{\alpha}^{\beta} \mu$ *for every fuzzy right ideal* λ *and every fuzzy left ideal* μ *with thresholds* (α, β) *of R.*

Proof. $(i) \Rightarrow (ii)$ Let λ be a fuzzy right ideal and μ be a fuzzy left ideal with thresholds (α, β) of R. Then by Lemma 8.4, $\lambda \circ_{\alpha}^{\beta} \mu \leq \lambda \wedge_{\alpha}^{\beta} \mu$. Let $a \in R$. Then there exists $x \in R$ such that $a = axa$. Thus

$$\begin{aligned}
\left(\lambda \circ_{\alpha}^{\beta} \mu\right)(a) &= \left\{\left(\sup_{a=\Sigma_{i=1}^{m} a_i b_i} \left\{\bigwedge_{i=1}^{m} \{\lambda(a_i) \wedge \mu(b_i)\}\right\}\right) \wedge \beta\right\} \vee \alpha \\
&\geq \{\{\lambda(a) \wedge \mu(xa)\} \wedge \beta\} \vee \alpha \\
&= \{\{\lambda(a) \wedge (\mu(xa) \vee \alpha)\} \wedge \beta\} \vee \alpha \\
&\geq \{\{\lambda(a) \wedge (\mu(a) \wedge \beta)\} \wedge \beta\} \vee \alpha = \{\{\lambda(a) \wedge \mu(a)\} \wedge \beta\} \vee \alpha \\
&= \left(\lambda \wedge_{\alpha}^{\beta} \mu\right)(a).
\end{aligned}$$

Thus $\left(\lambda \circ_{\alpha}^{\beta} \mu\right) \geq \left(\lambda \wedge_{\alpha}^{\beta} \mu\right)$. Hence $\left(\lambda \circ_{\alpha}^{\beta} \mu\right) = \left(\lambda \wedge_{\alpha}^{\beta} \mu\right)$.

$(ii) \Rightarrow (i)$ Let A and B be right and left ideals of R, respectively. Then χ_A and χ_B are fuzzy right and fuzzy left ideals with thresholds (α, β) of R. Hence by hypothesis $\chi_A \circ_{\alpha}^{\beta} \chi_B = \chi_A \wedge_{\alpha}^{\beta} \chi_B$. Thus $((\chi_A \circ \chi_B) \wedge \beta) \vee \alpha = ((\chi_A \wedge \chi_B) \wedge \beta) \vee \alpha$. This implies $(\chi_{AB} \wedge \beta) \vee \alpha = (\chi_{A \cap B} \wedge \beta) \vee \alpha$. Hence $AB = A \cap B$. This shows that R is regular.

Theorem 8.13. *For a semiring R the following conditions are equivalent.*

(i) R is regular.

(ii) $\{\lambda \wedge \beta\} \vee \alpha \leq (\lambda \circ_{\alpha}^{\beta} \mathscr{R} \circ_{\alpha}^{\beta} \lambda)$ *for every fuzzy bi-ideal* λ *with thresholds* (α, β) *of R.*

(iii) $\{\lambda \wedge \beta\} \vee \alpha \leq (\lambda \circ_{\alpha}^{\beta} \mathscr{R} \circ_{\alpha}^{\beta} \lambda)$ *for every fuzzy quasi-ideal* λ *with thresholds* (α, β) *of R.*

Proof. $(i) \Rightarrow (ii)$ Let λ be a fuzzy bi-ideal with thresholds (α, β) of R and $a \in R$. Then there exists $x \in R$ such that $a = axa$. Thus

$$\begin{aligned}
&(\lambda \circ_{\alpha}^{\beta} \mathscr{R} \circ_{\alpha}^{\beta} \lambda)(a) \\
&= \left\{\left(\sup_{a=\Sigma_{i=1}^{m} a_i b_i} \left\{\bigwedge_{i=1}^{m} \left\{\left(\lambda \circ_{\alpha}^{\beta} \mathscr{R}\right)(a_i) \wedge \lambda(b_i)\right\}\right\}\right) \wedge \beta\right\} \vee \alpha \\
&\geq \left\{\left\{\left(\lambda \circ_{\alpha}^{\beta} \mathscr{R}\right)(ax) \wedge \lambda(a)\right\} \wedge \beta\right\} \vee \alpha \\
&= \left\{\left\{\left(\left\{\left(\sup_{ax=\Sigma_{j=1}^{n} c_j d_j} \left\{\bigwedge_{j=1}^{n} \{\lambda(c_j) \wedge \mathscr{R}(d_j)\}\right\}\right) \wedge \beta\right\} \vee \alpha\right) \wedge \lambda(a)\right\} \wedge \beta\right\} \vee \\
&\alpha \\
&\geq \{\{(\{\{\lambda(axa) \wedge \mathscr{R}(x)\} \wedge \beta\} \vee \alpha) \wedge \lambda(a)\} \wedge \beta\} \vee \alpha \qquad \text{(because } ax = axax\text{)} \\
&\geq \{\lambda(a) \wedge \beta\} \vee \alpha.
\end{aligned}$$

$(ii) \Rightarrow (iii)$ This is straightforward.

$(iii) \Rightarrow (i)$ Let Q be a quasi-ideal of R. Then χ_Q is a fuzzy quasi-ideal with thresholds (α,β) of R. Thus by hypothesis $(\chi_Q \wedge \beta) \vee \alpha \leq (\chi_Q \circ_\alpha^\beta \mathscr{R} \circ_\alpha^\beta \chi_Q) = (\chi_{QRQ} \wedge \beta) \vee \alpha$. Hence $Q \subseteq QRQ$, but $QRQ \subseteq Q$ always holds. Therefore $Q = QRQ$. Thus R is regular.

Theorem 8.14. *For a semiring R, the following conditions are equivalent.*

(i) R is regular.

(ii) $\left(\lambda \wedge_\alpha^\beta \mu\right) \leq \left(\lambda \circ_\alpha^\beta \mu \circ_\alpha^\beta \lambda\right)$ for every fuzzy bi-ideal λ and every fuzzy ideal μ with thresholds (α,β) of R.

(iii) $\left(\lambda \wedge_\alpha^\beta \mu\right) \leq \left(\lambda \circ_\alpha^\beta \mu \circ_\alpha^\beta \lambda\right)$ for every fuzzy quasi-ideal λ and every fuzzy ideal μ with thresholds (α,β) of R.

Proof. $(i) \Rightarrow (ii)$ Let λ be a fuzzy bi-ideal and μ be a fuzzy ideal with thresholds (α,β) of R. Let $x \in R$. Then there exists $a \in R$ such that $x = xax$. Thus

$$
\begin{aligned}
&(\lambda \circ_\alpha^\beta \mu \circ_\alpha^\beta \lambda)(x)\\
&= \left\{\left(\sup_{x=\Sigma_{i=1}^m a_i b_i} \left\{\bigwedge_{i=1}^m \left\{\left(\lambda \circ_\alpha^\beta \mu\right)(a_i) \wedge \lambda(b_i)\right\}\right\}\right) \wedge \beta\right\} \vee \alpha\\
&\geq \left\{\left\{\left(\lambda \circ_\alpha^\beta \mu\right)(xa) \wedge \lambda(x)\right\} \wedge \beta\right\} \vee \alpha\\
&= \left\{\left\{\left(\left\{\left(\sup_{xa=\Sigma_{j=1}^n c_j d_j} \left\{\bigwedge_{j=1}^n \{\lambda(c_j) \wedge \mu(d_j)\}\right\}\right) \wedge \beta\right\} \vee \alpha\right) \wedge \lambda(a)\right\} \wedge \beta\right\} \vee \alpha\\
&\geq \{\{\{\{\lambda(x) \wedge \mu(axa)\} \wedge \beta\} \vee \alpha \wedge \lambda(a)\} \wedge \beta\} \vee \alpha \quad \text{(because } xa = xaxa)\\
&\geq \{\{\lambda(x) \wedge \mu(x)\} \wedge \beta\} \vee \alpha = \left(\lambda \wedge_\alpha^\beta \mu\right)(x).
\end{aligned}
$$

$(ii) \Rightarrow (iii)$ Straightforward.

$(iii) \Rightarrow (i)$ Let λ be a fuzzy quasi-ideal with thresholds (α,β) of R. Since $\mathscr{R}$ is a fuzzy ideal with thresholds (α,β) of R, so by hypothesis $\left(\lambda \wedge_\alpha^\beta \mathscr{R}\right) \leq \left(\lambda \circ_\alpha^\beta \mathscr{R} \circ_\alpha^\beta \lambda\right)$. Thus by Theorem 8.13, R is regular.

Theorem 8.15. *For a semiring R, the following conditions are equivalent.*

(i) R is regular.

(ii) $\left(\lambda \wedge_\alpha^\beta \mu\right) \leq \left(\lambda \circ_\alpha^\beta \mu\right)$ for every fuzzy bi-ideal λ and every fuzzy left ideal μ with thresholds (α,β) of R.

(iii) $\left(\lambda \wedge_\alpha^\beta \mu\right) \leq \left(\lambda \circ_\alpha^\beta \mu\right)$ for every fuzzy quasi-ideal λ and every fuzzy left ideal μ with thresholds (α,β) of R.

(iv) $\left(\lambda \wedge_\alpha^\beta \mu\right) \leq \left(\lambda \circ_\alpha^\beta \mu\right)$ for every fuzzy right ideal λ and every fuzzy bi-ideal μ with thresholds (α,β) of R.

(v) $\left(\lambda \wedge_\alpha^\beta \mu\right) \leq \left(\lambda \circ_\alpha^\beta \mu\right)$ for every fuzzy right ideal λ and every fuzzy quasi-ideal μ with thresholds (α,β) of R.

(vi) $\left(\lambda \wedge_\alpha^\beta \mu \wedge_\alpha^\beta \nu\right) \leq \left(\lambda \circ_\alpha^\beta \mu \circ_\alpha^\beta \nu\right)$ for every fuzzy right ideal λ, every fuzzy bi-ideal μ and every fuzzy left ideal ν with thresholds (α,β) of R.

(vii) $\left(\lambda \wedge_{\alpha}^{\beta} \mu \wedge_{\alpha}^{\beta} \nu\right) \leq \left(\lambda \circ_{\alpha}^{\beta} \mu \circ_{\alpha}^{\beta} \nu\right)$ *for every fuzzy right ideal* λ*, every fuzzy quasi-ideal* μ *and every fuzzy right ideal* ν *with thresholds* (α,β) *of* R.

Proof. $(i) \Rightarrow (ii)$ Let λ be a fuzzy bi-ideal and μ a fuzzy left ideal with thresholds (α,β) of R. Now for any $a \in R$ there exists $x \in R$ such that $a = axa$. Thus

$$\begin{aligned}
\left(\lambda \circ_{\alpha}^{\beta} \mu\right)(a) &= \left\{\left(\sup_{a=\Sigma_{i=1}^{m} a_i b_i} \left\{\bigwedge_{i=1}^{m} \{\lambda(a_i) \wedge \mu(b_i)\}\right\}\right) \wedge \beta\right\} \vee \alpha \\
&\geq \{\{\lambda(a) \wedge \mu(xa)\} \wedge \beta\} \vee \alpha \\
&= \{\{\lambda(a) \wedge (\mu(xa) \vee \alpha)\} \wedge \beta\} \vee \alpha \\
&\geq \{\{\lambda(a) \wedge (\mu(a) \wedge \beta)\} \wedge \beta\} \vee \alpha = \{\{\lambda(a) \wedge \mu(a)\} \wedge \beta\} \vee \alpha \\
&= \left(\lambda \wedge_{\alpha}^{\beta} \mu\right)(a).
\end{aligned}$$

So $\left(\lambda \wedge_{\alpha}^{\beta} \mu\right) \leq \left(\lambda \circ_{\alpha}^{\beta} \mu\right)$.

$(ii) \Rightarrow (iii)$ Straightforward.

$(iii) \Rightarrow (i)$ Let λ be a fuzzy right ideal and μ be a fuzzy left ideal with thresholds (α,β) of R. Since every fuzzy right ideal with thresholds (α,β) is a fuzzy quasi-ideal with thresholds (α,β), we have by (iii) that $\left(\lambda \wedge_{\alpha}^{\beta} \mu\right) \leq \left(\lambda \circ_{\alpha}^{\beta} \mu\right)$. But by Lemma 8.4, $\left(\lambda \wedge_{\alpha}^{\beta} \mu\right) \geq \left(\lambda \circ_{\alpha}^{\beta} \mu\right)$. Hence $\left(\lambda \wedge_{\alpha}^{\beta} \mu\right) = \left(\lambda \circ_{\alpha}^{\beta} \mu\right)$ for every fuzzy right ideal λ and every fuzzy left ideal μ with thresholds (α,β) of R. Thus by Theorem 8.12, R is regular.

Similarly we can show that $(i) \Leftrightarrow (iv) \Leftrightarrow (v)$.

$(i) \Rightarrow (vi)$ Let λ be a fuzzy right ideal, μ be a fuzzy bi-ideal and ν be a fuzzy left ideal with thresholds (α,β) of R. Now for any $a \in R$ there exists $x \in R$ such that $a = axa$. Hence

$$\begin{aligned}
&(\lambda \circ_{\alpha}^{\beta} \mu \circ_{\alpha}^{\beta} \nu)(a) \\
&= \left\{\left(\sup_{x=\Sigma_{i=1}^{m} a_i b_i} \left\{\bigwedge_{i=1}^{m} \left\{\left(\lambda \circ_{\alpha}^{\beta} \mu\right)(a_i) \wedge \nu(b_i)\right\}\right\}\right) \wedge \beta\right\} \vee \alpha \\
&\geq \left\{\left\{\left(\lambda \circ_{\alpha}^{\beta} \mu\right)(a) \wedge \nu(xa)\right\} \wedge \beta\right\} \vee \alpha \\
&= \left\{\left\{\left(\left\{\left(\sup_{a=\Sigma_{j=1}^{n} c_j d_j} \left\{\bigwedge_{j=1}^{n} \{\lambda(c_j) \wedge \mu(d_j)\}\right\}\right) \wedge \beta\right\} \vee \alpha\right) \wedge \nu(a)\right\} \wedge \beta\right\} \vee \\
&\alpha \\
&\geq \{\{(\{\{\lambda(ax) \wedge \mu(a)\} \wedge \beta\} \vee \alpha) \wedge \nu(a)\} \wedge \beta\} \vee \alpha \\
&\geq \{\{(\{\{\lambda(a) \wedge \mu(a)\} \wedge \beta\} \vee \alpha) \wedge \nu(a)\} \wedge \beta\} \vee \alpha \\
&= \left(\lambda \wedge_{\alpha}^{\beta} \mu \wedge_{\alpha}^{\beta} \nu\right)(a).
\end{aligned}$$

$(vi) \Rightarrow (vii)$ Obvious.

$(vii) \Rightarrow (i)$ Let λ be a fuzzy right ideal and ν be a fuzzy left ideal with thresholds (α,β) of R. Then

$$\left(\lambda \wedge_{\alpha}^{\beta} \nu\right) = \left(\lambda \wedge_{\alpha}^{\beta} \mathscr{R} \wedge_{\alpha}^{\beta} \nu\right) \leq \left(\lambda \circ_{\alpha}^{\beta} \mathscr{R} \circ_{\alpha}^{\beta} \nu\right) \leq \left(\lambda \circ_{\alpha}^{\beta} \nu\right).$$

But by Lemma 8.4, $\left(\lambda \circ_{\alpha}^{\beta} \nu\right) \leq \left(\lambda \wedge_{\alpha}^{\beta} \nu\right)$. Hence $\left(\lambda \circ_{\alpha}^{\beta} \nu\right) = \left(\lambda \wedge_{\alpha}^{\beta} \nu\right)$ for every fuzzy right ideal λ and for every fuzzy left ideal ν with thresholds (α, β) of R. Thus by Theorem 8.12, R is regular.

8.3 Intra-regular Semirings

In this section we characterize intra-regular and regular and intra-regular semirings by the properties of their fuzzy left (right) ideals, fuzzy bi-ideals and fuzzy quasi-ideals with thresholds (α, β).

Lemma 8.5. *A semiring R is intra-regular if and only if $\left(\lambda \wedge_{\alpha}^{\beta} \nu\right) \leq \left(\lambda \circ_{\alpha}^{\beta} \nu\right)$ for every fuzzy left ideal λ and for every fuzzy right ideal μ with thresholds (α, β) of R.*

Proof. Let λ and μ be fuzzy left ideal and fuzzy right ideal with thresholds (α, β) of R, respectively. Let $a \in R$. Then there exist $x_i, x_i' \in R$ such that $a = \sum_{i=1}^{m} x_i a^2 x_i'$. Thus we have

$$\begin{aligned}
&\left(\lambda \circ_{\alpha}^{\beta} \mu\right)(a) \\
&= \left\{ \left(\sup_{a=\sum_{i=1}^{m} a_i b_i} \left\{ \bigwedge_{i=1}^{m} \{\lambda(a_i) \wedge \mu(b_i)\} \right\} \right) \wedge \beta \right\} \vee \alpha \\
&\geq \left\{ \left\{ \bigwedge_{i=1}^{m} \{\lambda(x_i a) \wedge \mu(a x_i')\} \right\} \wedge \beta \right\} \vee \alpha \\
&= \left\{ \left\{ \bigwedge_{i=1}^{m} \{(\lambda(x_i a) \vee \alpha) \wedge (\mu(a x_i') \vee \alpha)\} \right\} \wedge \beta \right\} \vee \alpha \\
&\geq \left\{ \left\{ \bigwedge_{i=1}^{m} \{(\lambda(a) \wedge \beta) \wedge (\mu(a) \wedge \beta)\} \right\} \wedge \beta \right\} \vee \alpha \\
&= \{\{\lambda(a) \wedge \mu(a)\} \wedge \beta\} \vee \alpha = \left(\lambda \wedge_{\alpha}^{\beta} \nu\right)(a).
\end{aligned}$$

Conversely, assume that A and B are left and right ideals of R, respectively. Then χ_A and χ_B are fuzzy left and fuzzy right ideals with thresholds (α, β) of R, respectively. Thus by hypothesis $\chi_A \wedge_{\alpha}^{\beta} \chi_B \leq \chi_A \circ_{\alpha}^{\beta} \chi_B \Rightarrow (\chi_{A \cap B} \wedge \beta) \vee \alpha \leq (\chi_{AB} \wedge \beta) \vee \alpha \Rightarrow A \cap B \subseteq AB$. Thus by Theorem 5.7, R is intra-regular.

Theorem 8.16. *The following conditions are equivalent for a semiring R:*
(i) R is both regular and intra-regular.
(ii) $(\lambda \wedge \beta) \vee \alpha = \lambda \circ_{\alpha}^{\beta} \lambda$ for every fuzzy bi-ideal λ with thresholds (α, β) of R.
(iii) $(\lambda \wedge \beta) \vee \alpha = \lambda \circ_{\alpha}^{\beta} \lambda$ for every fuzzy quasi-ideal λ with thresholds (α, β) of R.

Proof. $(i) \Rightarrow (ii)$ Let λ be a fuzzy bi-ideal with thresholds (α,β) of R and $x \in R$. Since R is both regular and intra-regular, there exist elements $a, p_i, p_i' \in R$ such that $x = xax$ and $x = \sum_{i=1}^m p_i x x p_i'$.

Thus $x = xax = xaxax = xa\left(\sum_{i=1}^m p_i x x p_i'\right) ax = \sum_{i=1}^m (xap_i x)(xp_i' ax)$.

Now

$$
\begin{aligned}
&\left(\lambda \circ_\alpha^\beta \lambda\right)(x)\\
&= \left\{\left(\sup_{x=\sum_{i=1}^m a_i b_i} \left\{\bigwedge_{i=1}^m \{\lambda(a_i) \wedge \lambda(b_i)\}\right\}\right) \wedge \beta\right\} \vee \alpha\\
&\geq \left\{\left\{\bigwedge_{i=1}^m \{\lambda(xap_i x) \wedge \lambda(xp_i' ax)\}\right\} \wedge \beta\right\} \vee \alpha\\
&= \left\{\left\{\bigwedge_{i=1}^m \{(\lambda(xap_i x) \vee \alpha) \wedge (\lambda(xp_i' ax) \vee \alpha)\}\right\} \wedge \beta\right\} \vee \alpha\\
&\geq \left\{\left\{\bigwedge_{i=1}^m \{(\lambda(x) \wedge \lambda(x) \wedge \beta) \wedge (\lambda(x) \wedge \lambda(x) \wedge \beta)\}\right\} \wedge \beta\right\} \vee \alpha\\
&= \{\lambda(x) \wedge \beta\} \vee \alpha.
\end{aligned}
$$

On the other hand $\lambda \circ_\alpha^\beta \lambda \leq (\lambda \wedge \beta) \vee \alpha$.

Hence $\lambda \circ_\alpha^\beta \lambda = (\lambda \wedge \beta) \vee \alpha$.

$(ii) \Rightarrow (iii)$ Obvious.

$(iii) \Rightarrow (i)$ Let Q be a quasi-ideal of R. Then χ_Q is a fuzzy quasi-ideal with thresholds (α,β) of R. Thus by hypothesis,

$(\chi_Q \wedge \beta) \vee \alpha = \chi_Q \circ_\alpha^\beta \chi_Q = ((\chi_Q \circ \chi_Q) \wedge \beta) \vee \alpha = (\chi_{Q^2} \wedge \beta) \vee \alpha$.

$\Rightarrow Q = Q^2$. Hence by Theorem 5.9, R is both regular and intra-regular.

Theorem 8.17. *The following conditions are equivalent for a semiring R:*

(i) R is both regular and intra-regular.

(ii) $\lambda \wedge_\alpha^\beta \mu \leq \lambda \circ_\alpha^\beta \mu$ for all fuzzy bi-ideals λ and μ with thresholds (α,β) of R.

(iii) $\lambda \wedge_\alpha^\beta \mu \leq \lambda \circ_\alpha^\beta \mu$ for every fuzzy bi-ideal λ and every fuzzy quasi-ideal μ with thresholds (α,β) of R.

(iv) $\lambda \wedge_\alpha^\beta \mu \leq \lambda \circ_\alpha^\beta \mu$ for every fuzzy quasi-ideal λ and every fuzzy bi-ideal μ with thresholds (α,β) of R.

(v) $\lambda \wedge_\alpha^\beta \mu \leq \lambda \circ_\alpha^\beta \mu$ for all fuzzy quasi-ideals λ and μ with thresholds (α,β) of R.

Proof. $(i) \Rightarrow (ii)$ Let λ and μ be fuzzy bi-ideals with thresholds (α,β) of R and $x \in R$. Since R is both regular and intra-regular, there exist elements $a, p_i, p_i' \in R$ such that

$x = xax$ and $x = \sum_{i=1}^m p_i x x p_i'$.

Thus $x = xax = xaxax = xa\left(\sum_{i=1}^m p_i x x p_i'\right) ax = \sum_{i=1}^m (xap_i x)(xp_i' ax)$.

Now

$$\begin{aligned}
&\left(\lambda \circ_{\alpha}^{\beta} \mu\right)(x) \\
&= \left\{ \left(\sup_{x=\Sigma_{i=1}^{m} a_i b_i} \left\{ \bigwedge_{i=1}^{m} \{\lambda(a_i) \wedge \mu(b_i)\} \right\} \right) \wedge \beta \right\} \vee \alpha \\
&\geq \left\{ \left\{ \bigwedge_{i=1}^{m} \{\lambda(xap_i x) \wedge \mu(xp_i' ax)\} \right\} \wedge \beta \right\} \vee \alpha \\
&= \left\{ \left\{ \bigwedge_{i=1}^{m} \{(\lambda(xap_i x) \vee \alpha) \wedge (\mu(xp_i' ax) \vee \alpha)\} \right\} \wedge \beta \right\} \vee \alpha \\
&\geq \left\{ \left\{ \bigwedge_{i=1}^{m} \{(\lambda(x) \wedge \lambda(x) \wedge \beta) \wedge (\mu(x) \wedge \mu(x) \wedge \beta)\} \right\} \wedge \beta \right\} \vee \alpha \\
&= \{\lambda(x) \wedge \mu(x) \wedge \beta\} \vee \alpha = \left(\lambda \wedge_{\alpha}^{\beta} \mu\right)(x).
\end{aligned}$$

$(ii) \Rightarrow (iii) \Rightarrow (v)$ and $(ii) \Rightarrow (iv) \Rightarrow (v)$ are clear.

$(v) \Rightarrow (i)$ Let λ be a fuzzy left ideal and μ be a fuzzy right ideal with thresholds (α,β) of R. Then λ and μ are fuzzy bi-ideals with thresholds (α,β) of R. So by hypothesis $\lambda \wedge_{\alpha}^{\beta} \mu \leq \lambda \circ_{\alpha}^{\beta} \mu$. But by Lemma 8.4, $\lambda \wedge_{\alpha}^{\beta} \mu \geq \lambda \circ_{\alpha}^{\beta} \mu$ Thus $\lambda \wedge_{\alpha}^{\beta} \mu = \lambda \circ_{\alpha}^{\beta} \mu$. Hence by Theorem 8.12, R is regular. On the other hand by hypothesis we also have $\lambda \wedge_{\alpha}^{\beta} \mu \leq \lambda \circ_{\alpha}^{\beta} \mu$. By Lemma 8.5, R is intra-regular.

8.4 Fuzzy k-Ideals with Thresholds (α,β)

Definition 8.6. A fuzzy subset λ of a semiring R is said to be a *fuzzy left (resp. right) k-ideal with thresholds* (α,β) of R if it satisfies $(1),(3)$ and (10), where

(10) for all $x,a,b \in R$ such that $x+a=b$
$\max\{\lambda(x),\alpha\} \geq \min\{\lambda(a),\lambda(b),\beta\}$.

Definition 8.7. A fuzzy subset λ of a semiring R is said to be a *fuzzy k -bi-ideal with thresholds*(α,β) of R if it satisfies $(1),(2),(4)$ and (10).

Definition 8.8. A fuzzy subset λ of a semiring R is said to be a *fuzzy k-quasi-ideal with thresholds* (α,β) of R if it satisfies $(1),(10)$ and (11), where

(11) $\max\{\lambda(x),\alpha\} \geq \min\{\lambda \odot \mathscr{R}(x), \mathscr{R} \odot \lambda(z), \beta\}$.

The proof of the following Lemma is straightforward and hence omitted.

Lemma 8.6. *A nonempty subset A of R is a left k-ideal (right k-ideal, k-ideal, k-bi-ideal, k-quasi-ideal) of R if and only if* χ_A *is a fuzzy left k-ideal (right k-ideal, k-ideal, k-bi-ideal, k-quasi-ideal) with thresholds* (α,β) *of R.*

Theorem 8.18. *(i) If* λ *is a fuzzy left (right) k-ideal with thresholds* (α,β) *of a semiring R, then* $(\lambda \wedge \beta)$ *is a fuzzy left (right) k-ideal with thresholds* (α,β) *of R.*

(ii) If λ is a fuzzy k-bi-ideal of a semiring R, then $\lambda \wedge \beta$ is a fuzzy k-bi-ideal with thresholds (α, β) of R.

Proof. The proof is similar to the proof of Theorem 8.3.

Lemma 8.7. *Let A, B be subsets of R. Then*

$$C_A \oplus_\alpha^\beta C_B = (C_{\overline{A+B}} \wedge \beta) \vee \alpha.$$

Proof. Let A, B be subsets of R and $x \in R$. If $x \in \overline{A+B}$, then there exist $a_1, a_2 \in A$ and $b_1, b_2 \in B$ such that $x + (a_1 + b_1) = (a_2 + b_2)$. Thus

$$(\chi_A \oplus_\alpha^\beta \chi_B)(x)$$
$$= \left(\sup_{x+(a_1'+b_1')=(a_2'+b_2')} \min\{\chi_A(a_1'), \chi_A(a_2'), \chi_B(b_1'), \chi_B(b_2')\} \wedge \beta \right) \vee \alpha$$
$$= (1 \wedge \beta) \vee \alpha = (\chi_{\overline{A+B}}(x) \wedge \beta) \vee \alpha.$$

If $x \notin \overline{A+B}$, then there do not exist $a_1, a_2 \in A$ and $b_1, b_2 \in B$ such that $x + (a_1 + b_1) = (a_2 + b_2)$. Thus

$$(\chi_A \oplus_\alpha^\beta \chi_B)(x) = (0 \wedge \beta) \vee \alpha = (\chi_{\overline{A+B}}(x) \wedge \beta) \vee \alpha.$$

Hence $\chi_A \oplus_\alpha^\beta \chi_B = (\chi_{\overline{A+B}} \wedge \beta) \vee \alpha$.

Lemma 8.8. *A fuzzy subset λ of a semiring R satisfies conditions (1) and (10) if and only if it satisfies condition*

(12) $\lambda \oplus_\alpha^\beta \lambda \le (\lambda \wedge \beta) \vee \alpha$.

Proof. Suppose λ satisfies conditions (1) and (10). Let $x \in R$. Then

$$(\lambda \oplus_\alpha^\beta \lambda)(x) = \left\{ \left(\sup_{x+(a_1+b_1)=(a_2+b_2)} \{\min\{\lambda(a_1), \lambda(a_2), \lambda(b_1), \lambda(b_2)\}\} \right) \wedge \beta \right\} \vee \alpha$$
$$= \left\{ \left(\sup_{x+(a_1+b_1)=(a_2+b_2)} \min\{\min\{\lambda(a_1), \lambda(b_1), \wedge\beta\}, \min\{\lambda(a_2), \lambda(b_2), \wedge\beta\}\} \right) \wedge \beta \right\} \vee \alpha$$
$$\le \left\{ \left(\sup_{x+(a_1+b_1)=(a_2+b_2)} \min\{\{\max(\lambda(a_1+b_1), \alpha), \max(\lambda(a_2+b_2), \alpha)\}\} \right) \wedge \beta \right\} \vee \alpha$$
$$\le \left\{ \left(\sup_{x+(a_1+b_1)=(a_2+b_2)} \{\min\{\lambda(a_1+b_1), \lambda(a_2+b_2)\} \vee \alpha\} \right) \wedge \beta \right\} \vee \alpha$$
$$\le \left\{ \left(\sup_{x+(a_1+b_1)=(a_2+b_2)} \{\min\{\lambda(a_1+b_1), \lambda(a_2+b_2), \wedge\beta\}\} \right) \wedge \beta \right\} \vee \alpha$$
$$\le \{(\lambda(x) \vee \alpha) \wedge \beta\} \vee \alpha = \{\lambda(x) \wedge \beta\} \vee \alpha.$$

Thus $\lambda \oplus_\alpha^\beta \lambda \le (\lambda \wedge \beta) \vee \alpha$.

Conversely, assume that $\lambda \oplus_\alpha^\beta \lambda \le (\lambda \wedge \beta) \vee \alpha$ and $x, y \in R$. Then

$\lambda(x+y)\vee\alpha \geq (\lambda(x+y)\wedge\beta)\vee\alpha \geq (\lambda\oplus_\alpha^\beta\lambda)(x+y)$
$=\left\{\left(\sup_{x+(a_1+b_1)=(a_2+b_2)}\{\min\{\lambda(a_1),\lambda(a_2),\lambda(b_1),\lambda(b_2)\}\}\right)\wedge\beta\right\}\vee\alpha$
$\geq (\min\{\lambda(0),\lambda(x),\lambda(0),\lambda(y)\}\wedge\beta)\vee\alpha$ because $(x+y)+(0+0)=(x+y)$
$\geq (\lambda(x)\wedge\lambda(y))\wedge\beta$.
Thus λ satisfies condition (1).
Let $a,b,x\in R$ such that $x+a=b$. Then
$\lambda(x)\vee\alpha \geq (\lambda(x)\wedge\beta)\vee\alpha \geq (\lambda\oplus_\alpha^\beta\lambda)(x)$
$=\left\{\left(\sup_{x+(a_1+b_1)=(a_2+b_2)}\{\min\{\lambda(a_1),\lambda(a_2),\lambda(b_1),\lambda(b_2)\}\}\right)\wedge\beta\right\}\vee\alpha$
$\geq (\min\{\lambda(a),\lambda(0),\lambda(b)\}\wedge\beta)\vee\alpha$ because $x+(a+0)=(b+0)$
$=\min\{\lambda(a),\lambda(b),\beta\}$.
This shows that λ satisfies (10).

Theorem 8.19. *A fuzzy subset λ of a semiring R is a fuzzy left (resp. right) k-ideal with thresholds (α,β) of R if and only if λ satisfies conditions (12) and (13), where*
(13) $\mathscr{R}\odot_\alpha^\beta\lambda \leq (\lambda\wedge\beta)\vee\alpha$ $\left(\textit{resp. } \lambda\odot_\alpha^\beta\mathscr{R}\leq(\lambda\wedge\beta)\vee\alpha\right)$.

Proof. Suppose λ be a fuzzy left k-ideal with thresholds (α,β) of R. Then by Lemma 8.8, λ satisfies condition (12). Now we show that λ satisfies condition (13). Let $x\in R$. If $(\mathscr{R}\odot_\alpha^\beta\lambda)(x)=(0\wedge\beta)\vee\alpha$, then $(\mathscr{R}\odot_\alpha^\beta\lambda)(x)\leq(\lambda(x)\wedge\beta)\vee\alpha$. Otherwise, there exist elements $a_i,b_i,c_j,d_j\in R$ such that $x+\sum_{i=1}^m a_ib_i=\sum_{j=1}^n c_jd_j$. Then we have
$(\mathscr{R}\odot_\alpha^\beta\lambda)(x)$

$$=\left(\sup_{x+\Sigma_{i=1}^m a_ib_i=\Sigma_{j=1}^n a'_jb'_j}\left\{\min\left\{\mathscr{R}(a_i),\lambda(b_i),\mathscr{R}\left(a'_j\right),\lambda\left(b'_j\right)\right\}\right\}\wedge\beta\right)\vee\alpha$$

$$=\left(\sup_{x+\Sigma_{i=1}^m a_ib_i=\Sigma_{j=1}^n a'_jb'_j}\left\{\min\left\{\lambda(b_i),\lambda\left(b'_j\right)\right\}\right\}\wedge\beta\right)\vee\alpha$$

$$=\left(\sup_{x+\Sigma_{i=1}^m a_ib_i=\Sigma_{j=1}^n a'_jb'_j}\left\{\min\left\{\lambda(b_i)\wedge\beta,\lambda\left(b'_j\right)\wedge\beta\right\}\right\}\wedge\beta\right)\vee\alpha$$

$$\leq\left(\sup_{x+\Sigma_{i=1}^m a_ib_i=\Sigma_{j=1}^n a'_jb'_j}\left\{\min\left\{\lambda(a_ib_i)\vee\alpha,\lambda\left(a'_jb'_j\right)\vee\alpha\right\}\right\}\wedge\beta\right)\vee\alpha$$

$$\leq\left(\sup_{x+\Sigma_{i=1}^m a_ib_i=\Sigma_{j=1}^n a'_jb'_j}\left(\min\left\{\lambda\left(\Sigma_{i=1}^m a_ib_i\right),\lambda\left(\Sigma_{j=1}^n a'_jb'_j\right)\right\}\right)\wedge\beta\right)\vee\alpha$$

$\leq(\lambda(x)\wedge\beta)\vee\alpha$.
$\Rightarrow\mathscr{R}\odot_\alpha^\beta\lambda\leq(\lambda\wedge\beta)\vee\alpha$.

Conversely, assume that λ satisfies conditions (12) and (13). Then by Lemma 8.8, λ satisfies conditions (1) and (10). We show that λ satisfies condition (3). Let $x,y\in R$. Then we have

$$\begin{aligned}\lambda(xy)\vee\alpha &\geq (\lambda(xy)\wedge\beta)\vee\alpha \geq \left(\mathscr{R}\odot_{\alpha}^{\beta}\lambda\right)(xy)\\ &= \left(\sup_{xy+\Sigma_{i=1}^{m}a_ib_i=\Sigma_{j=1}^{n}a'_jb'_j}\left\{\min\left\{\mathscr{R}(a_i),\lambda(b_i),\mathscr{R}\left(a'_j\right),\lambda\left(b'_j\right)\right\}\right\}\wedge\beta\right)\vee\alpha\\ &= \left(\sup_{xy+\Sigma_{i=1}^{m}a_ib_i=\Sigma_{j=1}^{n}a'_jb'_j}\left\{\min\left\{\lambda(b_i),\lambda\left(b'_j\right)\right\}\right\}\wedge\beta\right)\vee\alpha\\ &\geq (\lambda(y)\wedge\beta)\vee\alpha \qquad \text{because } xy+0y=xy.\end{aligned}$$

This shows that λ satisfies condition (3). So λ is a fuzzy left k-ideal with thresholds (α,β) of R.

Similarly we can prove the following results:

Lemma 8.9. *A fuzzy subset λ of a semiring R satisfies conditions $(1),(2)$ and (10) if and only if it satisfies (12) and (14), where*
(14) $\lambda\odot_{\alpha}^{\beta}\lambda\leq(\lambda\wedge\beta)\vee\alpha.$

Lemma 8.10. *A fuzzy subset λ of a semiring R satisfies conditions $(1),(4)$ and (10) if and only if it satisfies (12) and (15), where*
(15) $\lambda\odot_{\alpha}^{\beta}\mathscr{R}\odot_{\alpha}^{\beta}\lambda\leq(\lambda\wedge\beta)\vee\alpha.$

By using the above Lemmas, we can prove the following Theorems.

Theorem 8.20. *A fuzzy subset λ of a semiring R is a fuzzy k-bi-ideal with thresholds (α,β) of R if and only if λ satisfies conditions $(12),(14)$ and (15).*

Theorem 8.21. *A fuzzy subset λ of a semiring R is a fuzzy k-quasi-ideal with thresholds (α,β) of R if and only if λ satisfies conditions (11) and (12).*

The proofs of the following results are straight forward.

Theorem 8.22. *Every fuzzy left (right) k-ideal with thresholds (α,β) of a semiring R is a fuzzy k-quasi-ideal with thresholds (α,β) of R.*

Theorem 8.23. *Every fuzzy k-quasi-ideal with thresholds (α,β) of a semiring R is a fuzzy k-bi-ideal with thresholds (α,β) of R.*

Theorem 8.24. *If λ and μ are fuzzy right and left k-ideals with thresholds (α,β) of R, respectively, then $\lambda\odot_{\alpha}^{\beta}\mu\leq\lambda\wedge_{\alpha}^{\beta}\mu$.*

Proof. Let λ and μ be fuzzy right and left k-ideals with thresholds (α,β) of R, respectively. Then

$$\begin{aligned}&\left(\lambda\odot_{\alpha}^{\beta}\mu\right)(x)\\ &= \left(\sup_{x+\Sigma_{i=1}^{m}a_ib_i=\Sigma_{j=1}^{n}c_jd_j}\{\min\{\lambda(a_i),\mu(b_i),\lambda(c_j),\mu(d_j)\}\wedge\beta\right)\vee\alpha\end{aligned}$$

$$= \left(\sup_{x+\sum_{i=1}^{m} a_i b_i = \sum_{j=1}^{n} c_j d_j} \{\min\{\lambda(a_i) \wedge \beta, \mu(b_i) \wedge \beta, \lambda(c_j) \wedge \beta, \mu(d_j) \wedge \beta\} \wedge \beta \right) \vee \alpha$$

$$\leq \left(\sup_{x+\sum_{i=1}^{m} a_i b_i = \sum_{j=1}^{n} c_j d_j} \left\{ \begin{array}{l} \min\{\lambda(a_i b_i) \vee \alpha, \mu(a_i b_i) \vee \alpha, \\ \lambda(c_j d_j) \vee \alpha, \mu(c_j d_j) \vee \alpha \end{array} \right\} \wedge \beta \right) \vee \alpha$$

$$\leq \left(\sup_{x+\sum_{i=1}^{m} a_i b_i = \sum_{j=1}^{n} c_j d_j} \left\{ \begin{array}{l} \min\{\lambda(a_i b_i), \mu(a_i b_i), \\ \lambda(c_j d_j), \mu(c_j d_j) \end{array} \right\} \wedge \beta \right) \vee \alpha$$

$$\leq \left(\sup_{x+\sum_{i=1}^{m} a_i b_i = \sum_{j=1}^{n} c_j d_j} \left\{ \min \left\{ \begin{array}{l} (\lambda(\sum_{i=1}^{m} a_i b_i) \vee \alpha), (\mu(\sum_{i=1}^{m} a_i b_i) \vee \alpha), \\ (\lambda(\sum_{j=1}^{n} c_j d_j) \vee \alpha), (\mu(\sum_{j=1}^{n} c_j d_j) \vee \alpha) \end{array} \right\} \wedge \beta \right\} \right) \vee \alpha$$

$$\leq \{\lambda(x) \wedge \mu(x) \wedge \beta\} \vee \alpha = \left(\lambda \wedge_{\alpha}^{\beta} \mu\right)(x).$$

Thus $\lambda \odot_{\alpha}^{\beta} \mu \leq \lambda \wedge_{\alpha}^{\beta} \mu$.

8.5 *k*-Regular Semirings

In this section we characterize *k*-regular semirings by the properties of their fuzzy left (right) *k*-ideals, fuzzy *k*-bi-ideals and fuzzy *k*-quasi-ideals with thresholds (α, β).

Theorem 8.25. *For a semiring R the following conditions are equivalent.*

(i) R is k-regular.

(ii) $(\lambda \wedge_{\alpha}^{\beta} \mu) = (\lambda \odot_{\alpha}^{\beta} \mu)$ *for every fuzzy right k-ideal* λ *and every fuzzy left k-ideal* μ *with thresholds* (α, β) *of R.*

Proof. The proof is similar to the proof of Theorem 8.15.

Theorem 8.26. *For a semiring R, the following conditions are equivalent.*

(i) R is k-regular.

(ii) $(\lambda \wedge \beta) \vee \alpha \leq (\lambda \odot_{\alpha}^{\beta} \mathscr{R} \odot_{\alpha}^{\beta} \lambda)$ *for every fuzzy k-bi-ideal* λ *with thresholds* (α, β) *of R.*

(iii) $(\lambda \wedge \beta) \vee \alpha \leq (\lambda \odot_{\alpha}^{\beta} \mathscr{R} \odot_{\alpha}^{\beta} \lambda)$ *for every fuzzy k-quasi-ideal* λ *with thresholds* (α, β) *of R.*

Proof. The proof is similar to the proof of Theorem 8.13.

Theorem 8.27. *For a semiring R, the following conditions are equivalent.*

(i) R is k-regular.

(ii) $\left(\lambda \wedge_{\alpha}^{\beta} \mu\right) \leq \left(\lambda \odot_{\alpha}^{\beta} \mu \odot_{\alpha}^{\beta} \lambda\right)$ *for every fuzzy k-bi-ideal* λ *and every fuzzy k-ideal* μ *with thresholds* (α, β) *of R.*

(iii) $\left(\lambda \wedge_{\alpha}^{\beta} \mu\right) \leq \left(\lambda \odot_{\alpha}^{\beta} \mu \odot_{\alpha}^{\beta} \lambda\right)$ *for every fuzzy k-quasi-ideal* λ *and every fuzzy k-ideal* μ *with thresholds* (α, β) *of R.*

Proof. The proof is similar to the proof of Theorem 8.14.

Theorem 8.28. *For a semiring R, the following conditions are equivalent.*

(i) R is k-regular.

(ii) $\left(\lambda \wedge_{\alpha}^{\beta} \mu\right) \leq \left(\lambda \odot_{\alpha}^{\beta} \mu\right)$ *for every fuzzy k-bi-ideal* λ *and every fuzzy left k-ideal* μ *with thresholds* (α, β) *of R.*

(iii) $\left(\lambda \wedge_{\alpha}^{\beta} \mu\right) \leq \left(\lambda \odot_{\alpha}^{\beta} \mu\right)$ *for every fuzzy k-quasi-ideal* λ *and every fuzzy left k-ideal* μ *with thresholds* (α, β) *of R.*

(iv) $\left(\lambda \wedge_{\alpha}^{\beta} \mu\right) \leq \left(\lambda \odot_{\alpha}^{\beta} \mu\right)$ *for every fuzzy right k-ideal* λ *and every fuzzy k-bi-ideal* μ *with thresholds* (α, β) *of R.*

(v) $\left(\lambda \wedge_{\alpha}^{\beta} \mu\right) \leq \left(\lambda \odot_{\alpha}^{\beta} \mu\right)$ *for every fuzzy right k-ideal* λ *and every fuzzy k-quasi-ideal* μ *with thresholds* (α, β) *of R.*

(vi) $\left(\lambda \wedge_{\alpha}^{\beta} \mu \wedge_{\alpha}^{\beta} \nu\right) \leq \left(\lambda \odot_{\alpha}^{\beta} \mu \odot_{\alpha}^{\beta} \nu\right)$ *for every fuzzy right k-ideal* λ*, every fuzzy k-bi-ideal* μ *and every fuzzy left k-ideal* ν *with thresholds* (α, β) *of R.*

(vii) $\left(\lambda \wedge_{\alpha}^{\beta} \mu \wedge_{\alpha}^{\beta} \nu\right) \leq \left(\lambda \odot_{\alpha}^{\beta} \mu \odot_{\alpha}^{\beta} \nu\right)$ *for every fuzzy right k-ideal* λ*, every fuzzy k-quasi-ideal* μ *and every fuzzy left k-ideal* ν *with thresholds* (α, β) *of R.*

Proof. The proof is similar to the proof of Theorem 8.15.

8.6 *k*-Intra-regular Semirings

In this section we characterize *k*-intra-regular and *k*-regular and *k*-intra-regular semirings by the properties of their fuzzy left (right) *k*-ideals, fuzzy *k*-bi-ideals and fuzzy *k*-quasi-ideals with thresholds (α, β).

Lemma 8.11. *A semiring R is k-intra-regular if and only if* $\lambda \wedge_{\alpha}^{\beta} \mu \leq \lambda \odot_{\alpha}^{\beta} \mu$ *for every fuzzy left k-ideal* λ *and for every fuzzy right k-ideal* μ *with thresholds* (α, β) *of R.*

Proof. The proof is similar to the proof of Lemma 8.5.

Theorem 8.29. *The following conditions are equivalent for a semiring R*

(i) R is both k-regular and k-intra-regular.

(ii) $(\lambda \wedge \beta) \vee \alpha = \lambda \odot_{\alpha}^{\beta} \lambda$ *for every fuzzy k-bi-ideal* λ *with thresholds* (α, β) *of R.*

(iii) $(\lambda \wedge \beta) \vee \alpha = \lambda \odot_{\alpha}^{\beta} \lambda$ *for every fuzzy k-quasi-ideal* λ *with thresholds* (α, β) *of R.*

Proof. The proof is similar to the proof of Theorem 8.16.

Theorem 8.30. *The following conditions are equivalent for a semiring R*

(i) R is both k-regular and k-intra-regular.

(ii) $\lambda \wedge_{\alpha}^{\beta} \mu \leq \lambda \odot_{\alpha}^{\beta} \mu$ *for all fuzzy k-bi-ideals* λ *and* μ *with thresholds* (α, β) *of R.*

(*iii*) $\lambda \wedge_{\alpha}^{\beta} \mu \leq \lambda \odot_{\alpha}^{\beta} \mu$ *for every fuzzy k-bi-ideal* λ *and every fuzzy k-quasi-ideals* μ *with thresholds* (α, β) *of* R.

(*iv*) $\lambda \wedge_{\alpha}^{\beta} \mu \leq \lambda \odot_{\alpha}^{\beta} \mu$ *for every fuzzy k-quasi-ideal* λ *and every fuzzy k-bi-ideals* μ *with thresholds* (α, β) *of* R.

(*v*) $\lambda \wedge_{\alpha}^{\beta} \mu \leq \lambda \odot_{\alpha}^{\beta} \mu$ *for all fuzzy k-quasi-ideals* λ *and* μ *with thresholds* (α, β) *of* R.

Proof. The proof is similar to the proof of Theorem 8.17.

8.7 Ternary Semirings

In 1932, Lehmer introduced ternary algebraic systems [103]. Dutta and Kar [50] introduced ternary semirings in 2003. Many papers are available on ternary semirings [see [50, 51, 52, 85]]. Fuzzy ideals and fuzzy bi-ideals in ternary semirings are studied in [93, 110].

Part II
Invited Chapters

Chapter 9
On Fuzzy *LD*-Bigroupoids

Hee Sik Kim[1] and J. Neggers[2]

[1] Department of Mathematics, Hanyang University, Seoul, 131 – 791, Korea
heekim@hanyang.ac.kr
[2] Department of Mathematics, University of Alabama, Tuscaloosa, AL 35487 – 0350, U.S.A
jneggers@as.ua.edu

Abstract. In this paper, we define a generalization of what it means to be a semiring, i.e., the class of *LD*-bigroupoids with companion classes *RD*-bigroupoids and *D*-bigroupoids. After development of several basic ideas we consider the fuzzified versions of these algebraic systems and we investigate how theideas developed for *LD*-bigroupoids carry over into the realm of fuzzy *LD*-bigroupoids, yielding a generalization of the theory of fuzzy semirings. The results obtained demonstrate that it is quite possible to take these ideas much further as we expect with happen in the future.

Keywords and phrases: (Fuzzy) *LD*-bigroupoid, (star, dot)-prime, barrier value, dot-left-ideal, null element, (star, dot)-rising, μ-compatible, specialization, duality.

2000 Mathematics Subject Classification. 03E72, 16Y60.

9.1 Preliminaries

The notion of semiring was first introduced by H. S. Vandiver in 1934, and since then many other researchers also developed the theory of semirings as a generalization of rings. Semirings occur in different mathematical fields, e.g., as ideals of a ring, as positive cones of partially ordered rings and fields, in the context of topological considerations, and in the foundations of arithmetic, including questions raised by school education ([5]). In the 1980's the theory of semirings contributed to computer science, since the rapid development of computer science need additional theoretical mathematical background. The semiring structure does not contain an additive inverse, and this point is very helpful in developing the theoretical structure of computer science. For example, hemirings, as a semiring with zero and commutative addition, appeared in a natural manner in some applications to the theory of automata and formal languages. We refer to J. S. Golan's remarkable book for general reference ([4]).

J. Ahsan et al.: Fuzzy Semirings with Applications, STUDFUZZ 278, pp. 163–174.
springerlink.com

L. A. Zadeh ([9]) introduced the notion of fuzzy sets and fuzzy set operations. Since then, fuzzy set theory developed by Zadeh and others has evoked great interest among researchers working in different branches of mathematics. The concept of fuzzy set has been applied by many authors to generalize some of the basic notions of algebra. J. N. Mordeson and D. S. Malik ([7]) achieved a synthesis of fuzzy algebras. Fuzzy semirings were first investigated by J. Ahsan et al. ([1]), and have been studied by many researchers ([2, 3, 8]).

In this paper, we consider properties of the set $F(X,*,\cdot)$, where $(X,*,\cdot)$ is a left-distributive bigroupoid(an *LD*-bigroupoid), i.e., a set X equipped with two binary operations $(x,y) \to x*y$ and $(x,y) \to x \cdot y$ such that $x \cdot (y*z) = (x \cdot y)*(x \cdot z)$, and where $\mu \in F(X,*,\cdot)$ provided $\mu : (X,*,\cdot) \to [0,1]$ is a fuzzy subset of X for which $\mu(x*y) \geq \min\{\mu(x),\mu(y)\}$ and $\mu(x \cdot y) \geq \mu(y)$. This usage follows the standard usage in the theory of fuzzy semirings where these mappings will be referred to as fuzzy left-ideals as well. The difference here is that the class of *LD*-bigroupoids is of course vastly larger than the class of semirings and that therefore the classes of fuzzy *LD*-bigroupoids in other cases can exhibit very different behaviors. Interesting classes of mappings which are new also obtained in natural ways. We have concentrated our attention on *LD*-bigroupoids, realizing that there are corresponding theories for *RD*-bigroupoids which are right-distributive and *D*-bigroupoids which are distributive. As it stands, we believe we have accomplished a good bit in providing foundations for an area of study for which much more can be accomplished.

9.2 *LD*-Bigroupoids

An *LD-bigroupoid* is a nonempty set X with two binary operations "$*$" and "$\cdot$" satisfying the following axioms:

(I) $(X,*),(X,\cdot)$ are groupoids,
(LD) $x \cdot (y*z) = (x \cdot y)*(x \cdot z)$, for all $x,y,z \in X$.

In that case, $(X,*,\cdot)$ is an *LD-bigroupoid.* If $(X,\cdot,*)$ is an *LD*-bigroupoid, then

$$x*(y \cdot z) = (x*y) \cdot (x*z)$$

for all $x,y,z \in X$. On the other hand, if

$$(x*y) \cdot z = (x \cdot z)*(y \cdot z)$$

for all $x,y,z \in X$ (right distributive las holds.), then $(X,*,\cdot)$ is an *RD-bigroupoid.* If $(X,*,\cdot)$ is both an *LD*-bigroupoid and *RD*-bigroupoid, then it is a *D-bigroupoid.*

If $(X,*)$ and $(X,\cdot)$ are both semigroups, then if $(X,*,\cdot)$ is an *LD*-bigroupoid, then $(X,*,\cdot)$ is an *LD-bisemigroup*. Similarly, if $(X,*,\cdot)$ is an *RD*-bigroupoid, then it is an *RD-bisemigroup*. If it is both, then it is a *D-bisemigroup*.

If $(X,*)$ is a commutative semigroup and if $(X,\cdot)$ is a semigroup such that $(X,*,\cdot)$ is a *D*-bisemigroup, then $(X,*,\cdot)$ is a *semiring*.

Example 9.1. 1. Every ring is a semiring; every semiring is a D-bigroupoid; evey D-bisemirgroup is a D-bigroupoid.

2. Every distributive lattice, incline algebra is a D-bigroupoid.

3. Let $(X,*,f)$ be a leftoid, i.e., $x*y=f(x)$, where $f:X\to X$ is a map. Then $(X,*,*)$ is an LD-bigroupoid if $f(f(x))=f(x)$ for any $x\in X$.

Example 9.2. Let X be the set of all real numbers. If we define binary operations "$*$" and "$\cdot$" on X by $x*y:=3+2x-3y, x\cdot y:=3/2$ for any $x,y\in X$, then $(X,*,\cdot)$ is a D-bigroupoid, but not a semiring, since $(X,*)$ is not a semigroup.

9.3 Fuzzy *LD*-Bigroupoids

In this section, we introduce the notion of fuzzy LD-bigroupoid and discuss star prime and (open) barrier value for a fuzzy LD-bigroupoid.

Definition 9.1. Let $(X,*,\cdot)$ be an LD-bigroupoid. A map $\mu:X\to[0,1]$ is called a *fuzzy LD-bigroupoid* of X if, for any $x,y\in X$,

(i) $\mu(x*y)\geq\min\{\mu(x),\mu(y)\}$,
(ii) $\mu(x\cdot y)\geq\mu(y)$.

This follows the approach used in the discussion of fuzzy semirings([1]). Given a fuzzy set μ on X, we define useful subsets of X.

1. $\mu_t^{OC}:=\{x\in X|\mu(x)\leq t\}$, $\mu_t^{C}:=\{x\in X|\mu(x)<t\}$,
2. $\mu_t:=\{x\in X|\mu(x)\geq t\}$, $\mu_t^{O}:=\{x\in X|\mu(x)>t\}$.

Definition 9.2. Let $(X,*,\cdot)$ be an LD-bigroupoid. A nonempty subset S of X is said to be *star-prime* if for any $a,b\in X$ with $a*b\in S$, $\{a,b\}\cap S\neq\emptyset$.

Proposition 9.1. *Let $(X,*,\cdot)$ be an LD-bigroupoid. If μ is a fuzzy LD-bigroupoid of X, then μ_t^{OC} is star-prime if it is a nonempty set.*

Proof. Let $a,b\in X$ such that $a*b\in\mu_t^{OC}$. Then $t\geq\mu(a*b)\geq\min\{\mu(a),\mu(b)\}$. This means that at least one of $\mu(a),\mu(b)$ is less than or equal to t. Hence $\{a,b\}\cap S\neq\emptyset$. □

Corollary 9.1. *Let $(X,*,\cdot)$ be an LD-bigroupoid. If μ is a fuzzy LD-bigroupoid of X, then μ_t^{C} is star-prime if it is a nonempty set.*

Proof. Similar to Proposition 9.1 □

Definition 9.3. Let μ be a fuzzy LD-bigroupoid of X and let $t\in[0,1]$ with $\mu_t\neq\emptyset$. t is said to be a *barrier value* for μ if $\mu(a*b)\geq t$ implies $\max\{\mu(a),\mu(b)\}\geq t$.

With this notion we obtain the following:

Proposition 9.2. *Let μ be a fuzzy LD-bigroupoid of X and let t be a barrier for μ. Then μ_t is star-prime if $\mu_t\neq\emptyset$.*

If we apply this concept to the set μ_t^O, then we obtain the following:

Definition 9.4. Let μ be a fuzzy *LD*-bigroupoid of X and let $t \in [0,1]$ with $\mu_t^O \neq \emptyset$. t is said to be an *open barrier value* for μ if $\mu(a*b) > t$ implies $\max\{\mu(a), \mu(b)\} > t$.

Using Definition 3.7 we obtain an exact analog of Proposition 3.6.

Proposition 9.3. *Let μ be a fuzzy LD-bigroupoid of X and let t be an open barrier for μ. Then μ_t^O is star-prime if $\mu_t^O \neq \emptyset$.*

The notion of a star-prime subset of an *LD*-bigroupoid X has the following relations with a subgroupoid of X.

Proposition 9.4. *Let $(X, *, \cdot)$ be an LD-bigroupoid and let S be a star-prime subset of X. Then $(X \setminus S, *)$ is a subgroupoid of $(X, *)$.*

Proof. Given $a, b \in X \setminus S$, we have $a*b \notin S$, since S is star-prime, proving the proposition.

Proposition 9.5. *Let $(X, *, \cdot)$ be an LD-bigroupoid and let $(S, *)$ be a subgroupoid of $(X, *)$. Then $(X \setminus S, *)$ is a star-prime subset of $(X, *)$.*

Proof. Straightforward.

9.4 Mean Value Property

In this section we introduce the notion of mean value property and discuss some relations with (open) barrier values for fuzzy *LD*-bigroupoids.

Definition 9.5. Let $(X, *, \cdot)$ be an *LD*-bigroupoid. A fuzzy *LD*-bigroupoid μ is said to have the *mean value property* if $\mu(a*b) \leq \max\{\mu(a), \mu(b)\}$ for any $a, b \in X$.

If μ has the mean value property, then $\min\{\mu(a), \mu(b)\} \leq \mu(a*b) \leq \max\{\mu(a), \mu(b)\}$ for any $a, b \in X$. If we consider a and b to be components of a "system" $a*b$, and if $\mu(x)$ measures the "quality/reliability" of x, then the mean value property merely reflects the fact that usually the quality/reliability of a system is no better than the reliability of the best component and at least as good as the reliability of the worst component.

Proposition 9.6. *Let $(X, *, \cdot)$ be an LD-bigroupoid. If a fuzzy LD-bigroupoid μ has the mean value property, then t is a barrier value for μ, for any $t \in [0,1]$ with $\mu_t \neq \emptyset$.*

Proof. Given $a, b \in X$, if $\mu(a*b) \geq t$ for some $t \in [0,1]$, then $\max\{\mu(a), \mu(b)\} \geq \mu(a*b) \geq t$, since μ has the mean value property, proving the proposition.

Corollary 9.2. *Let $(X, *, \cdot)$ be an LD-bigroupoid. If a fuzzy LD-bigroupoid μ has the mean value property, then t is an open barrier value for μ, for any $t \in [0,1]$ with $\mu_t^O \neq \emptyset$.*

Proposition 9.7. *Let μ be a fuzzy LD-bigroupoid of X and let $t \in [0,1]$ with $\mu_t \neq \emptyset$. If t is a barrier value for μ, then it is an open barrier value for μ.*

Proof. Assume that there exist $a, b \in X$ such that $\mu(a * b) > t$ and $\max\{\mu(a), \mu(b)\} \leq t$. Let $\varepsilon > 0$ such that $t + \varepsilon \leq \mu(a * b)$. Since t is a barrier value for μ, we obtain $t < t + \varepsilon \leq \max\{\mu(a), \mu(b)\}$, a contradiction.

Proposition 9.8. *If t is an open barrier value and* $\max\{\mu(a), \mu(b)\} \leq t \leq \mu(a * b)$, *then $\mu(a * b) = t$.*

Proof. Assume that $\mu(a * b) > t$. Since t is an open barrier value for μ, $\max\{\mu(a), \mu(b)\} > t$, a contradiction.

9.5 Dot-Prime and Scalar Barrier Value

In this section, we introduce the notions of dot-prime subset and dot-left-ideal of an *LD*-bigroupoid and discuss some relations with (open) scalar barrier values for the fuzzy *LD*-bigroupoid.

Definition 9.6. Let $(X, *, \cdot)$ be an *LD*-bigroupoid and $\emptyset \neq S \subseteq X$. S is said to be a *dot-prime-subset* of X if $a \cdot b \in S$, then $b \in S$. S is said to be a *dot-left-ideal* of X if $X \cdot S \subseteq S$, i.e., if $x \in X, a \in S$, then $x \cdot a \in S$.

Proposition 9.9. *Let $(X, *, \cdot)$ be an LD-bigroupoid and let S be a nonempty subset of X. Then S is a dot-prime subset of X if and only if $X \setminus S$ is a dot-left-ideal of X.*

Proof

$$\begin{aligned}
&S \text{ is a dot-prime-subset of } X \\
\Leftrightarrow\ &\text{for any } a, b \in X \text{ with } a \cdot b \in S \Rightarrow b \in S \\
\Leftrightarrow\ &\text{for any } a, b \in X, b \notin S \Rightarrow a \cdot b \notin S \\
\Leftrightarrow\ &\text{for any } b \in X \setminus S \Rightarrow a \cdot b \in X \setminus S \\
\Leftrightarrow\ &X \cdot (X \setminus S) \subseteq (X \setminus S) \\
\Leftrightarrow\ &X \setminus S \text{ is a dot-left-ideal of } X
\end{aligned}$$

Proposition 9.10. *Let $(X, *, \cdot)$ be an LD-bigroupoid and μ be a fuzzy LD-bigroupoid of X. Then μ_t^O is a dot-left-ideal of X for any $t \in [0,1]$ with $\mu_t^O \neq \emptyset$.*

Proof. If $b \in \mu_t^O$, then $\mu(b) > t$. Since μ is a fuzzy *LD*-bigroupoid, $\mu(a \cdot b) \geq \mu(b) > t$, for any $a \in X$, proving that $a \cdot b \in \mu_t^O$.

Corollary 9.3. *Let $(X, *, \cdot)$ be an LD-bigroupoid and μ be a fuzzy LD-bigroupoid of X. Then μ_t is a dot-left-ideal of X for any $t \in [0,1]$ with $\mu_t \neq \emptyset$.*

Corollary 9.4. *Let $(X, *, \cdot)$ be an LD-bigroupoid and μ be a fuzzy LD-bigroupoid of X. Then μ_t^C is a dot-prime subset of X for any $t \in [0,1]$ with $\mu_t \neq \emptyset$.*

Proof. It follows from Proposition 9.9 and Corollary 9.3

Using Corollary 9.1 and Corollary 9.4 we can see that μ_t^C is both a star-prime and a dot-prime subset of an *LD*-bigroupoid X. Similarly, μ_t^{OC} is also both a star-prime and a dot-prime subset of an *LD*-bigroupoid X, if it is not empty.

Definition 9.7. Let μ be a fuzzy *LD*-bigroupoid of X and let $t \in [0,1]$ with $\mu_t^{OC} \neq \emptyset$. t is said to be a *scalar barrier value* for μ if for any $a,b \in X$ with $\mu(b) \leq t$, then $\mu(a \cdot b) \leq t$, and t is said to be an *open barrier value* for μ if for any $a,b \in X$ with $\mu(b) < t$, then $\mu(a \cdot b) < t$.

Proposition 9.11. *Let $(X,*,\cdot)$ be an LD-bigroupoid and μ be a fuzzy LD-bigroupoid of X. If t is a scalar barrier value, then $\mu(b) = t$ implies $\mu(a \cdot b) = t$ for any $a,b \in X$.*

Proof. Assume that $\mu(b) = t$. Since t is a scalar barrier value, we have $\mu(a \cdot b) \leq t$. We claim that $\mu(a \cdot b) = t$. If $\mu(a \cdot b) < t$, then $a \cdot b \in \mu_t^C$. It follows from Corollary 9.4 that $b \in \mu_t^C$, i.e., $\mu(b) < t$, a contradiction.

Proposition 9.12. *Let $(X,*,\cdot)$ be an LD-bigroupoid and μ be a fuzzy LD-bigroupoid of X. If t is a scalar barrier value for any $t \in [0,1]$ with $\mu_t \neq \emptyset$, then $\mu(a \cdot b) = \mu(b)$.*

Definition 9.8. Let $(X,*,\cdot)$ be an *LD*-bigroupoid and $\emptyset \neq S \subseteq X$. An algebraic system $(S,*,\cdot)$ is said to be a *star-left-ideal* if (i) $(S,*)$ is a subgroupoid of $(X,*)$, (ii) $(S,\cdot)$ is a dot-left-ideal of $(X,\cdot)$

Proposition 9.13. *Let $(X,*,\cdot)$ be an LD-bigroupoid and μ be a fuzzy LD-bigroupoid of X. If $t \in [0,1]$ with $\mu_t \neq \emptyset$, then μ_t is a star-left-ideal of X.*

Proof. Given $a,b \in \mu_t$, we have $\mu(a) \geq t, \mu(b) \geq t$. Since μ is a fuzzy *LD*-bigroupoid of X, we obtain $\mu(a*b) \geq \min\{\mu(a),\mu(b)\} \geq t$, proving $a*b \in \mu_t$. Given $a \in X$ and $b \in \mu_t$, since μ is a fuzzy *LD*-bigroupoid, we have $\mu(a \cdot b) \geq \mu(b) \geq t$ and hence $a \cdot b \in \mu_t$, proving the proposition.

Proposition 9.14. *Let $(X,*,\cdot)$ be an LD-bigroupoid and μ be a fuzzy LD-bigroupoid of X. If $t \in [0,1]$ is an open barrier for μ, then μ_t^O is a star-left-ideal of X if $\mu_t^O \neq \emptyset$.*

Proof. It follows immediately from Propositions 9.3 and 9.10

Consider now a set $\mu_s \cap \mu_t^{OC}$. If $x \in \mu_s \cap \mu_t^{OC}$, then $\mu(x) \geq s$ and $\mu(x) \not> t$, i.e., $\mu_s \cap \mu_t^{OC} = \mu^{-1}([s,t])$, the closed interval $[s,t]$ being $\{\alpha | s \leq \alpha \leq t\}$. Hence, using these descriptions we may also consider such pre-images by μ of various types of intervals of $[0,1]$.

9.6 Null-Elements

In this section, we discuss a strong-dot-null-element and a strong-star-null-element of an *LD*-bigroupoid.

Definition 9.9. Let $(X,*,\cdot)$ be an *LD*-bigroupoid. An element $n \in X$ is called a *strong-dot-null-element* of X if $x \cdot x = n$ for any $x \in X$. An element ν of X is called a *strong-star-null-element* of X if $x * x = \nu$ for any $x \in X$.

Proposition 9.15. *Let $(X,*,\cdot)$ be an LD-bigroupoid and μ be a fuzzy LD-bigroupoid of X. If n is a strong-dot-null-element of X, then $\mu(n) \geq \mu(x)$ for any $x \in X$.*

Proof. For any $x \in X$, we have $\mu(n) = \mu(x \cdot x) \geq \mu(x)$.

Proposition 9.16. *Let $(X,*,\cdot)$ be an LD-bigroupoid and μ be a fuzzy LD-bigroupoid of X. If ν is a strong-star-null-element of X, then $\mu(\nu) \geq \mu(x)$ for any $x \in X$.*

Proof. For any $x \in X$, we have $\mu(\nu) = \mu(x * x) > \mu(x)$.

Proposition 9.17. *Let $(X,*,\cdot)$ be an LD-bigroupoid. If n is a strong-dot-null-element and ν is a strong-star-null-element of X, then $n = \nu$.*

Proof. For any $x,y \in X$, $x \cdot \nu = x \cdot (y * y) = (x \cdot y) * (x \cdot y) = \nu$. If we let $x := \nu$, then $\nu \cdot \nu = \nu$. Since n is a strong-dot-null-element, we obtain $n = \nu \cdot \nu = \nu$, proving the proposition.

Let $(X,*,\cdot)$ be an *LD*-bigroupoid. If n_1, n_2 are strong-dot-null-elements of X, then it is easy to show that $n_1 = n_2$. Similarly, a strong-star-null-element is unique if it exists. An element $n \in X$ is said to be a *zero* of X if $x \cdot x = x * x = n$ for any $x \in X$.

9.7 Star-rising and Dot-rising

In this section, we introduce the notion of star-rising and dot-rising to fuzzy *LD*-bigroupoids.

Definition 9.10. Let $(X,*,\cdot)$ be an *LD*-bigroupoid. A fuzzy subset μ of X is said to be *star-rising* if $\mu(x * y) \geq \max\{\mu(x), \mu(y)\}$ for any $x,y \in X$, and μ is said to be *dot-rising* if $\mu(x \cdot y) \geq \max\{\mu(x), \mu(y)\}$ for any $x,y \in X$. A fuzzy subset μ of X is said to be *rising* if it is both star-rising and dot-rising.

Proposition 9.18. *Let $(X,*,\cdot)$ be an LD-bigroupoid and μ is star-rising. If either $e * x = x$ for any $x \in X$ or $x * e = x$ for any $x \in X$, then $\mu(x) \geq \mu(e)$ for any $x \in X$.*

Proposition 9.19. *Let $(X,*,\cdot)$ be an LD-bigroupoid and μ is star-rising. If either $e * x = e$ for any $x \in X$ or $x * e = e$for any $x \in X$, then $\mu(e) \geq \mu(x)$ for any $x \in X$.*

Y. B. Jun et. al ([6]) defined the notion of a *KS-algebra*, i.e., an algebra $(X,*,\cdot,0)$ satisfying the conditions: (i) $(X,*,0)$ is a *BCK*-algebra, (ii) $(X,\cdot)$ is a semigroup, (iii) "$\cdot$" is distributive (on both sides) over "$*$".

Corollary 9.5. *Let $(X,*,\cdot,0)$ be a KS-algebra. If μ is star-rising, then $\mu(x) = \mu(0)$ for any $x \in X$, i.e., μ is a constant function on X.*

Proof. If $(X, *, \cdot, 0)$ is a *KS*-algebra, then $x * 0 = x, 0 * x = 0$ for any $x \in X$. By Propositions 9.18 and 9.19, we obtain $\mu(0) = \mu(x)$.

Example 9.3. Let $(X := [0, \infty), +, \cdot)$ with usual addition and multiplication on the real numbers. Then it is an *LD*-bigroupoid. Define a map μ on X by

$$\mu(x) := 1 - \exp^{-\lambda(x-1)}, \lambda > 0$$

Then $\mu(1) = 0$ and $\lim_{x \to \infty} \mu(x) = 1$. Since $\mu'(x) = \lambda \exp^{-\lambda(x-1)} > 0$, μ is monotonically increasing and thus $\mu(x+y) \geq \max\{\mu(x), \mu(y)\}$ and $\mu(x \cdot y) \geq \max\{\mu(x), \mu(y)\}$ for any $x, y \in X$. This proves that μ is a fuzzy *LD*-bigroupoid and rising.

Proposition 9.20. *Let $(X, *, \cdot)$ be an LD-bigroupoid. If μ is rising and $t \in [0, 1]$ is a barrier value for μ, then $\mu(x * y) = \max\{\mu(x), \mu(y)\}$ for any $x, y \in X$.*

Proof. Since μ is star-rising, $\mu(x * y) \geq \max\{\mu(x), \mu(y)\}$ for any $x, y \in X$. If we let $t := \mu(x * y)$, then $t \leq t = \mu(x * y)$. Since t is a barrier value, we obtain $\mu(x * y) \leq \max\{\mu(x), \mu(y)\}$ for any $x, y \in X$, which shows that $\mu(x * y) = \max\{\mu(x), \mu(y)\}$.

An *LD*-bigroupoid $(X, *, \cdot)$ is said to be an *LD-rising-bigroupoid* if it has an associated non-constant rising fuzzy *LD*-bigroupoid. The bigroupoid $([0, \infty), +, \cdot)$ discussed in Example 9.3 is an *LD*-rising-bigroupoid.

Proposition 9.21. *Let $(X, *, \cdot)$ be an LD-rising-bigroupoid. Then it has no element $e \in X$ such that either $e * x = e, x * e = x$ for any $x \in X$ or $e * x = e, x * e = x$ for any $x \in X$.*

Proof. Assume that there exists $e \in X$ such that either $e * x = e, x * e = x$ for any $x \in X$ or $e \cdot x = e, x \cdot e = x$ for any $x \in X$. Since $(X, *, \cdot)$ is an *LD*-rising-bigroupoid, there exists a non-constant rising fuzzy *LD*-bigroupoid of X. Hence $\mu(e) = \mu(e * x) \geq \max\{\mu(e), \mu(x)\}$ and $\mu(x) = \mu(x * e) \geq \max\{\mu(x), \mu(e)\}$ for any $x \in X$, which proves that $\mu(e) = \mu(x)$ for any $x \in X$, i.e., μ is a constant function, a contradiction.

9.8 Specializations of fuzzy *LD*-Bigroupoids.

In this section, we introduce the notion of a μ-compatible fuzzy *LD*-bigroupoid and discuss several properties of specializations of fuzzy *LD*-bigroupoids.

Let $(X, *, \cdot)$ be an *LD*-bigroupoid. If we define a binary operation "$\circledast$" on X by $x \circledast y := y * x$ for any $x, y \in X$, then it is easy to see that $(X, \circledast, \cdot)$ is an *LD*-bigroupoid, called a *star-opposite-LD-bigroupoid*.

Proposition 9.22. *Let $(X, *, \cdot)$ be an LD-bigroupoid and its star-opposite-LD-bigroupoid $(X, \circledast, \cdot)$. Then*

(i) if μ is a fuzzy LD-bigroupoid of $(X, *, \cdot)$, then μ is also a fuzzy LD-bigroupoid of $(X, \circledast, \cdot)$,
(ii) if μ is (star, dot) rising on $(X, *, \cdot)$, then it is also (star, dot) rising on $(X, \circledast, \cdot)$.

Proof. Straightforward.

Proposition 9.23. *If μ has the mean value property on $(X,*,\cdot)$, then μ has also the mean value property on $(X,\circledast,\cdot)$.*

If μ has the mean value property on $(X,*,\cdot)$, i.e., $(1-\mu)(x*y) = 1-\mu(x*y) \geq 1-\max\{\mu(x),\mu(y)\} = \min\{(1-\mu)(x),(1-\mu)(y)\}$ for any $x,y \in X$, then $1-\mu$ is a fuzzy subgroupoid of $(X,*)$, but $1-\mu$ is not usually a fuzzy *LD*-bigroupoid of $(X,*,\cdot)$. In fact, if μ has the mean value property, then it is a fuzzy *LD*-bigroupoid of X so that $\mu(x\cdot y) \geq \mu(y)$, which implies that $(1-\mu)(x\cdot y) \leq (1-\mu)(y)$ where $x,y \in X$.

The following propositions can easily be obtained, and we omit the proofs.

Proposition 9.24. *Let $(X,*,\cdot)$ be an LD-bigroupoid and let $\alpha \in [0,1]$. If we define a map $\mu_\alpha(x) := \alpha\mu(x)$ for any $x \in X$, then μ_α is a fuzzy LD-bigroupoid of $(X,*,\cdot)$.*

Proposition 9.25. *Let $(X,*,\cdot)$ be an LD-bigroupoid and let $\alpha \in [0,1]$. If μ has the mean value property on $(X,*,\cdot)$, then μ_α has also the mean value property on $(X,*,0)$.*

Proposition 9.26. *Let $(X,*,\cdot)$ be an LD-bigroupoid and let $\alpha \in [0,1]$. If μ is (star, dot) rising on $(X,*,\cdot)$, then μ_α (star, dot) rising on $(X,*,0)$.*

Given a function $f: X \to X$, and a fuzzy *LD*-bigroupoid μ on $(X,*,\cdot)$, let $\mu_f(x) := \mu(f(x))$. Notice that if $f(x*y) = f(x)*f(y)$ and if $f(x\cdot y) = f(x)\cdot f(y)$, then $\mu(f(x*y)) = \mu(f(x)*f(y)) \geq \min\{\mu(f(x)),\mu(f(y))\}$ and if $f(x\cdot y) = f(x)\cdot f(y)$, then $\mu(f(x*y)) = \mu(f(x)*f(y)) \geq \min\{\mu(f(x)),\mu(f(y))\}$ and $\mu(f(x\cdot y)) = \mu(f(x)\cdot f(y)) \geq \mu(f(y))$, so that μ_f is a fuzzy *LD*-bigroupoid. We summarize:

Proposition 9.27. *Let $(X,*,\cdot)$ be an LD-bigroupoid and μ be a fuzzy LD-bigroupoid of X. If $f: X \to X$ be a homomorphism, then the map $\mu_f: (X,*,\cdot) \to [0,1]$ defined by $\mu_f(x) := \mu(f(x))$ is a fuzzy LD-bigroupoid of X.*

Definition 9.11. Let $(X,*,\cdot)$ be an *LD*-bigroupoid and μ be a fuzzy *LD*-bigroupoid of X. A map $f: X \to X$ is said to be *μ-compatible* if μ_f is a fuzzy *LD*-bigroupoid.

Note that every homomorphism $f: (X,*,\cdot) \to (X,*,\cdot)$ is μ -compatible.

Proposition 9.28. *Let $(X,*,\cdot)$ be an LD-bigroupoid and μ be a fuzzy LD-bigroupoid of X. Let $g: X \to X$ be a map and let $f: X \to X$ be map defined by $f(x*y) = f(x)*f(y), f(x\cdot y) = g(x)\cdot f(y)$ for any $x,y \in X$. Then $\mu_f: X \to X$ is a fuzzy LD-bigroupoid of X.*

Proof. Straightforward.

The map f defined in Proposition 9.28 is μ-compatible.

Proposition 9.29. *Let $(X,*,\cdot)$ be an LD-bigroupoid and μ be a fuzzy LD-bigroupoid of X. If we let $\mu C := \{g | g : \mu\text{-compatible on } X\}$, then $(\mu C, \circ)$ is a subsemigroup of $(X^X, \circ)$ where "$\circ$" is the composition of mappings.*

Proof. If $f,g \in \mu C$, then μ_f and μ_g are fuzzy *LD*-bigroupoids of X. For any $x \in X$, $\mu_{gof}(x) = \mu((g \circ f)(x)) = \mu(g(f(x))) = \mu_g(f(x)) = (\mu_g)_f(x)$ is a fuzzy *LD*-bigroupoid of X, proving that $g \circ f \in \mu C$.

Definition 9.12. Let $(X,*,\cdot)$ be an *LD*-bigroupoid and μ be a fuzzy *LD*-bigroupoid of X. Define $< \mu >:= \{\mu_{f,\alpha} | f \in \mu C, \alpha \in [0,1]\}$ where $\mu_{f,\alpha}(x) := \alpha\mu_f(x)$ for any $x \in X$. We call $\mu_{f,\alpha}$ a *specialization* of μ.

Definition 9.13. Let $(X,*,\cdot)$ be an *LD*-bigroupoid and μ and δ be a fuzzy *LD*-bigroupoids of X. δ is said to be a *specialization* of μ, and denoted by $\mu \rightarrow \delta$, if there exists $\alpha \in [0,1]$ and $f \in \mu C$ such that $\delta = \mu_{f,\alpha}$. We denote $\mu \equiv \delta$ if $\mu \rightarrow \delta, \delta \rightarrow \mu$. In this case, we say that μ and δ are *mutual specializations*. In fact, if $\mu \rightarrow \delta, \delta \rightarrow \mu$, then $\delta = \mu_{f,\alpha}, \mu = \delta_{g,\beta}$ for some $\alpha, \beta \in [0,1]$ and $f,g \in \mu C$, which means that $\mu(x) = \beta\delta(g(x))$ and $\delta(x) = \alpha\mu(f(x))$ for any $x \in X$. It follows that, for any $x \in X$,

$$\begin{aligned}\mu(x) =&\beta\delta(g(x))\\ =&\beta\alpha\mu(f(g(x)))\\ =&(\beta\alpha)\mu_{fog}(x).\end{aligned}$$

It is easy to see that the relation $\equiv$ is an equivalence relation on the class of fuzzy *LD*-bigroupoids defined on $(X,*,\cdot)$. This yields a new class of objects, viz., the equivalence classes $[\mu]$ consisting of all δ such that $\mu \equiv \delta$.

Note that a specialization of a constant mapping is a constant mapping. In fact, if $\mu : X \rightarrow [0,1]$ is a constant mapping, i.e., $\mu(x) = c$ for some $c \in [0,1]$ for any $x \in X$, then it is a fuzzy *LD*-bigroupoid of X. Hence $\mu_{f,\alpha}(x) = \alpha(\mu(f(x))) = \alpha c$ for any $f \in \mu C, \alpha \in [0,1]$, which proves that $< \mu >:= \{\mu_{f,\alpha} | \mu_{f,\alpha} : X \rightarrow [0,1] :$ constant, $f \in \mu C, \alpha \in [0,1]\}$.

Lemma 9.1. *If* $\mu \rightarrow \nu$ *and* $\nu \rightarrow \delta$, *then* $\mu \rightarrow \delta$.

Proof. If $\mu \rightarrow \nu$ and $\nu \rightarrow \delta$, then $\nu = \mu_{f,\alpha}, \delta = \nu_{g,\beta}$ for some $\alpha, \beta \in [0,1]$ and for some $f,g \in \mu C$, which means that $\nu(x) = \mu_{f,\alpha}(x) = \alpha\mu_f(x) = \alpha\mu(f(x))$ and hence $\delta(x) = \beta\nu(g(x)) = \beta\alpha\mu(f(g(x))) = (\beta\alpha)\mu_{fog}(x) = \mu_{fog,\beta\alpha}(x)$, proving that $\mu \rightarrow \delta$. □

Proposition 9.30. *If* $\mu \rightarrow \nu$, *then* $[\nu] \subseteq < \mu >$.

Proof. If $\delta \in [\nu]$, then $\delta \equiv \nu$ and hence $\nu \rightarrow \delta$. Since $\mu \rightarrow \nu$, by Lemma 8.12, we obtain that $\mu \rightarrow \delta$. Hence $\delta = \mu_{f,\alpha}$ for some $\alpha \in [0,1]$ and $f \in \mu C$, which proves that $\delta \in < \mu >$.

Proposition 9.31. *Let* $\mu \rightarrow \nu$, $\nu = \mu_{f,\alpha}$ *and* $f(x*y) = f(x)*f(y)$ *for any* $x,y \in X$. *If* μ *is star-rising, then* ν *is star-rising.*

Proof. For any $x,y \in X$, we have

$$\begin{aligned}\nu(x*y) =&\mu_{f,\alpha}(x*y)\\ =&\alpha\mu(f(x*y))\\ =&\alpha\mu(f(x)*f(y))\\ \geq&\alpha\max\{\mu(f(x)),\mu(f(y))\}\\ =&\max\{\alpha\mu(f(x)),\alpha\mu(f(y))\}\\ =&\max\{\nu(x),\nu(y)\},\end{aligned}$$

proving that ν is star-rising.

Proposition 9.32. *Let μ_i be fuzzy LD-bigroupoids of $(X,*,\cdot)$ $(i=1,2)$. Define $\mu := \mu_1 \vee \mu_2$, i.e., $\mu(x) = \max\{\mu_1(x),\mu_2(x)\}$. Then*

(i) μ is a fuzzy LD-bigroupoid of X,
(ii) if μ_1 and μ_2 have the mean value property, then μ has also the mean value property,
(iii) if μ_1 and μ_2 are star-rising(dot-rising, resp.), then μ is also star-rising(dot-rising, resp.).

Proof. Straightforward.

9.9 Duality

In this section, we define a fuzzy *LD*-bigroupoid on the collection of fuzzy *LD*-bigroupoids defined on $(X,*,\cdot)$, and show that it is isomorphic to $(X,*,\cdot)$. Let $(X,*,\cdot)$ be an *LD*-bigroupoid. Suppose that $F(X,*,\cdot)$ denotes the collection of all fuzzy *LD*-bigroupoids defined on $(X,*,\cdot)$. Given $a \in X$, let $a^\diamond : F(X,*,\cdot) \to [0,1]$ be defined by $a^\diamond(\mu) := \mu(a)$. We denote $X^\diamond := \{a^\diamond \mid a \in X\}$.

Proposition 9.33. *Let $(X,*,\cdot)$ be an LD-bigroupoid. Define two binary operations "$\bigtriangledown$" and "$\bigtriangleup$" on $X^\diamond$ by*

$$a^\diamond \bigtriangledown b^\diamond := (a*b)^\diamond,$$

$$a^\diamond \bigtriangleup b^\diamond := (a\cdot b)^\diamond$$

for any $a,b \in X$. Then $(X^\diamond,\bigtriangledown,\bigtriangleup)$is an LD-bigroupoid and isomorphic to $(X,*,\cdot)$.

Proof. Given $\mu \in F(X,*,\cdot)$, we have

$$\begin{aligned}(a^\diamond \bigtriangleup (b^\diamond \bigtriangledown c^\diamond))(\mu) =&(a^\diamond \bigtriangleup (b*c)^\diamond)(\mu)\\ =&(a\cdot(b*c))^\diamond(\mu)\\ =&((a\cdot b)*(a\cdot c))^\diamond(\mu)\\ =&[(a\cdot b)^\diamond \bigtriangledown (a\cdot c)^\diamond](\mu)\\ =&[(a^\diamond \bigtriangleup b^\diamond) \bigtriangledown (a^\diamond \bigtriangleup c^\diamond)](\mu),\end{aligned}$$

proving that $a^\diamond \bigtriangleup (b^\diamond \bigtriangledown c^\diamond) = (a^\diamond \bigtriangleup b^\diamond) \bigtriangledown (a^\diamond \bigtriangleup c^\diamond)$. Hence $(X^\diamond, \bigtriangledown, \bigtriangleup)$ is an *LD*-bigroupoid. If we define a map $\varphi : (X, *, \cdot) \to (X^\diamond, \bigtriangledown, \bigtriangleup)$ by $\varphi(a) := a^\diamond$, then it is easy to see that it is an isomorphism.

Proposition 9.34. *Let $(X, *, \cdot)$ be an LD-bigroupoid. For any $\mu \in F(X, *, \cdot)$, if we define $\mu^\diamond : (X^\diamond, \bigtriangledown, \bigtriangleup) \to [0,1]$, then it is a fuzzy LD -bigroupoid of $(X^\diamond, \bigtriangledown, \bigtriangleup)$.*

Proof. For any $a^\diamond, b^\diamond \in X^\diamond$, we have $\mu^\diamond(a^\diamond \bigtriangledown b^\diamond) = \mu^\diamond((a * b)^\diamond) = \mu(a * b) \geq \min\{\mu(a), \mu(b)\} = \min\{\mu^\diamond(a^\diamond), \mu^\diamond(b^\diamond)\}$ and $\mu^\diamond(a^\diamond \bigtriangleup b^\diamond) = \mu^\diamond((a \cdot b)^\diamond) = \mu(a \cdot b) \geq \mu(b) = \mu^\diamond(b^\diamond)$, proving the proposition.

If we let $F(X^\diamond, \bigtriangledown, \bigtriangleup) := \{\mu^\diamond | \mu \in F(X, *, \cdot)\}$, then by Proposition 9.2, it is in one-one correspondence with $F(X, *, \cdot)$ via the mapping $\Phi(\mu) = \mu^\diamond$. $X^\diamond$ has an order relation $\leq$ via $a^\diamond \leq b^\diamond$ if $a^\diamond(\mu) \leq b^\diamond(\mu)$ for all $\mu \in F(X, *, \cdot)$. From the above it follows that $\min\{a^\diamond, b^\diamond\} \leq a^\diamond \bigtriangledown b^\diamond$ and $b^\diamond \leq (a^\diamond \bigtriangleup b^\diamond)$ where $(\min\{a^\diamond, b^\diamond\})(\mu) = \min\{a^\diamond(\mu), b^\diamond(\mu)\}$.

References

1. Ahsan, J., Saifullah, K., Khan, M.F.: Fuzzy semirings. Fuzzy Sets and Sys. 60, 309–320 (1993)
2. Dudek, W.A.: Special types of intuitionistic fuzzy left *h*-ideals of hemirings. Soft Computing 12, 359–364 (2008)
3. Ghosh, S.: Fuzzy *k*-ideals of semirings. Fuzzy Sets and Sys. 95, 103–108 (1998)
4. Golan, J.S.: Semirings and their applications. Kluwer Academic, Boston (1999)
5. Hebisch, U., Weinert, H.J.: Semirings, algebraic theory and applications in computer science. World Scientific, Singapore (1993)
6. Jun, Y.B., Xin, X.L., Roh, E.H.: A class of algebras related to *BCI*-algebras and semigroups. 24, 309–321 (1998)
7. Mordeson, J.N., Malik, D.S.: Fuzzy commutative algebra. World Scientific, Singapore (1998)
8. Neggers, J., Jun, Y.B., Kim, H.S.: On *L*-fuzzy ideals in semirings II. Czech. Math. J. 49, 127–133 (1999)
9. Zadeh, L.A.: Fuzzy sets. Inf. Contr. 8, 338–353 (1965)

Chapter 10
Semiring Parsing

Yudong Liu

Computer Science Department,
Western Washington University,
Bellinig ham, Washington
yudong.liu@wwu.edu

Syntactic parsing is an important task in natural language processing (NLP). In this chapter an application of semiring theory in parsing (a.k.a."semiring parsing") will be introduced. A semiring parsing framework is proposed and studied in [6].

10.1 Introduction

Statistical parsing algorithms are useful for structure predictions in many diverse application areas, ranging from natural language processing to biological sequence analysis. Natural language is highly prone to ambiguities, which means that there may be hundreds, even thousands, of syntactic parse trees for certain natural sentences. Such a fact has been a major challenge in natural language processing, given that syntactic analysis is an important intermediate step in many natural language applications. [3] proposed some methods for dealing with such syntactic ambiguity in ways that exploit certain *regularities* among alternative parse trees. A special case is that ambiguity coefficients follow a well-known combinatorial series called the *Catalan Numbers* [8]. Theoretically, such encoding of ambiguity indicates that parsing can be brought into an *algebraic power series* framework and there exists a solid algebraic foundation to define parsing as the algebraic operations in this framework. In fact, formal power series are a well-known device in the formal language literature for developing the algebraic properties of context-free grammars [2, 9]. From the application point of view, parsing algorithms can be used to compute many interesting quantities, such as Viterbi value (the value of the best derivation), Viterbi derivation (the parse tree corresponding to the Viterbi value), n-best derivations, derivation forests, derivation count, inside and outside probabilities of components of the input string. Currently there are a variety of efficient parsing algorithms available for different grammar formalisms, from context-free grammars (CFGs) to tree adjoining grammars [7]. Conventionally, different parsing descriptions are needed for different tasks; a fair amount of work can be required to construct each one. Therefore it is preferable to unify all these parsers into a general framework to make it work across diverse tasks and application areas.

J. Ahsan et al.: Fuzzy Semirings with Applications, STUDFUZZ 278, pp. 175–192.
springerlink.com

Semiring parsing [6] is motivated by the discussions above. This framework is based on the theory of power series, and takes statistical parsing to be the computation of the *coefficients* for the elements of a free monoid. The monoid typically represents a set of languages. In [6] a semiring parsing system consists of a deductive component and a semiring interpreter component. A deductive system provides us a unified way to represent a variety of parsing algorithms and particularly suitable for rapid prototyping new parsing strategies. And a semiring interpreter component assigns the corresponding semantics to the result of the deductive system by a task-specific semiring specialization.

Therefore, by separation of the algebra and the parsing algorithms, a semiring parsing system provides a generalized and modularized framework to unify a variety of parsing algorithms across a variety of tasks.

10.2 Context-Free Parsing Algorithms

There exist efficient recognition and parsing algorithms for arbitrary context-free languages. Compared with a pushdown automaton which recognizes a particular context-free language, these algorithms are more general and can be used for any context-free language. In this section, we are going to introduce some fundamental recognition and parsing algorithms, and how they work for one particular weighted context-free language – probabilistic context-free languages.

10.2.1 Recognition and Parsing Algorithms for Context-Free Languages

The recognition problem for context-free languages is essentially a decision problem: "Given a context-free grammar G and a string w, is $w \in L(G)$"? A parsing process is a recognition process which additionally outputs a parse or derivation of each acceptable input. So if we can effectively recognize a string, then the parsing is trivial. Once recognition is accomplished, we can use backtracking strategy to retrieve the derivations in linear time.

In the following, we give three classic context-free parsing algorithms: CKY algorithm, Earley's algorithm and GHR parsing algorithm.

CKY (Cocke-Kasami-Younger) Parsing Algorithm

The CKY algorithm is a tabular dynamic programming method. It works bottom-up by finding all derivations for a string by using previously computed derivations for all possible substrings. The remarkable property of CKY algorithm is that it has the ability to parse arbitrarily ambiguous CFGs and find all possible parse trees (exponentially many) for a given input in polynomial time and space. As we will see, its time complexity is $O(|G|^2n^3)$ and its space complexity is $O(|G|n^2)$ where $|G|$ is the size of grammar and n is the length of input string.

The following shows the basic CKY algorithm:

Input string w **of size** n
Initialize $2D$ **chart of size** n^2 //each cell of chart can be viewed as a set
for $i = 0$ **to** $n-1$
add A into chart$[i][i+1]$ **if** $w[i] = a$ **and** rule $A \to a \in G$
for $j = 2$ **to** N
    for $i = j-2$ **downto 0**
        for $k = i+1$ **to** $j-1$
        add A into chart$[i][j]$ **if** $B \in$ chart$[i][k]$ **and** $C \in$ chart$[k][j]$ **and** rule $A \to BC \in G$
return "yes" **if** $S \in$ chart$[0][n]$
else return "no"

To apply the CKY algorithm, the given grammar has to be transformed into Chomsky Normal Form (CNF). As one of the most-used algorithms for parsing natural languages, CKY parsing can be found in most of the literature on parsing.

Earley's Parsing Algorithm

Earley's algorithm is proposed by [4]. It is also a tabular parsing method based on a dynamic programming strategy. However, it avoids the main limitation of CKY algorithm on grammar and therefore applicable to arbitrary context-free grammars without relying on any grammar transformations. It uses a top-down parsing strategy with a bottom-up filter in a way that it initializes the top-down rule in form of $S \to \alpha$ and scan the string from left to right with applying three kinds of rules: *predictor, scanner* and *completer*. Earley's algorithm uses "lookahead" by using "dotted rules" to avoid lots of useless matchings, which is superior to CKY again. In some cases, lookahead may lead to things more complicated and increases the number of states. More details on this algorithm can also be found on most of the literature on parsing.

GHR (Graham-Harrison-Ruzzo) Algorithm

GHR (Graham-Harrison-Ruzzo) algorithm [11] is another improved recognition algorithm working on arbitrary context-free grammars. One principle drawback of the CKY method is that the algorithm may find a lot of "useless" matches which cannot lead to derivations to S. Compared with CKY, GHR algorithm improves the average case performance by considering only those matches that are consistent with the left context, which is in the same spirit of Earley's algorithm. Specifically:

CKY : *Add A to chart*$[i,j]$ *if and only if* $A \Rightarrow^* w_{i,j}$
GHR : *Add* $A \to \alpha \cdot \beta$ *to chart*$[i,j]$ *if and only if* $\alpha \Rightarrow^* w_{i,j}$ *and* $S \Rightarrow^* w_i A \delta$ *for some* $\delta \in \Sigma^*$.

Another trick in GHR is to pre-compute the "prediction set", which saves time by such off-line computation. GHR seems to do as well as one could expect in an *on-line* recognizer that recognizes each prefix of w without lookahead. Still the algorithm runs in time $O(|G^2|n^3)$ and space $O(|G|n^2)$ for arbitrary context-free grammar. And it is also possible to implement the algorithm in time asymptotically less than $O(|G^2|n^3)$.

10.2.1.1 Summary of CKY, Earley's and GHR Algorithms

The algorithms of CKY, Earley and GHR are different styles of parsing for context-free grammars (CFGs). All of them ensure that, given a set of grammar rules and an input sentence, all possible parses of that sentence are produced. The dynamic programming strategy makes the results not only include all the possible parse trees but all of the subtrees that were found during the generation of complete parse trees. In particular, there are some close connections between CKY and Earley, which is shown by relating both to GHR ([11] section 5). The unification of superficially dissimilar methods may give us some clue to represent these different-appearance algorithms in a more general way. It turns out that it is feasible to achieve. More details can be found in the next Section.

10.2.2 Probabilistic Context-Free Grammars (PCFGs)

Ambiguity is a fundamental problem in natural language processing. For the parsing task, one possible solution is to find the most likely parse tree among the potentially exponentially many parses. Probabilistic context free grammar is proposed based on this motivation.

Definition 10.1. A *probabilistic context-free grammar* (or PCFG) is a context-free grammar that associates a probability to each of its production rules. Formally, a PCFG G consists of (Σ, N, P, S) and a corresponding set of probabilities on rules such that:

$$\forall N^i \in N \ \sum_{\alpha} P(N^i \to \alpha) = 1$$

The following grammar is a well-defined probabilistic context-free grammar:

$S \to XX$	0.7
$S \to YY$	0.3
$X \to AA$	0.4
$X \to B$	0.6
$Y \to AA$	0.5
$Y \to B$	0.5
$A \to a$	1.0
$B \to b$	1.0

A PCFG generates the same set of parses for a given input that the corresponding CFG does, and assigns a probability to each parse tree. The probability of a parse tree is the product of the probabilities of all rules applied in the parse tree. The probability of a sentence is the sum of the probabilities of all parse trees.

Example 10.1. Use the above grammar to parse the input *aab*, and get two parse trees: The probabilities for these two trees are:

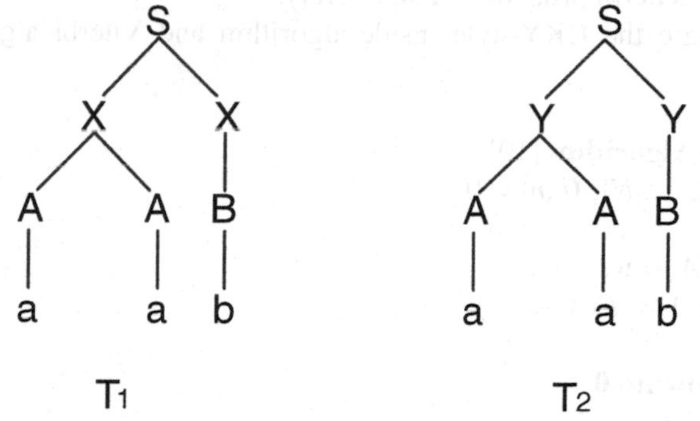

Fig. 10.1 *parse trees of aab.*

$$\begin{aligned} P(T_1) &= P(S \to XX) \times P(X \to AA) \times P(X \to B) \times P(A \to a)^2 \times P(B \to b) \\ &= 0.7 \times 0.4 \times 0.6 \times 1.0^2 \times 1.0 = 0.168 \\ P(T_2) &= P(S \to YY) \times P(Y \to AA) \times P(Y \to B) \times P(A \to a)^2 \times P(B \to b) \\ &= 0.3 \times 0.5 \times 0.5 \times 1.0^2 \times 1.0 = 0.075 \end{aligned}$$

Since T_1 and T_2 are all possible parse trees for input *aab*, then the entire probability of generating input *aab* is : $P(aab) = P(T_1) + P(T_2) = 0.168 + 0.075 = 0.243$. And between T_1 and T_2, T_1 is the more likely parse tree for input *aab* based on the given grammar.

Inside Algorithm and Viterbi Algorithm

As illustrated in the example, generally we are more interested in two types of probabilities based on PCFGs:

- What is the probability of a sentence w_{1m} according to the grammar G: $P(w_{1m} \mid G)$?
- What is the most likely parse tree for a sentence: $\text{argmax}_t P(t \mid w_{1m}, G)$?

Typically, the first probability is called *inside probability*[1] of the input sentence; and the second *Viterbi probability*[2] for the input sentence. Interested readers can refer to [12] for more introduction.

Intuitively, if simply enumerate all possible parse trees of the string, just as the example illustrated, the answers to both probabilities are pretty straightforward. However, generally it is impractical as there might be exponentially many number of parse trees. A more efficient way has to be explored to calculate these values. Inside algorithm and Viterbi algorithm use dynamic programming to compute the inside probability and the Viterbi probability respectively.

The following are the CKY-style inside algorithm and Viterbi algorithm for PCFGs:

CKY-style Inside Algorithm [10]
float chart$[0..n-1,\ 1..|N|,\ 0..n] = 0$
for $i = 0$ **to** $n-1$
  for each rule $A \rightarrow w_s \in P$
    chart$[i, A, i+1] = P(A \rightarrow w_s)$
for $j = 2$ **to** N
  for $i = j-2$ **downto 0**
    for $k = i+1$ **to** $j-1$
      for each rule $A \rightarrow BC \in P$
        chart$[i,A,j]$ = chart$[i,A,j]$+(chart$[i,B,k]\times$ chart$[k,C,j] \times P(A \rightarrow BC)$)
return chart$[0,S,n]$

CKY-style Viterbi Algorithm
float chart$[0..n-1, 1..|N|, 0..n] = 0$
for $i = 0$ **to** $n-1$
  for each rule $A \rightarrow w_s \in P$
    chart$[i, A, i+1] = P(A \rightarrow w_s)$
for $j = 2$ **to** N
  for $i = j-2$ **downto 0**
    for $k = i+1$ **to** $j-1$
      for each rule $A \rightarrow BC \in P$
        chart$[i,A,j]$ = max(chart$[i,A,j]$,(chart$[i,B,k]\times$ chart$[k,C,j] \times P(A \rightarrow$
BC)))
return chart$[0,S,n]$

[1] Formally, the inside probability $\beta_j(p,q)$, $\beta_j(p,q) = P(w_{pq} \mid N^j_{pq}, G)$, is the total probability of generating words $w_p \cdots w_q$ given that one is starting off with the nonterminal N^j.

[2] By the notation $\delta_i(p,q)$ = the highest inside probability parse of a subtree N^i_{pq}, the Viterbi probability for generating words $w_p \cdots w_q$ starting off with the nonterminal N^i is $\delta_i(p,q) = \max_{1 \le j,k \le n, p \le r < q} P(N^i \rightarrow N^j N^k)\delta_j(p,r)\delta_k(r+1,q)$

Comments:

1. chart$[i,A,j]$ is used to denote different element in the same cell of the table, which is different from the notation in basic CKY algorithm.
2. Comparing these two algorithms, the only difference is that Viterbi algorithm finds the most probable parse by taking the maximum operation instead of summing over all parse trees as in inside algorithm. The notation of semirings can be used to unify these two algorithms. And Inside algorithm corresponds to the semiring $(R_0^1,+,\times,0,1)$ and Viterbi algorithm corresponds to the semiring $(R_0^1,\max,\times,0,1)$.

These two algorithms can be rewritten into one general form as follows, by using the semiring notation:

chart$[0..n-1,1..|N|,0..n] = \bar{0}$
for $i = 0$ **to** $n-1$
  for each rule $A \to w_s \in P$
    chart$[i,A,i+1] = P(A \to w_s)$
for $j = 2$ **to** N
  for $i = j-2$ **downto 0**
    for $k = i+1$ **to** $j-1$
      for each rule $A \to BC \in P$
        chart$[i,A,j] = \bigoplus($chart$[i,A,j]$,chart$[i,B,k] \otimes$ chart$[k,C,j] \otimes P(A \to BC))$
return chart$[0,S,n]$

Similarly, a generic Earley or GHR parsing algorithm can be used to compute both probabilities; furthermore, the n most likely parse trees can be calculated for a given input using the same generic algorithm. All the computations can be generalized into the same framework based on the concept of power series and the notation of semirings.

10.2.2.1 Summary of This Section

In this section, we introduced the basic parsing algorithms of context free grammars (CFGs) and probabilistic context free grammars (PCFGs). Parsing algorithms play a core role both in the syntactic structure analysis and value computations related to syntactic structures. Combining the semiring notation, parsing algorithms across tasks can be unified into a general form.

10.3 Deductive Parsing

From the previous discussion, we see that over different semirings, a single parsing algorithm can be applied across various tasks. From the introduction of the parsing algorithms, we can see the certain similarity between CKY, Earley and GHR algorithms. At this point, we may ask if there is a unified way to represent all these different parsing strategies. Deductive parsing [15] provides such a general framework by taking parsing as a deductive process.

10.3.1 Components of a Deductive System

In [15] a deductive parsing system consists of an *item-based description* component and a *control engine* component. Item-based descriptions are basically used to describe parsing algorithms by treating parsing as deduction, and the control engine component interprets the deduction to implement the corresponding parser. In the following, we will introduce these components and use some examples to illustrate the idea of deduction.

Item-Based Description: Items and Inference Rules

In most parsers, there is at least one chart of some form. In this general description, a corresponding concept *item* is proposed with the form $[i,A,j]$. For example, in CKY parsing, item is of the form $[i,A,j]$ and states that the nonterminal A derives the substring between index i and j in the string. In [15] this form is used to denote every chart element.

The general form for an inference rule will be

$$\frac{A_1 \cdots A_k}{B} \langle \text{side conditions on } A_1 \cdots A_k, B \rangle$$

where $A_1,\ldots,A_k$ are the *antecedents* and B the *consequent* of the inference rule. The existence of side conditions depends on the parser and its value is only true or false. When the inference rule is used, the metavariables it contains will be instantiated by appropriate *terms*. *Terms* refer to grammar rules and items.

An *axiom* in a deductive parsing refers to a sound item. For example, for each word w_i in the string, it is clear that the item $[i,w_i,i+1]$ makes a true claim, so that such item can be taken as axiomatic. Generally the parsing procedure is started with axioms.

Item-Based Description: Examples of CKY and Earley Parsing

The following gives item-based descriptions of a CKY-style parser and Earley-style parser respectively.

Item form: $[i,A,j]$

Goal: $[0,S,n]$

Axioms: $[i,w_i,i+1] \qquad 0 \le i < n$

Inference rules: $\frac{R(A \to w_i)}{[i,A,i+1]}\ [i, w_i, i+1]$ Prediction

$\frac{R(A \to BC)\quad [i,B,k]\quad [k,C,j]}{[i,A,j]}$ Completion

Item-based description of a CKY parser

Item form: $[i,A \to \alpha \bullet \beta, j]$

Goal: $[0,S' \to S\bullet, n]$

Axioms: $[0, S' \to \bullet S, 0]$

Inference rules: $\frac{[i,A \to \alpha \bullet w_{j+1} \beta, j]}{[i,A \to \alpha w_{j+1} \bullet \beta, j+1]}$ Scanning

$\frac{R(B \to \gamma)}{[j,B \to \bullet\gamma, j]}\ [i, A \to \alpha \bullet B\beta, j]$ Prediction

$\frac{[i,A \to \alpha \bullet B\beta, k]\quad [k,B \to \gamma\bullet, j]}{[i,A \to \alpha B \bullet \beta, j]}$ Completion

Item-based description of an Earley's parser

In this representation, some conventional notations are used for metavariables ranging over the objects under discussion: n for the length of the object language string to be parsed; $A,B,C,\ldots$ for arbitrary formulas or symbols such as grammar nonterminals; $a,b,c,\ldots$ for arbitrary terminal symbols; $i,j,k,\ldots$ for indices into various strings, especially the string w; $\alpha,\beta,\gamma,\ldots$ for strings or terminal and nonterminal symbols.

In [15] the context-free grammars (CFGs) are used to illustrate the description format even though it can similarly be extended to other formalisms.

Interpretation Engine – Control Strategy

The specification of inference rules only partially characterizes a parsing algorithm, in that it provides for what items are to be computed, but not the order. So a further control strategy is needed to operate over the inference rules. One crucial concern is that an item should not be enumerated more than once. To prevent this, it is standard to maintain a cache of lemmas, adding to the cache only those items that have not been seen so far. This cache plays the same role as the *chart* in the chart-parsing algorithms, the *well-formed substring table* in CYK parsing, and the *state sets* in Earley's algorithm.

As a reasoning engine, the soundness and completeness of this strategy should be proved when in use.

In general, item should be added to the chart as they are proved. However, each new item may itself generate new consequences. The issue as to when to compute the consequences of a new item is quite subtle. So a standard solution is to keep a separate *agenda* of items that have been proved but whose consequences have not been computed. When an item is removed from the agenda and added to the chart, its consequences are computed and themselves added to the agenda for later consideration.

10.3.2 Item-Based Descriptions

Item-based descriptions in deductive parsing provides a facility to represent a variety of parsing algorithms in a unified way. An item-based description provides the flexibility and generality of the deductive parsing. Some related work are listed as follows:

Prolog Implementation of Deductive Parsing

In [15], a deductive parsing system is implemented in Prolog. In this system, a series of parsing algorithms are literally represented by *inference rules*, and a uniform *deduction engine* is given by parameterized with the inference rules so that it can be used to parse according to any of the associated algorithms.

Dyna

Dyna (*http://www.dyna.org*) is a Turing-complete programming language for specifying dynamic programs and training their weights. You write a short declarative specification in Dyna, and the Dyna optimizing compiler produces efficient C++ classes that form the core of your C++ application. For example, for CKY inside parsing algorithm

:- item(item, double, 0).
constit(X,I,J) += rewrite(X,W) $*$ word(W,I,J).
constit(X,I,K) += rewrite(X,Y,Z)$*$constit(Y,I,J)$*$ constit(Z,J,K).
goal += constit($s,0,N$) $*$ end(N).

Dyna is a declarative language like Prolog. It can express the abstract structure of an algorithm.

10.3.2.1 Summary of This Section

Deductive parsing provides a general way to represent different parsing algorithms. By taking parsing as deduction, it uses item-based description to describe the parsing algorithm directly and a single control strategy to implement the corresponding parser. The modularity of deductive parsing is especially useful for rapid prototyping of an experimentation with new parsing algorithms. It can be easily extended to develop algorithms for parsing with various grammars.

10.4 Semiring Parsing

From the discussion before, we know that a variety of parsing algorithms can be represented as a deductive procedure based on the generalized item-based descriptions. As we have seen, parsing algorithms can be used to compute many interesting quantities in practice, such as *Viterbi value (the value of the best parse tree), Viterbi derivation (the parse tree with Viterbi value), n-best derivations, derivation forests, derivation count, inside and outside probabilities* of components of the string. Since deductive parsing has already provided a unified way to represent different parsing algorithms, and parsing has solid algebraic foundation in the *formal power series* framework, all these provide a probability of unifying statistical NLP into a general framework by taking advantage of the algebraic properties of parsing algorithms. A semiring parsing framework is proposed and studied in [6].

10.4.1 Principles of Semiring Parsing

As aforementioned, parsing can be brought into an algebraic power series framework. It generalizes the notion of recognition, parsing, statistical parsing, etc. by describing the computations executed from an input to a domain with a semiring structure. An abstract semiring plays a central role in a power series framework on related domain of grammars and parses. In algebraic power series framework, each production of G is associated with an element of a semiring A, and parsing amounts to finding the element of A associated with x (the *coefficient* of x) with respect to the grammar. When one deals with the boolean semiring, the power series represents recognition; when one uses the semiring of parse forests, the power series represents parsing.

Semiring parsing system is essentially an implementation of an algebraic power series framework of parsing. It associates the grammar with specific semiring in terms of particular task. In addition, it simplifies the representations of a variety of parsers with a unified *item-based description*; in particular, such unification makes it easier to generalize parsers across tasks: a single item-based description can be used to compute values for a variety of applications, simply by changing semirings.

In the following part, we will use examples to make the discussion more transparent.

10.4.2 Examples: Inside Semiring and Viterbi Semiring

A semiring parser requires a grammar, a string to be parsed and an item-based description of the parsing algorithm as input. An example is given to illustrate how semiring parsing works. For simplicity, we will use CKY parsing algorithm working on Inside semiring and Viterbi semiring.

The Viterbi semiring computes the probability of the most probable derivation of the input sentence, given a probabilistic grammar. It is formalized as: $(R_0^1, \max, \times, 0, 1)$. R_0^1 indicates the set of real numbers from 0 to 1 inclusive, which denotes the domain of grammar rules and parses values. And $\oplus$ in abstract semiring is specialized into *max*, $\otimes$ into $\times$, identity $\bar{0}$ into 0 and identity $\bar{1}$ into 1 with general semantics. Since we handle *probabilities* in this task, it is easy to understand the definition of the semiring. Similarly, inside semiring is $(R_0^1, +, \times, 0, 1)$.

The input string is *xxx* and the grammar is as follows:

$$\begin{array}{ll} S \rightarrow XX & 1.0 \\ X \rightarrow XX & 0.2 \\ X \rightarrow x & 0.8 \end{array}$$

Now use item-based description of a CKY parser described in the previous section. Starting from axioms, which are $[0,x,1],[1,x,2],[2,x,3]$ respectively, prediction inference rule is firstly triggered and instantiated as

$$\frac{R(X \rightarrow x)}{[0,X,1]}[0,x,1], \quad \frac{R(X \rightarrow x)}{[1,X,2]}[1,x,2], \quad \frac{R(X \rightarrow x)}{[2,X,3]}[2,x,3].$$

Axioms are obviously true, so we compute the following items values:

$$\begin{aligned} [0,X,1] &= 0.8 \\ [1,X,2] &= 0.8 \\ [2,X,3] &= 0.8 \end{aligned}$$

Next, completion inference rule can be used. Consider the instantiation $i = 0, k = 1, j = 2, A = X, B = X, C = X$:

$$\frac{R(X \rightarrow XX) \quad [0,X,1] \quad [1,X,2]}{[0,X,2]}$$

We use the multiplicative operator of the Viterbi semiring to multiply together the values of the top line, deducing that $[0,X,3] = 0.2 \times 0.8 \times 0.8 = 0.128$. Similarly,

$$\begin{aligned} [0,X,2] &= 0.128 \\ [1,X,3] &= 0.128 \\ [0,S,2] &= 0.64 \\ [1,S,3] &= 0.64 \end{aligned}$$

There are two more ways to instantiate the conditions of the completion rule:

$$\frac{R(S \rightarrow XX) \quad [0,X,1] \quad [1,X,3]}{[0,S,3]}$$

$$\frac{R(S \rightarrow XX) \quad [0,X,2] \quad [2,X,3]}{[0,S,3]}$$

The first has the value of $1 \times 0.8 \times 0.128 = 0.1024$, and the second has the value 0.1024. For the ambiguities, we typically use additive operator of semiring to "sum" them up. In particular, for Viterbi semiring, additive operator is *max*. So the Viterbi value for the goal item is $[0,S,3] = \max(0.1024, 0.1024) = 0.1024$. For inside semiring, the inside value for the goal item is $[0,S,3] = 0.1024 + 0.1024 = 0.2048$.

Practically, we use semiring to compute the item values. There are some important detail for semiring parser that cannot be skipped – the order of item value to be computed. Specifically, the value of a consequent item cannot be computed until the values of all its antecedents have been computed. If an order on all derivable items can be given, the values of all items can be calculated one by one in this order. As for the proof of correctness of semiring parsing and more detailed discussion, refer to [6].

10.4.3 Some Classic Semirings in Semiring Parsing

The following are some classic semirings in semiring parsing [6].

Boolean Semiring (TRUE, FALSE, $\vee, \wedge$, FALSE, TRUE) recognition

Inside Semiring $(R_0^1, +, \times, 0, 1)$ string probability

Counting Semiring $(N_0^\infty, +, \times, 0, 1)$ number of derivations

Derivation Forests Semiring $(2^E, \cup, \cdot, \emptyset, \{\langle\rangle\})$ set of derivations

Viterbi Derivation Semiring
$(R_0^1 \times 2^E, \max_{Vit}, \times_{Vit}, \langle 0, \emptyset \rangle, \langle 1, \{\langle\rangle\} \rangle)$ best derivation

Viterbi-n-best Derivation Semiring
$(\left\{ \mathrm{topn}(X) | X \in 2^{R_0^1 \times E} \right\}, \max_{Vit-n}, \times_{Vit-n}, \emptyset, \langle 1, \{\langle\rangle\} \rangle)$ best n derivations

E is the set of all derivations in some canonical form, and 2^E to indicate the set of all sets of derivations in canonical form.

10.4.4 Semiring Parsing – A Prototype Implementation

Figure 10.2 shows the communications between the components in a semiring parsing prototype system. In the following sections, we will describe each component in detail.

Item-Based Description Component

In last section, an *item-based description* has been exemplified with a CKY parser. In a working system, this component is initialized with a particular parsing algorithm. That is the user will represent the parser into this form and initialize the component in his own right, which adds more flexibility to the system. Essentially

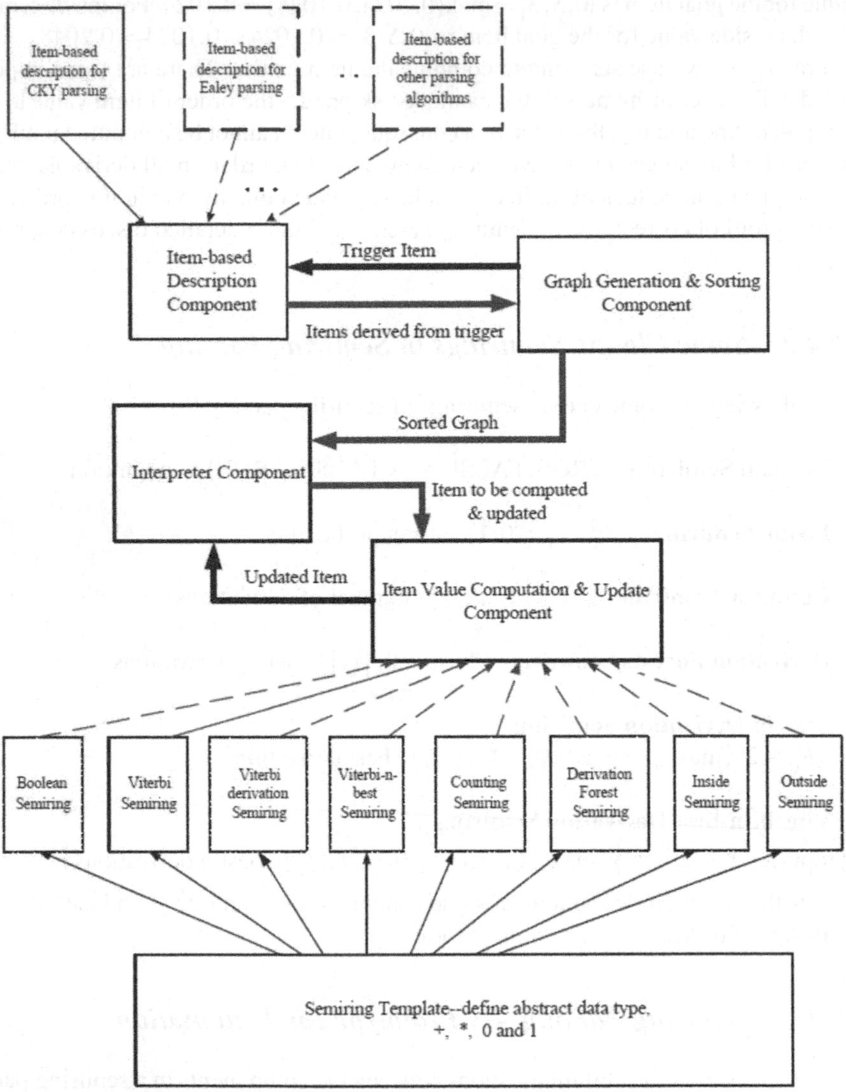

Fig. 10.2 An Implementation of a Semiring Parsing Prototype

this component takes a *deductive parsing-style* strategy and the inference rules play a central role in it.

Graph Generation and Sorting Component

As pointed out in [6], "an order needs to be imposed on the items, in such a way that no item precedes any item on which it depends". To this end, in [6], each item

x has been assigned to a *bucket* B, writing *bucket*$(x) = B$ and saying x is *associated* with B. So buckets are ordered in such a way that if item y depends on item x, then *bucket*$(x) \leq$ *bucket*(y). For those items that depend on themselves or depend on each other, they are associated with special *looping buckets*, whose value may require infinite sums to compute. More discussion sees [6].

Overview of Bucket Sorting

[6] suggests to organize the buckets with the required ordering constraints by creating a graph of the dependencies, with a node for each item, and an edge from each item x to each item b that depends on it. Thus a a topological sort can be performed on the graph.

[6] discusses the following three approaches to determine the buckets and ordering.

The first one is simply a brute-force enumeration of all items, derivable or not, followed by a topological sort. Obviously, this approach will have suboptimal time and space complexity for most item-based description.

The second approach is to use a deductive parser to determine the derivable items and their dependencies, and then to perform a topological sort. Theoretically, the time complexity of deductive parsing can achieve optimal ($O(n^3)$) and the time complexity of topological sort is bounded by $O(|V| + |E|)$, where $|V|$ is the number of items that can be bounded by $O(n^2)$ and E is the number of dependencies of the graph, which is bounded by a cubic complexity [1]. So the entire time complexity can achieve optimal; Unfortunately, for memory, there are several $O(E)$ data structures holding edges. Therefore, the space complexity of this approach will not achieve $O(n^2)$, which denotes optimal space complexity.

The third approach to bucketing is to create algorithm-specific bucketing code; this results in a parser with both optimal time and optimal space complexity. For instance, in a CKY-style parser, we can simply create one bucket for all items with same span of length. Essentially, to achieve optimal performance, this approach is trying to put the items with no any dependencies in between into the same bucket. It helps maximize sharing of structure and reduce the number of dependencies, which is a key factor in the space complexity.

Example of the Second Approach

The second approach was implemented in the prototype system. In this subsection, we will still use the example in Section 10.4.2 to illustrate this process. Figure 10.3 shows the dependency graph generated by approach 2. In this graph, each node denotes a derived item and a directed edge from node i to node j denotes the dependency relation of item i to item j. The graph is an *AND-OR graph*; *AND-nodes* correspond to the usual derivable items, which can be derived from some unique set of antecedents. For instance, all nodes in the graph but $[0,X,3]$*up and*$[0,S,3]$ are AND-nodes; while *OR-nodes* correspond to those items that can be derived from multiple sets of antecedents, i.e. such items can be deducted from different paths

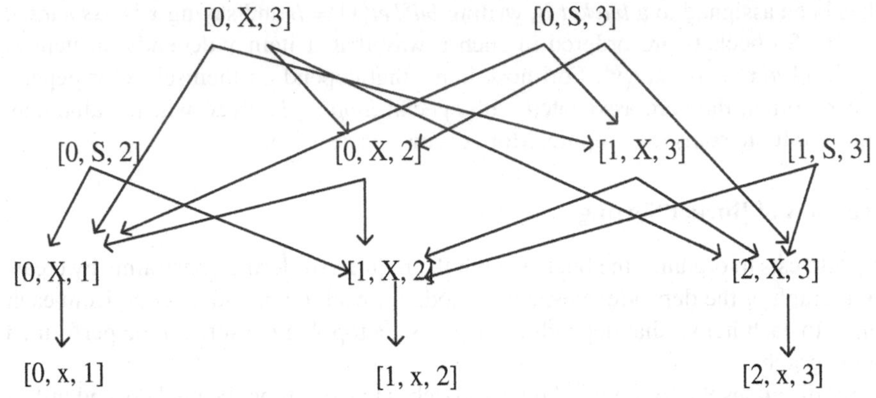

Fig. 10.3 dependency graph by CKY parsing

that represent ambiguities. The nodes $[0,X,3] up\ and [0,S,3]$ in the graph are the examples of OR-nodes, which have two different derivation paths respectively.

Note that in this graph every item is derivable and associated with one bucket implicitly. A *chart-based, agenda-driven* strategy is used to control the item derivation by iteratively invoking the item-based description component.

Agenda is essentially a cache for items that are newly produced and remain to be processed. During each iteration, one item is pulled off from the agenda and added to the chart (on condition that no its duplicate in the chart). Take this item as a trigger, more items may be created by invoking the inference rules in the item-based description component and added into agenda. Repeat such iterations until agenda gets exhausted and all derivable items will be stored in the chart.

During this process, every time a new item is derived, it is inserted into the graph and its dependency relation is recorded as well. In our implementation, we sort the graph during the graph construction. Every time an OR-node is created, a partial sort is taken to keep the graph in a sorted state.

Try this example on the first approach, all possible items, such as $[0,S,1]$ would be organized into the graph, whether the item is derivable or not. For the third approach, those items with the same length such as $[0,X,1]$, $[1,X,2]$ and $[2,X,3]$ will be put into the same bucket and presents one node in the graph.

Dependency Graph and Shared Forests Structure

In [1], the structure of shared forests in ambiguous parsing is discussed in detail. The structure of shared forests is precisely an AND-OR shared forest graph. One experimenting result in that paper has shown that *sophistication* may have a negative effect on the efficiency of all-path parsing. This result is verified in the prototype system.

It is also mentioned that one essential guideline to achieve better sharing (and often also reduced computation time) is to try to *recognize every grammar rule in only one place of the generated chart parser code, even at the cost of increasing non-determinism.* This guideline has the same idea with the third approach. In the prototype system, the number of items and the dependencies between items are the key factors in improving the time and space efficiency. In other words, in order to improve the efficiency, we have to reduce the number of *buckets*, which equals to increasing the sharing of the structure in the graph on some level. However, this method will sacrifice the generality of the system and take more programming efforts.

Interpreter Component

The interpreter component executes the process by receiving a sorted item list from graph generation and sorting component and repeatedly invokes the corresponding item value computation and update component, which is essentially a specialized semiring, to compute and update the item values in the specified order.

Item Value Computation and Update Component

The definition of *item value* depends on a specific semiring. This component specializes the semirings. For example, in the Viterbi semiring, the item value is the largest probability of deriving the item; in Viterbi derivation semiring, item value is a pair of Viterbi value and a derivation list that can derive this item. So in this component, multiple semirings will be defined and a particular semiring will be invoked according to the particular task. In the implementation of the prototype system, a template semiring is defined first, including the definition of abstract multiplicative identity, additive identity, multiplication and addition operation. Every specific semiring is defined based on the template. Specifically, data type, multiplication, addition, multiplicative identity and additive identity are defined in each specific semiring. So the value computation and update gets easier when all these elements are specialized.

10.5 Summary

In this chapter, we first reviewed some basics parsing algorithms for context-free languages and probabilistic context-free grammars. The use of algebraic properties helps us view the statistical parsing in a more general and unified perspective. In the framework of power series, parsing can be taken to be the computation of the coefficients for the elements of a free monoid. Semiring parsing is based on the general view and by assigning different semantics to the operations in parsing to achieve the unified representation of parsing across diverse tasks. Semiring parsing turns out to have a solid theoretical foundation and can provide the basis for a fast and extensible toolkit for many applications.

In practice, we have to consider how to efficiently organize the exponentially many parses and how to effectively introduce fast search strategies in deduction without loss of system generality. Some research endeavour has been made along this line [5, 14, 13].

A more in-depth introduction on the theoretical foundation and detailed implementation of a semring parsing system can be found in the referenced literatures and [6].

References

1. Billot, S., Lang, B.: The structure of shared forests in ambiguous parsing. In: Proceedings of ACL 1989 (1989)
2. Chomsky, N., Schutzenberger, M.P.: The algebraic theory of context-free languages. In: Braffort, P., Hirschberg, D. (eds.) Computer Programming and Formal Systems, pp. 118–161. North-Holland, Amsterdam (1963)
3. Church, K., Patil, R.: Coping with syntactic ambiguity or how to put the block in the box on the table. America Journal of Computational Linguistics 8(3-4) (1982)
4. Earley, J.: An efficient context-free parsing algorithm. Communications of the ACM 13(2), 94–102 (1970)
5. Eisner, J.: Expectation semirings: Flexible EM for finite-state transducers. In: van Noord, G. (ed.) Proceedings of the ESSLLI Workshop on Finite-State Methods in Natural Language Processing (2001)
6. Goodman, J.: Semiring parsing. Computational Linguistics 25(4), 573–605 (1999)
7. Joshi, A., Schabes, Y.: Tree-adjoining grammars. Handbook of Formal Languages, 3 (1997)
8. Knuth, D.E.: The Art of Computer Programming, Volume I: Fundamental Algorithms, 2nd edn. Addison-Wesley (1973)
9. Kuich, W., Salomaa, A.: Semirings, Automata, Languages. Springer (1985)
10. Lari, K., Young, S.J.: The estimation of stochastic context-free grammars using the insideoutside algorithm. Computer Speech and Language 4, 35–56 (1990)
11. Graham, S.L., Harrison, M.A., Ruzzo, W.L.: An improved context-free recognizer. ACM Transactions on Programming Languages and Systems 2(3), 415–462 (1980)
12. Manning, C.D., Schütze, H.: Foundations of Statistical Natural Language Processing. MIT Press, Cambridge (1999)
13. McAllester, D., Collins, M., Pereira, F.: Case-factor diagrams for structured probabilistic modeling. In: UAI 2004 (2004)
14. Miyao, Y., Tsujii, J.: Maximum entropy estimation for feature forests. In: Proceedings of HLT 2002 (2002)
15. Pereira, F.C.N., Shieber, S.M., Schabes, Y.: Principles and implementations of deductive parsing. Journal of Logic Programming 24(1-2), 3–36 (1995)

Chapter 11
Coverings and Decompositions of Semiring-Weighted Finite Transition Systems*,**

Manfred Droste[1], Ingmar Meinecke[1], Branimir Šešelja[2], and Andreja Tepavčević[2]

[1] Institut für Informatik, Universität Leipzig, D-04109 Leipzig, Germany
{droste,meinecke}@informatik.uni-leipzig.de

[2] Department of Mathematics and Informatics, University of Novi Sad,
Trg Dositeja Obradovica 4, 21000 Novi Sad, Serbia
{seselja,andreja}@dmi.uns.ac.rs

Abstract. We consider weighted finite transition systems (WTS) with weights from naturally ordered semirings. Such semirings comprise the natural numbers with ordinary addition and multiplication as well as distributive lattices and the max-plus-semiring. For these systems we explore the concepts of covering and cascade product. We show a cascade decomposition result for such WTS using special partitions of the state set of the system. This extends a classical result of automata theory to the weighted setting.

11.1 Introduction

Synthesis and analysis of automata have been a central topic of computer science since its beginning. By Kleene's famous result [16], combinations of very simple automata have the full power of all automata. On a related strand, cascade and wreath products of transition systems and semigroups were investigated, leading to the fundamental theorem of Krohn-Rhodes [18]. For the several variations of the Krohn-Rhodes decomposition theorem see the books by Ginzburg [13], Eilenberg [11], Holcombe [14], and Straubing [28]. More recent books on the topic are those by Dömösi and Nehaniv [7], by Rhodes and Steinberg [26], and a new edition of a book by Rhodes [25]. The proof of this deep result is still the interest of on-going research, see for instance [12]. However, there are also other decomposition results than the one by Krohn and Rhodes. One possibility is to obtain simpler systems by an admissible partition of the state space of the machine [13, 14]. Decompositions using admissible partitions were applied e.g. in [17] in order to decompose a controller (obtained from a model of a system and a specification of a property) into

* This research was partially supported by the DAAD-Serbia project "Weighted Automata over Semirings and Lattices" and the DFG-project "Gewichtete Automaten und gewichtete Logiken für diskrete Strukturen", DR 202 / 11-1.

** The research of the last two authors is supported by Serbian Ministry of Science and Tech. Develop., Grant No. 174013.

J. Ahsan et al.: Fuzzy Semirings with Applications, STUDFUZZ 278, pp. 193–216.
springerlink.com

smaller distributed controllers; see also Holcombe [14] for various further applications of cascade products. In our work, we have investigated coverings by cascade products for the class of weighted automata.

In weighted automata, the transitions may carry weights modeling, e.g., the resources used during the execution of the transition. Already Schützenberger [27] extended Kleene's fundamental result to this model of quantitative automata whose theory quickly developed, cf. [10] and recently [8]. Here, the weights are taken from a semiring; multiplication is used to define the weights of paths and addition gives the total weight of all paths realizing a given word. However, decompositions of semiring weighted automata are missing. Only much more specialized settings have been considered like fuzzy finite state machines [5, 15, 21, 22, 23] where the weight structure is the interval $([0,1],\max,\min)$. Our concepts of covering and cascade product were somehow inspired by this work. Also a decomposition of probabilistic transition systems was given [20, 24], but there decomposition means only that the probabilistic part was separated from the underlying boolean transitional structure.

In our work, assuming that the semiring carries a partial order compatible with the semiring operations, we introduce a notion of *covering* for weighted finite transition systems (WTS); if a WTS M' covers M, then a subset of the states of M' can be mapped onto the states of M such that the weights of the transitions of M' bound the weight of the corresponding transition of M. This extends the corresponding notion for classical (unweighted) transition systems, and it is related to (but different from) other notions of coverings from the literature, cf., e.g. [1, 2].

We also extend the classical notion of cascade products to weighted transition systems, and we derive that cascade products preserve the covering relation. Next, we turn to admissible partitions for WTS; they naturally lead to a quotient WTS covered by the original system. These admissible partitions appeared in the literature, also for weighted automata, under several names: blockstochastic matrices [19], simulation [3], bisimulation [4], and out-licit equivalences, quotient, covering [1, 2].

For our main results, we assume that the naturally ordered semiring S satisfies the condition that $s \le s^2$ for all $s \in S$. This assumption is satisfied by many important semirings including the semiring of natural numbers, the max-plus-semiring, all distributive lattices, and non-commutative semirings of formal languages, but not by e.g. the min-plus-semiring. Examples show that our main results do not hold without this condition on S. Assuming the condition, we show that we can cover any given WTS by a sequence of cascade products of smaller irreducible WTS. These smaller systems result from the original one by using admissible partitions. We can guarantee that the weights of the simpler systems have an upper bound depending on the weights and the size of the original system. This resembles somehow the idea of Colcombet [6] to relate two cost functions by a correction function. Furthermore, in case there are two admissible equivalence relations which are orthogonal, the original WTS can be covered by a direct product of the two corresponding quotient WTS. We obtain the classical results for unweighted transition systems [13, 14] as an immediate consequence by choosing S as the Boolean semiring $(\{0,1\},\vee,\wedge)$.

In Mordeson-Malik [23], such a cascade decomposition result was claimed for the fuzzy semiring $([0,1],\max,\min)$ and a stronger notion of covering with equality of weights (instead of bounds by weights). We give a counter-example showing that the proposed proof method does not work for such *strong* coverings. We also show that our condition that $s \leq s^2$ for the elements of the semiring is necessary for the approach used here. In case of the Boolean semiring $(\{0,1\},\vee,\wedge)$, our result translates precisely to the classical one for unweighted finite transition systems.

To the best of our knowledge, this is the first paper dealing with cascade products of general semiring weighted transition systems. It remains open whether suitable wreath products could also be defined leading (ideally) to a weighted Krohn-Rhodes theory, and whether the applications of cascade products for e.g. neural networks (cf. Holcombe [14]) could also be developed for a quantitative setting.

An extended abstract of this chapter appeared in [9].

11.2 Semirings and Weighted Finite Transition Systems

We recall the classical notion of transition systems. A structure $M = (Q,A,T)$ where Q (the set of states) and A (the input alphabet) are finite nonempty sets and $T \subseteq Q \times A \times Q$ is called *finite transition system* (*TS* for short). Next, transition systems will be enriched with weights from a semiring.

A *semiring* $(S,+,\cdot,0,1)$ consists of a set S together with two binary operations $+$ (called addition) and $\cdot$ (called multiplication) on S such that

- $(S,+,0)$ is a commutative monoid,
- $(S,\cdot,1)$ is a monoid,
- $\cdot$ distributes over $+$, i.e., for all $x,y,z \in S$ we have $x \cdot (y+z) = x \cdot y + x \cdot z$ and $(y+z) \cdot x = y \cdot x + z \cdot x$, and
- $s \cdot 0 = 0 \cdot s = 0$ for all $s \in S$.

A semiring S is *commutative* whenever the multiplication is commutative, i.e., $s \cdot s' = s' \cdot s$ for all $s,s' \in S$. A semiring S is *naturally ordered* if the relation $\leq$ defined by

$$s_1 \leq s_2 :\Longleftrightarrow \exists s \in S:\ s_1 + s = s_2$$

is a partial order. Then $\leq$ is compatible with multiplication, i.e., for all $s,s_1,s_2 \in S$:

$$s_1 \leq s_2 \Longrightarrow s_1 \cdot s \leq s_2 \cdot s \text{ and } s \cdot s_1 \leq s \cdot s_2 .$$

Note that in a naturally ordered semiring $0 \leq s$ for all $s \in S$.

Distributive lattices provide examples of commutative naturally ordered semirings. One instance is the semiring $F = ([0,1],\vee,\wedge,0,1)$ of the interval $[0,1]$ with supremum $\vee$ as sum and infimum $\wedge$ as product. Other examples which are not distributive lattices are given by the semiring of natural numbers $(\mathbb{N},+,\cdot,0,1)$, semirings of formal languages together with union and concatenation, and the max-plus-semiring $(\mathbb{N} \cup \{-\infty\},\max,+,-\infty,0)$.

Definition 11.1. A *weighted finite transition system (WTS) M* over the semiring S is a triple $M = (Q,A,\mu)$, where Q (the *set of states*) and A (the *set of input symbols*) are nonempty finite sets and $\mu : Q \times A \times Q \to S$ is a mapping.

Note that we defined μ as a total function. As usual, those triples $(q,a,p) \in Q \times A \times Q$ for which $\mu(q,a,p) \neq 0$ are called transitions.

Remark 11.1. To fix our notation, we recall the basic connection between unweighted TS and WTS over the Boolean semiring $S = \mathbb{B} = (\{0,1\}, \vee, \wedge, 0, 1)$. Let $M = (Q,A,\mu)$ be a WTS over $\mathbb{B}$. Then we can associate to M an unweighted TS $ts(M) = (Q,A,T)$ with $(q,a,p) \in T$ if and only if $\mu(q,a,p) = 1$. Vice versa, if $M = (Q,A,T)$ is a TS, then we associate to M the WTS $wts(M) = (Q,A,\mu)$ over $\mathbb{B}$ where $\mu(q,a,p) = 1$ if $(q,a,p) \in T$ and $\mu(q,a,p) = 0$ otherwise. Note that for a WTS M over $\mathbb{B}$ the following holds: $wts(ts(M)) = M$. And for a TS M we have $ts(wts(M)) = M$.

Let A^* be the set of all finite words over A (including the empty word ε). Together with concatenation, A^* is a monoid. For $w \in A^*$, $|w|$ denotes the length of w.

Given a function $\xi : A_1 \to A_2$ between two alphabets, let $\xi^* : A_1^* \to A_2^*$ be its homomorphic extension. $S^{Q \times Q}$ is the monoid of $Q \times Q$-matrices over S together with matrix multiplication. Now $\mu : Q \times A \times Q \to S$ can be understood as a function $\mu : A \to S^{Q \times Q}$. Then $\mu^* : A^* \to S^{Q \times Q}$ is the homomorphic extension of μ. If we write μ^* as a function from $Q \times A^* \times Q$ to S, then we have:

$$\mu^*(q,\varepsilon,p) = \begin{cases} 1 & \text{if } p = q, \\ 0 & \text{if } p \neq q, \end{cases}$$
$$\mu^*(q,wa,p) = \sum_{r \in Q} (\mu^*(q,w,r) \cdot \mu(r,a,p))$$

for all $w \in A^*$ and $a \in A$. For notational simplicity, we will denote μ^* also by μ and ξ^* by ξ.

11.3 Coverings

From now on, we consider WTS over naturally ordered semirings $(S,+,\cdot,0,1)$. For a partial function $f : A \dashrightarrow B$ let $dom(f) = \{a \in A \mid f(a) \text{ is defined}\}$, the domain of f.

Definition 11.2. Let $M_1 = (Q_1,A_1,\mu_1)$ and $M_2 = (Q_2,A_2,\mu_2)$ be WTS. Let $\eta : Q_2 \dashrightarrow Q_1$ be a surjective partial function and $\xi : A_1 \to A_2$ a function.

M_2 *covers* M_1 *via* (η,ξ) and (η,ξ) is a *covering* of M_1 by M_2, written as $M_1 \leq_{(\eta,\xi)} M_2$, if for every $q_2 \in dom(\eta)$, $q_1 \in Q_1$, and $a \in A_1$,

$$\mu_1(\eta(q_2),a,q_1) \leq \sum_{r_2 \in \eta^{-1}(q_1)} \mu_2(q_2,\xi(a),r_2). \quad (11.1)$$

We say that M_2 *covers* M_1 or M_1 *is covered by* M_2, written as $M_1 \leq M_2$, if there is a covering (η,ξ) of M_1 by M_2.

Whenever $\eta : Q_2 \to Q_1$ is total, i.e. a function from Q_2 to Q_1, then we say the covering (η, ξ) is *total*.

If (11.1) holds with equality instead of $\leq$, then the covering is called *strong*.

Remark 11.2. Suppose $M_1 = (Q_1, A_1, \mu_1)$ and $M_2 = (Q_2, A_2, \mu_2)$ are WTS over $\mathbb{B}$. Let $ts(M_1) = (Q_1, A_1, T_1)$ and $ts(M_2) = (Q_2, A_2, T_2)$ be the associated unweighted transition systems, respectively (cf. Remark 11.1). Suppose $M_1 \leq_{(\eta,\xi)} M_2$. Then for every $q_2 \in dom(\eta)$, $a \in A_1$, and $q_1 \in Q_1$ with $(\eta(q_2), a, q_1) \in T_1$ there is a transition $(q_2, \xi(a), r_2) \in T_2$ for some $r_2 \in \eta^{-1}(q_1)$. In this sense, we say that *$ts(M_1)$ is covered by $ts(M_2)$ via (η, ξ)*. By Proposition 11.1 (see below), also a path between $\eta(q_2)$ and q_1 on $w \in A_1^*$ in $ts(M_1)$ can be simulated by a path in $ts(M_2)$ between q_2 and some $r_2 \in \eta^{-1}(q_1)$ on $\xi(w) \in A_2^*$. Altogether we have: $M_1 \leq M_2$ iff $ts(M_1) \leq ts(M_2)$.

Whenever the systems are *deterministic*, then the state r_2 is uniquely determined. This coincides with the classical concept of covering for deterministic transition systems, cf. [14, p. 43].

Remark 11.3. The notion of covering used by Mordeson and Malik in [22, Def. 3.1] and [23, Def. 6.6.1] for fuzzy finite state machines is precisely the notion of strong and total covering of WTS over the semiring $F = ([0,1], \vee, \wedge, 0, 1)$ in our sense. As we will see later on, strong coverings are too restrictive to obtain good decomposition results.

In [15, Def. 4.1], a covering is defined for T-generalized state machines over the semiring $F = ([0,1], \vee, \wedge, 0, 1)$ by an inequality instead of an equality. But different from our definition, the inequality has to be satisfied for every single state $r_2 \in \eta^{-1}(q_1)$ (cf. (11.1) in Definition 11.2) and not for the sum over all $r_2 \in \eta^{-1}(q_1)$. This is a much stricter condition than the one we defined.

In [1, 2], Béal, Lombardy, and Sakarovitch use a notion of covering for weighted automata which is closer to what we call admissible partitions (see Def. 11.4). Moreover, they require the equivalence of the original and the covering automaton, i.e., they have the same behavior. Here, we have a weaker condition due to the inequality in (11.1).

The defining inequality of coverings transfers to words:

Proposition 11.1. *Let $M_1 \leq_{(\eta,\xi)} M_2$. Then we have for every $q_2 \in dom(\eta) \subseteq Q_2$, $q_1 \in Q_1$, and $w \in A_1^*$ that*

$$\mu_1(\eta(q_2), w, q_1) \leq \sum_{r_2 \in \eta^{-1}(q_1)} \mu_2(q_2, \xi(w), r_2). \tag{11.2}$$

If (η, ξ) is total and strong, then (11.2) holds with equality replacing $\leq$.

Proof. Let $w = \varepsilon$. Then

$$\mu_1(\eta(q_2), \varepsilon, q_1) = \begin{cases} 1 & \text{if } \eta(q_2) = q_1, \\ 0 & \text{otherwise} \end{cases} = \sum_{r_2 \in \eta^{-1}(q_1)} \mu_2(q_2, \varepsilon, r_2).$$

For $w = a \in A_1$ the claim holds true due to Definition 11.2. Now consider $w' = wa \in A_1^*$ for $w \in A_1^*$ and $a \in A_1$. Then we get

$$\begin{aligned}
&\mu_1(\eta(q_2), wa, q_1)\\
&= \sum_{p_1 \in Q_1} \mu_1(\eta(q_2), w, p_1) \cdot \mu_1(p_1, a, q_1)
\end{aligned}$$

and by induction hypothesis for w

$$\begin{aligned}
&\leq \sum_{p_1 \in Q_1} \left(\sum_{s_2 \in \eta^{-1}(p_1)} \mu_2(q_2, \xi(w), s_2) \right) \cdot \mu_1(p_1, a, q_1))\\
&= \sum_{p_1 \in Q_1} \left(\sum_{s_2 \in \eta^{-1}(p_1)} \mu_2(q_2, \xi(w), s_2) \cdot \mu_1(p_1, a, q_1) \right)
\end{aligned}$$

and by applying Definition 11.2 for $a \in A_1$ and every $s_2 \in \eta^{-1}(p_1)$

$$\leq \sum_{p_1 \in Q_1} \left(\sum_{s_2 \in \eta^{-1}(p_1)} \mu_2(q_2, \xi(w), s_2) \cdot \left[\sum_{r_2 \in \eta^{-1}(q_1)} \mu_2(s_2, \xi(a), r_2) \right] \right)$$

and now by distributivity

$$= \sum_{r_2 \in \eta^{-1}(q_1)} \sum_{p_1 \in Q_1} \left(\sum_{s_2 \in \eta^{-1}(p_1)} \mu_2(q_2, \xi(w), s_2) \cdot \mu_2(s_2, \xi(a), r_2) \right)$$

and finally by using that $s_2 \in dom(\eta) \subseteq Q_2$

$$\begin{aligned}
&\leq \sum_{r_2 \in \eta^{-1}(q_1)} \sum_{s_2 \in Q_2} \mu_2(q_2, \xi(w), s_2) \cdot \mu_2(s_2, \xi(a), r_2)\\
&= \sum_{r_2 \in \eta^{-1}(q_1)} \mu_2(q_2, \xi(wa), r_2)
\end{aligned}$$

which shows the claim for $w' = wa$.

Let (η, ξ) be strong and total. Due to the strongness of the covering, the induction hypothesis above is applied with equality instead of $\leq$. Since (η, ξ) is total, the last inequality turns also into an equality. This shows the second claim.

Note that for every WTS M, $M \leq M$ by the identity mappings (id_Q, id_A). Moreover, the covering relation is transitive as shown by the next proposition.

Proposition 11.2. *Let M_1, M_2, and M_3 be WTS over S. If $M_1 \leq_{(\eta_1,\xi_1)} M_2$ and $M_2 \leq_{(\eta_2,\xi_2)} M_3$, then $M_1 \leq_{(\eta_3,\xi_3)} M_3$ with $\eta_3 = \eta_1 \circ \eta_2$ and $\xi_3 = \xi_2 \circ \xi_1$.*[1]

If (η_1,ξ_1) and (η_2,ξ_2) are strong, then (η_3,ξ_3) is also strong. Moreover, if the first two coverings are total, then (η_3,ξ_3) is also total.

Proof. Let $M_i = (Q_i, A_i, \mu_i)$, for $i \in \{1,2,3\}$. Let $\eta_1 : Q_2 \dashrightarrow Q_1$ and $\eta_2 : Q_3 \dashrightarrow Q_2$ be surjective partial functions and let $\xi_1 : A_1 \to A_2$ and $\xi_2 : A_2 \to A_3$ be functions such that $M_1 \leq_{(\eta_1,\xi_1)} M_2$ and $M_2 \leq_{(\eta_2,\xi_2)} M_3$.

Now, we consider $\eta_3 = \eta_1 \circ \eta_2 : Q_3 \dashrightarrow Q_1$ and $\xi_3 = \xi_2 \circ \xi_1 : A_1 \to A_3$. Obviously, η_3 is surjective because η_1 and η_2 are surjective. Moreover, if η_1 and η_2 are total, then also η_3 is total. We prove that $M_1 \leq_{(\eta_3,\xi_3)} M_3$. Indeed, we have for all $q_3 \in dom(\eta_3)$, $p_1 \in Q_1$, and $a \in A_1$

$$\begin{aligned}
\mu_1(\eta_3(q_3), a, p_1) &= \mu_1(\eta_1(\eta_2(q_3)), a, p_1) \\
&\leq \sum_{r \in \eta_1^{-1}(p_1)} \mu_2(\eta_2(q_3), \xi_1(a), r) \\
&\leq \sum_{r \in \eta_1^{-1}(p_1)} \sum_{t \in \eta_2^{-1}(r)} \mu_3(q_3, \xi_2(\xi_1(a)), t) \\
&= \sum_{t \in \eta_2^{-1}(\eta_1^{-1}(p_1))} \mu_3(q_3, \xi_2(\xi_1(a)), t) \\
&= \sum_{t \in \eta_3^{-1}(p_1)} \mu_3(q_3, \xi_3(a), t)
\end{aligned}$$

which shows the assertion. Note that for strong coverings (η_1,ξ_1) and (η_2,ξ_2) the two inequalities above are replaced by equality. Hence, in this case also (η_3,ξ_3) is a strong covering.

11.4 Cascade Products

Now we introduce the central construction for decomposition: the cascade product. This concept is well-known for unweighted TS and was established also for WTS over the semiring $F = ([0,1], \vee, \wedge, 0, 1)$, cf. [5, 22, 23].

Definition 11.3. Let $M_1 = (Q_1, A_1, \mu_1)$ and $M_2 = (Q_2, A_2, \mu_2)$ be WTS over the semiring S. Let $\omega : Q_2 \times A_2 \to A_1$ be a mapping. The *cascade product* of M_1 and M_2 via ω is the WTS $M_1 \omega M_2 = (Q, A_2, \mu_\omega)$ where $Q = Q_1 \times Q_2$ and $\mu_\omega : Q \times A_2 \times Q \to S$ is the mapping defined by

$$\mu_\omega((q_1,q_2), b, (p_1,p_2)) = \mu_1(q_1, \omega(q_2,b), p_1) \cdot \mu_2(q_2, b, p_2)$$

for all $q_1, p_1 \in Q_1$, $q_2, p_2 \in Q_2$, and $b \in A_2$.

[1] Note that the composition of $f : A \to B$ and $g : B \to C$ is defined by $g \circ f : A \to C$ (first apply f and then g).

Note that the equation in the definition above cannot be generalized from letters $b \in A_2$ to words $w \in A_2^*$. Indeed, let $b, b' \in A_2$. Then we have for a commutative semiring

$$\begin{aligned}
&\mu_\omega((q_1,q_2),bb',(p_1,p_2)) \\
&= \sum_{(r_1,r_2)\in Q_1\times Q_2} \mu_\omega((q_1,q_2),b,(r_1,r_2))\cdot\mu_\omega((r_1,r_2),b',(p_1,p_2)) \\
&= \sum_{(r_1,r_2)\in Q_1\times Q_2} \mu_1(q_1,\omega(q_2,b),r_1)\cdot\mu_2(q_2,b,r_2)\cdot\mu_1(r_1,\omega(r_2,b'),p_1)\cdot\mu_2(r_2,b',p_2) \\
&= \sum_{(r_1,r_2)\in Q_1\times Q_2} \mu_1(q_1,\omega(q_2,b),r_1)\cdot\mu_1(r_1,\omega(r_2,b'),p_1)\cdot\mu_2(q_2,b,r_2)\cdot\mu_2(r_2,b',p_2)
\end{aligned}$$

The problem is now that the letter $\omega(r_2,b') \in A_1$ depends on the intermediate state $r_2 \in Q_2$. Since this state varies, we cannot dissolve the sum above into the product of two sums each only in M_1 and M_2, respectively.

Remark 11.4. Let $M_1 = (Q_1,A_1,\mu_1)$ and $M_2 = (Q_2,A_2,\mu_2)$ be two WTS over the Boolean semiring $\mathbb{B}$. Let $ts(M_1) = (Q_1,A_1,T_1)$ and $ts(M_2) = (Q_2,A_2,T_2)$ be their associated unweighted transition systems as defined in Remark 11.1. Let $\omega : Q_2 \times A_2 \to A_1$. Now the *cascade product* $ts(M_1)\,\omega\,ts(M_2)$ of $ts(M_1)$ and $ts(M_2)$ is the transition system $ts(M_1)\,\omega\,ts(M_2) = (Q_1 \times Q_2, A_2, T_\omega)$ with

$$((q_1,q_2),b,(p_1,p_2)) \in T_\omega \iff (q_1,\omega(q_2,b),p_1) \in T_1 \text{ and } (q_2,b,p_2) \in T_2$$

which is the classical notion of a cascade product of two transition systems, cf. [14, p. 52]. Then we can show easily that $ts(M_1\,\omega M_2) = ts(M_1)\,\omega\,ts(M_2)$, i.e., it does not matter if we build first the unweighted TS and take their cascade product or if we take the cascade product of the WTS and associate afterwards the respective TS.

We can also associate for a given unweighted TS M the WTS $wts(M)$ over $\mathbb{B}$. Then for two TS M and N we have $wts(M\omega N) = wts(M)\,\omega\,wts(N)$.

Next we show that the cascade product construction is compatible with the covering relation.

Proposition 11.3. *Let $M_i = (Q_i,A_i,\mu_i)$ for $i \in \{1,2,3\}$ be WTS over the semiring S. If $M_1 \le M_2$ via (η,ξ), then we have:*

(i) *For every cascade product $M_1\omega_{13}M_3$ there is a cascade product $M_2\omega_{23}M_3$ such that $M_1\omega_{13}M_3 \le M_2\omega_{23}M_3$.*
For every cascade product $M_2\omega_{23}M_3$ with $\omega_{23}(Q_3 \times A_3) \subseteq \xi(A_1)$ there is a cascade product $M_1\omega_{13}M_3$ such that $M_1\omega_{13}M_3 \le M_2\omega_{23}M_3$.

(ii) *For every cascade product $M_3\omega_{31}M_1$ with $\omega_{31}(q_1,a) = \omega_{31}(q_1,a')$ if $\xi(a) = \xi(a')$ there is a cascade product $M_3\omega_{32}M_2$ such that $M_3\omega_{31}M_1 \le M_3\omega_{32}M_2$.*
For every cascade product $M_3\omega_{32}M_2$ with $\omega_{32}(q_2,b) = \omega_{32}(q_2',b)$ if $\eta(q_2) = \eta(q_2')$ there is a cascade product $M_3\omega_{31}M_1$ such that $M_3\omega_{31}M_1 \le M_3\omega_{32}M_2$.

If the covering (η,ξ) of M_1 by M_2 is strong (total), then all the coverings above are also strong (total, respectively).

Proof. Let $M_i = (Q_i, A_i, \mu_i)$ for $i \in \{1,2,3\}$. From $M_1 \leq M_2$ it follows that there is a surjective partial function $\eta : Q_2 \dashrightarrow Q_1$ and a function $\xi : A_1 \to A_2$ such that for all $q_2 \in dom(\eta)$, $q_1 \in Q_1$, and $a \in A_1$

$$\mu_1(\eta(q_2), a, q_1) \leq \sum_{r_2 \in \eta^{-1}(q_1)} \mu_2(q_2, \xi(a), r_2).$$

If (η, ξ) is strong, then $\leq$ is replaced by equality.

(i) Let $M_1 \omega_{13} M_3 = (Q_1 \times Q_3, A_3, \mu_{\omega_{13}})$ be an arbitrary cascade product of M_1 and M_3 where $\omega_{13} : Q_3 \times A_3 \to A_1$ and $\mu_{\omega_{13}} : (Q_1 \times Q_3) \times A_3 \times (Q_1 \times Q_3) \to S$ is defined by

$$\mu_{\omega_{13}}((q_1, q_3), c, (p_1, p_3)) = \mu_1(q_1, \omega_{13}(q_3, c), p_1) \cdot \mu_3(q_3, c, p_3).$$

Now we define the cascade product $M_2 \omega_{23} M_3 = (Q_2 \times Q_3, A_3, \mu_{\omega_{23}})$ where $\omega_{23} = \xi \circ \omega_{13} : Q_3 \times A_3 \to A_2$. Then $\mu_{\omega_{23}} : (Q_2 \times Q_3) \times A_3 \times (Q_2 \times Q_3) \to S$ is defined by

$$\mu_{\omega_{23}}((q_2, q_3), c, (p_2, p_3)) = \mu_2(q_2, \omega_{23}(q_3, c), p_2) \cdot \mu_3(q_3, c, p_3).$$

We define $\eta' : Q_2 \times Q_3 \dashrightarrow Q_1 \times Q_3$ by $\eta'(q_2, q_3) = (\eta(q_2), q_3)$. Then η' is an onto partial function. Note that η' is total whenever η is total. Moreover, let $\xi_1 = id_{A_3} : A_3 \to A_3$ be the identity function.

We prove that $M_1 \omega_{13} M_3 \leq_{(\eta', \xi_1)} M_2 \omega_{23} M_3$, i.e.,

$$\mu_{\omega_{13}}(\eta'(q_2, q_3), c, (p_1, p_3)) \leq \sum_{(r_2, r_3) \in \eta'^{-1}(p_1, p_3)} \mu_{\omega_{23}}((q_2, q_3), c, (r_2, r_3))$$

for every $(q_2, q_3) \in dom(\eta')$, $(p_1, p_3) \in Q_1 \times Q_3$, and $c \in A_3$. Indeed, we have

$$\begin{aligned}
\mu_{\omega_{13}}(\eta'((q_2, q_3)), c, (p_1, p_3)) &= \mu_{\omega_{13}}((\eta(q_2), q_3), c, (p_1, p_3)) \\
&= \mu_1(\eta(q_2), \omega_{13}(q_3, c), p_1) \cdot \mu_3(q_3, c, p_3) \\
&\leq \sum_{r_2 \in \eta^{-1}(p_1)} \mu_2(q_2, \xi(\omega_{13}(q_3, c)), r_2) \cdot \mu_3(q_3, c, p_3) \\
&= \sum_{r_2 \in \eta^{-1}(p_1)} \mu_2(q_2, \omega_{23}(q_3, c), r_2) \cdot \mu_3(q_3, c, p_3) \\
&= \sum_{(r_2, r_3) \in \eta'^{-1}(p_1, p_3)} \mu_{\omega_{23}}((q_2, q_3)), c, (r_2, r_3)).
\end{aligned}$$

Note that if (η, ξ) is a strong covering of M_1 by M_2, then also (η', id_{A_3}) is a strong covering of $M_1 \omega_{13} M_3$ by $M_2 \omega_{23} M_3$.

If the cascade product $M_2 \omega_{23} M_3$ with $\omega_{23} : Q_3 \times A_3 \to A_2$ is given, then we can define $\omega_{13} : Q_3 \times A_3 \to A_1$ by choosing $\omega_{13}(q_3, c) \in \xi^{-1}(\omega_{23}(q_3, c))$. Note that $\xi^{-1}(\omega_{23}(q_3, c)) \neq \emptyset$ since $\omega_{23}(Q_3 \times A_3) \subseteq \xi(A_1)$. Then we have again $\omega_{23} = \xi \circ \omega_{13}$ and we proceed as above, thus showing $M_1 \omega_{13} M_3 \leq M_2 \omega_{23} M_3$.

(ii) Let $M_3\omega_{31}M_1 = (Q_3 \times Q_1, A_1, \mu_{\omega_{31}})$ be a cascade product with $\omega_{31} : Q_1 \times A_1 \to A_3$ where $\omega_{31}(q_1,a) = \omega_{31}(q_1,a')$ whenever $\xi(a) = \xi(a')$ and

$$\mu_{\omega_{31}}((q_3,q_1),a,(p_3,p_1)) = \mu_3(q_3,\omega_{31}(q_1,a),p_3) \cdot \mu_1(q_1,a,p_1).$$

Now we define $M_3\omega_{32}M_2 = (Q_3 \times Q_2, A_2, \mu_{\omega_{32}})$ with $\omega_{32} : Q_2 \times A_2 \to A_3$ as follows:

$$\omega_{32}(q_2,b) = \begin{cases} \omega_{31}(\eta(q_2),a) & \text{if } q_2 \in dom(\eta),\, b \in \xi(A_1), \text{ and } a \in \xi^{-1}(b) \text{ is fixed,} \\ c & \text{otherwise, where } c \in A_3 \text{ is fixed.} \end{cases}$$

Then $\omega_{31}(\eta(q_2),a) = \omega_{32}(q_2,\xi(a))$ for all $q_2 \in dom(\eta)$ and $a \in A_1$ because $\omega_{31}(q_1,a) = \omega_{31}(q_1,a')$ for all $a' \in \xi^{-1}(\xi(a))$. Moreover,

$$\mu_{\omega_{32}}((q_3,q_2),b,(p_3,p_2)) = \mu_3(q_3,\omega_{32}(q_2,b),p_3) \cdot \mu_2(q_2,b,p_2).$$

Let $\eta' : Q_3 \times Q_2 \dashrightarrow Q_3 \times Q_1$ be defined by $\eta'(q_3,q_2) = (q_3,\eta(q_2))$. Note that η' is onto. Moreover, if η is total, then η' is also total. Recall that $\xi : A_1 \to A_2$. Now we show that $M_3\omega_{31}M_1 \le_{(\eta',\xi)} M_3\omega_{32}M_2$, i.e.,

$$\mu_{\omega_{31}}(\eta'(q_3,q_2),a,(p_3,p_1)) \le \sum_{(r_3,r_2)\in\eta'^{-1}(p_3,p_1)} \mu_{\omega_{32}}((q_3,q_2),\xi(a),(r_3,r_2)),$$

for all $(q_3,q_2) \in dom(\eta')$, $(p_3,p_1) \in Q_3 \times Q_1$, and $a \in A_1$. Indeed, we get

$$\begin{aligned} \mu_{\omega_{31}}(\eta'(q_3,q_2),a,(p_3,p_1)) &= \mu_{\omega_{31}}((q_3,\eta(q_2)),a,(p_3,p_1)) \\ &= \mu_3(q_3,\omega_{31}(\eta(q_2),a),p_3) \cdot \mu_1(\eta(q_2),a,p_1) \\ &\le \mu_3(q_3,\omega_{31}(\eta(q_2),a),p_3) \cdot \sum_{r_2\in\eta^{-1}(p_1)} \mu_2(q_2,\xi(a),r_2) \end{aligned}$$

and since $\omega_{31}(\eta(q_2),a) = \omega_{32}(q_2,\xi(a))$ for all $q_2 \in dom(\eta)$ and $a \in A_1$

$$\begin{aligned} &= \sum_{r_2\in\eta^{-1}(p_1)} \mu_3(q_3,\omega_{32}(q_2,\xi(a)),p_3) \cdot \mu_2(q_2,\xi(a),r_2) \\ &= \sum_{(r_3,r_2)\in\eta'^{-1}(p_3,p_1)} \mu_{\omega_{32}}((q_3,q_2),\xi(a),(r_3,r_2)) \end{aligned}$$

which shows the assertion. Again, if (η,ξ) is a strong covering of M_1 by M_2, then (η',ξ) is a strong covering of $M_3\omega_{31}M_1$ by $M_3\omega_{32}M_2$.

If the cascade product $M_3\omega_{32}M_2 = (Q_3 \times Q_2, A_2, \mu_{\omega_{32}})$ with $\omega_{32} : Q_2 \times A_2 \to A_3$ is given, then we define a cascade product $M_3\omega_{31}M_1 = (Q_3 \times Q_1, A_1, \mu_{\omega_{31}})$ with $\omega_{31} : Q_1 \times A_1 \to A_3$ by $\omega_{31}(q_1,a) = \omega_{32}(q_2,\xi(a))$ where $q_2 \in \eta^{-1}(q_1) \ne \emptyset$ (recall that η is onto) is a fixed element. Again $\omega_{31}(\eta(q_2),a) = \omega_{32}(q_2,\xi(a))$ for all $q_2 \in dom(\eta)$ and $a \in A_1$ because $\omega_{32}(q_2',b) = \omega_{32}(q_2,b)$ whenever $\eta(q_2') = \eta(q_2)$. Now we proceed in the same way and with the same η' and ξ as above, thus, showing $M_3\omega_{31}M_1 \le M_3\omega_{32}M_2$ via (η',ξ).

11.5 Admissible Partitions

If $\pi = \{H_1, H_2, \ldots, H_k\}$ is a partition of a set Q, then we denote by $\sim_\pi$ the equivalence relation induced by π. The equivalence class of $q \in Q$ with respect to $\sim_\pi$ will be denoted by $[q]_\pi$.

Definition 11.4 (cf. [1, 3, 19]). Let $M = (Q, A, \mu)$ be a WTS over S. Let $\pi = \{H_1, H_2, ..., H_k\}$ be a partition of Q. Then π is an *admissible partition* of M (or: $\sim_\pi$ is *admissible*), if for every $a \in A$, for every $i \in \{1, 2, \ldots k\}$ and for all $p, q \in Q$ the following holds:

$$p \sim_\pi q \Longrightarrow \sum_{r \in H_i} \mu(p, a, r) = \sum_{r \in H_i} \mu(q, a, r).$$

Remark 11.5. Admissible partitions have appeared in the weighted automata literature several times under different names: Kuich and Salomaa call them *block-stochastic matrices* [19, Ex. 4.5], Bloom and Ésik speak about *simulations* [3], Buchholz calls them *bisimulations* [4, Def. 3.4], and Béal, Lombardy, and Sakarovitch use the terms *out-licit equivalences* and *coverings* [1, 2]. The requirements for these notions are sometimes stronger since also initial and final weights are considered.

In the literature about fuzzy finite state machines [21, 23] the notion *admissible partition* is used as for unweighted automata, cf. [14].

If $M = (Q, A, \mu)$ is a WTS over S and $\pi = \{H_i \mid i \in I\}$ is an admissible partition of M, then $M/\pi = (\pi, A, \mu_\pi)$ is a *quotient weighted transition system* with $\mu_\pi : \pi \times A \times \pi \to S$ being the mapping defined by

$$\mu_\pi(H_i, a, H_j) = \sum_{r \in H_j} \mu(q, a, r) \quad \text{for some } q \in H_i. \tag{11.3}$$

Note that μ_π is well defined. Indeed, for $q, q' \in H_i$ we have $\sum_{r \in H_j} \mu(q, a, r) = \sum_{r \in H_j} \mu(q', a, r)$ because $q \sim_\pi q'$ and $\sim_\pi$ is admissible.

Remark 11.6. If $M = (Q, A, \mu)$ is a WTS over the Boolean semiring $\mathbb{B}$, we can consider again the associated unweighted transition system $ts(M) = (Q, A, T)$, cf. Remark 11.1. Then $\pi = \{H_1, H_2, ..., H_k\}$ is an admissible partition of M if and only if for all $a \in A$, $i \in \{1, 2, \ldots k\}$ and $p, q \in Q$ the following holds:

$$p \sim_\pi q \Longrightarrow \Big(\exists r \in H_i : (p, a, r) \in T \iff \exists r \in H_i : (q, a, r) \in T\Big).$$

In this sense, we say that π is an admissible partition of the unweighted transition system $ts(M)$. This definition coincides with the notion of an *admissible partition* as defined in [14, p. 39].

The *quotient transition system* $ts(M)/\pi = (\pi, A, T_\pi)$ is defined by

$$(H_i, a, H_j) \in T_\pi \iff \exists q \in H_i \exists r \in H_j : (q, a, r) \in T.$$

If M is a WTS over $\mathbb{B}$ and N an unweighted TS, then we have

$$M/\pi = wts\big(ts(M)/\pi\big) \text{ and } N/\pi = ts\big(wts(N)/\pi\big)\,.$$

For $k \in \mathbb{N}$ and $s \in S$, let $ks = \underbrace{s + \ldots + s}_{k \text{ times}} \in S$ where $0s = 0$. For the set

$$weights(M) = \{s_1, \ldots, s_m\} = \{\mu(q,a,p) \mid q,p \in Q, a \in A\}$$

of weights appearing in $M = (Q,A,\mu)$, we put

$$\sum weights(M) = \{k_1 s_1 + \ldots + k_m s_m \mid 0 \le k_1 + \ldots + k_m \le |Q|\}\,.$$

Then we have $weights(M/\pi) \subseteq \sum weights(M)$. If $weights(M)$ is bounded from above by an element $s_M \in S$, then $weights(M/\pi)$ is bounded from above by $|Q|s_M$.

Proposition 11.4. *Let $M = (Q,A,\mu)$ be a WTS, π an admissible partition of M, and M/π the quotient WTS. Then $M/\pi \le_{(\eta_\pi, id_A)} M$ where $\eta_\pi : Q \dashrightarrow \pi : q \mapsto [q]_\pi$, the canonical projection. Moreover, the covering (η_π, id_A) is strong and total.*

Proof. Note that η_π is surjective and total. Moreover, for any $q \in Q$, $H \in \pi$, and $a \in A$ we have

$$\mu_\pi(\eta_\pi(q), a, H) = \sum_{r \in H} \mu(q,a,r) \qquad \text{(by definition of } \mu_\pi)$$

and $H = \eta_\pi^{-1}(H)$ which shows that (η_π, id_A) is a strong and total covering.

By Propositions 11.1 and 11.4, Equation (11.3) extends to words.

Proposition 11.5. *Let π be an admissible partition of a WTS M and let M/π be the quotient WTS of M with respect to π. Then $\mu_\pi(H_i, w, H_j) = \sum_{r \in H_j} \mu(q,w,r)$ for all $H_i, H_j \in \pi$, $w \in A^*$, and any $q \in H_i$.*

Proof. Let $M = (Q,A,\mu)$ and $\pi = \{H_i \mid i \in I\}$. Due to Proposition 11.4, (η_π, id_A) is a strong and total covering of M/π by M. Thus we have for all $H_i, H_j \in \pi$, $q \in H_i$, and $w \in A^*$ by using Proposition 11.1

$$\mu_\pi(H_i, w, H_j) = \mu_\pi(\eta_\pi(q), w, H_j) = \sum_{r \in H_j} \mu(q,w,r)$$

which shows the claim.

To obtain decomposition results, we will need an additional property of the underlying semiring S. Let S be naturally ordered by $\le$. We will demand that $s \le s^2$ for all $s \in S$. This is true for all distributive lattices because $s \wedge s = s$. But also for the semiring $(\mathbb{N}, +, \cdot, 0, 1)$ of natural numbers we have $s \le s \cdot s$ whereas in $(\mathbb{Q}_+, +, \cdot, 0, 1)$ with $\mathbb{Q}_+ = \{q \in \mathbb{Q} \mid q \ge 0\}$ this condition is not satisfied. Note that whenever $(S, +, \cdot, 0, 1)$ is a naturally ordered semiring where $1 \le s$ for every $s \neq 0$, then $s \le s^2$ because

$\leq$ is compatible with multiplication. Moreover, every naturally ordered semiring $(S,+,\cdot,0,1)$ contains the subsemiring $(\{s \in S \mid s \geq 1\} \cup \{0\},+,\cdot,0,1)$ and, thus, for this subsemiring the condition is fulfilled. Hence, also the max-plus-semiring $(\mathbb{N} \cup \{-\infty\},\max,+,-\infty,0)$ (here, the natural order is the usual one) and the non-commutative semiring $(\{L \subseteq A^* \mid \varepsilon \in L\} \cup \{\emptyset\},\cup,\cdot,\emptyset,\{\varepsilon\})$ where $\cdot$ denotes concatenation satisfy this condition. But the tropical semiring $(\mathbb{N} \cup \{\infty\},\min,+,\infty,0)$ does not satisfy the condition (note that this semiring is ordered by the opposite of the usual order).

Proposition 11.6. *Let S be a naturally ordered semiring with $s \leq s^2$ for every $s \in S$. Let $M = (Q,A,\mu)$ be a WTS, $H,K \subseteq Q$, $q_0 \in Q$, and $p_0 \in H \cap K$. Then $\mu(q_0,w,p_0) \leq \sum_{p\in H}\sum_{p'\in K}\mu(q_0,w,p)\cdot\mu(q_0,w,p')$ for all $w \in A^*$.*

Proof. We have for all $w \in A^*$

$$\begin{aligned}
&\mu(q_0,w,p_0)\\
\leq\,&\mu(q_0,w,p_0)\cdot\mu(q_0,w,p_0) && \text{(because of } s \leq s^2\text{)}\\
\leq\,&\sum_{p\in H}\sum_{p'\in K}\mu(q_0,w,p)\cdot\mu(q_0,w,p') && \text{(because } p_0 \in H \cap K\text{)}
\end{aligned}$$

which shows the claim.

Before we turn again to the cascade product, we show how admissible partitions imply a decomposition of a WTS using a direct product construction. Let $M_1 = (Q_1,A,\mu_1)$ and $M_2 = (Q_2,A,\mu_2)$ be WTS over the same alphabet A. The *product WTS* $M_1 \times M_2$ of M_1 and M_2 is defined by $M_1 \times M_2 = (Q_1 \times Q_2,A,\mu)$ with $\mu((q_1,q_2),a,(p_1,p_2)) = \mu_1(q_1,a,p_1)\cdot\mu_2(q_2,a,p_2)$. If the semiring S is commutative, then this equation lifts to words. Recall that this product construction is a synchronized product as used for the Hadamard product of two formal power series. For a set Q, let $\Delta_Q = \{(q,q) \mid q \in Q\}$, the diagonal of $Q \times Q$. Using Proposition 11.6, we can show:

Theorem 11.1. *Let S be a naturally ordered semiring with $s \leq s^2$ for all $s \in S$. Let $M = (Q,A,\mu)$ be a WTS over S and let π and τ be admissible partitions of Q. If $\sim_\pi \cap \sim_\tau = \Delta_Q$, then $M \leq M/\pi \times M/\tau$.*

This result is a natural extension of [14, Thm. 3.2.1].

Proof. Let $M/\pi = (\pi,A,\mu_\pi)$ and $M/\tau = (\tau,A,\mu_\tau)$ be the quotient WTS of M with respect to π and τ. The WTS $M/\pi \times M/\tau$ is defined as $M/\pi \times M/\tau = (\pi \times \tau,A,\mu_\bullet)$ with

$$\mu_\bullet((H_i,K_j),a,(H_u,K_v)) = \mu_\pi(H_i,a,H_u)\cdot\mu_\tau(K_j,a,K_v)$$

for all $(H_i,K_j),(H_u,K_v) \in \pi \times \tau$ and $a \in A$.

Let $\eta : \pi \times \tau \dashrightarrow Q$ be defined by $\eta(H_i,K_j) = q_0$ if $H_i \cap K_j = \{q_0\}$, otherwise η is undefined. Then η is onto. Moreover, put $\xi = id_A$.

Now we have for every $a \in A$, $(H_i, K_j) \in dom(\eta)$ (hence, $H_i \cap K_j \neq \emptyset$), and $q_1 \in Q$

$$\begin{aligned}
&\mu(\eta(H_i, K_j), a, q_1) \\
&= \mu(q_0, a, q_1) && (\{q_0\} = H_i \cap K_j) \\
&\leq \sum_{r \in H_u} \mu(q_0, a, r) \cdot \sum_{t \in K_v} \mu(q_0, a, t) && (\text{Prop. 11.6}, \{q_1\} = H_u \cap K_v) \\
&= \mu_\pi(H_i, a, H_u) \cdot \mu_\tau(K_j, a, K_v) && (\text{by definition of } \mu_\pi \text{ and } \mu_\tau) \\
&= \mu_\bullet((H_i, K_j), a, (H_u, K_v)) && (\text{by definition of } \mu_\bullet) \\
&= \sum_{(H_u, K_v) \in \eta^{-1}(q_1)} \mu_\bullet((H_i, K_j), a, (H_u, K_v)) && (\text{since } \sim_\pi \cap \sim_\tau = \Delta_Q)
\end{aligned}$$

which shows $M \leq_{(\eta, id_A)} M/\pi \times M/\tau$.

11.6 A Decomposition by Cascade Products

Now we turn again to the cascade product and show that the existence of admissible partitions implies a covering of the WTS by a cascade product of simpler transition systems. For a WTS $M = (Q, A, \mu)$, let $|M| = |Q|$ be the *size* of M.

Theorem 11.2. *Let S be a naturally ordered semiring with $s \leq s^2$ for all $s \in S$. Let $M = (Q, A, \mu)$ be a WTS over S. If π is an admissible partition of M, then there exists a WTS N such that $M \leq N\,\omega\,(M/\pi)$ where $weights(N) \subseteq \sum weights(M)$ and $weights(M/\pi) \subseteq \sum weights(M)$.*

Moreover, $|N| < |M|$ whenever $|\pi| \geq 2$.

Proof. Let π be an admissible partition of M. Then there is a partition τ of Q such that $\sim_\pi \cap \sim_\tau = \Delta_Q$ and $\sim_\tau \neq \Delta_Q$ if $\sim_\pi \neq Q \times Q$. Note that τ does not have to be admissible. We put $N = (\tau, \pi \times A, \mu')$ where

$$\mu'(K_j, (H_i, a), K_v) = \begin{cases} \sum_{p \in K_v} \mu(q_0, a, p) & \text{if } H_i \cap K_j = \{q_0\}, \\ 0 & \text{if } H_i \cap K_j = \emptyset. \end{cases}$$

If $|\pi| \geq 2$, we have $|N| < |M|$. Let $\omega = id_{\pi \times A}$ be the identity mapping on $\pi \times A$. Recall that $N\,\omega\,(M/\pi)$ is defined as $N\,\omega\,(M/\pi) = (\tau \times \pi, A, \mu_\omega)$ with $\mu_\omega((K_j, H_i), a, (K_v, H_u)) = \mu'(K_j, (H_i, a), K_v) \cdot \mu_\pi(H_i, a, H_u)$ for all $a \in A$.

Now, we define $\eta : \tau \times \pi \dashrightarrow Q$ by $\eta(K_j, H_i) = q_0$ if $H_i \cap K_j = \{q_0\}$ and which is not defined otherwise. Note that η is bijective on its domain. We show that M is covered via (η, id_A) by $N\,\omega\,(M/\pi)$, i.e.,

$$\mu(\eta(K_j, H_i), a, q) \leq \sum_{(K_{v'}, H_{u'}) \in \eta^{-1}(q)} \mu_\omega((K_j, H_i), a, (K_{v'}, H_{u'}))$$

for every $(K_j, H_i) \in dom(\eta)$, $a \in A$, and $q \in Q$.

Indeed, we get:

$$
\begin{aligned}
& \mu(\eta(K_j,H_i),a,q) \\
&= \mu(q_0,a,q) && \text{(where } K_j \cap H_i = \{q_o\}\text{)} \\
&\leq \sum_{p' \in K_v} \mu(q_0,a,p') \cdot \sum_{p \in H_u} \mu(q_0,a,p) && \text{(Prop. 11.6 with } K_v \cap H_u = \{q\}\text{)} \\
&= \mu'(K_j,(H_i,a),K_v) \cdot \mu_\pi(H_i,a,H_u) && \text{(definition of } \mu'\text{, } \pi \text{ admissible)} \\
&= \mu_\omega((K_j,H_i),a,(K_v,H_u)) && \text{(definition of } \mu_\omega\text{)} \\
&= \sum_{(K_{v'},H_{u'}) \in \eta^{-1}(q)} \mu_\omega((K_j,H_i),a,(K_{v'},H_{u'})) && \text{(bijectivity of } \eta\text{)}
\end{aligned}
$$

which shows $M \leq_{(\eta,id_A)} N\,\omega\,(M/\pi)$. Due to the definition of N and M/π, we have $weights(N) \subseteq \sum weights(M)$ and $weights(M/\pi) \subseteq \sum weights(M)$.

The fact that

$$weights(N) \subseteq \sum weights(M) \text{ and } weights(M/\pi) \subseteq \sum weights(M)$$

guarantees that the weights of the factor WTS N and M/π do not grow arbitrarily large. Otherwise, one could try to find a decomposition covering the original WTS M by putting the weights of N as large as necessary in order to obtain the required inequalities. But here we have a uniform bound on the weights of the factors depending on the size and the weights of M. This has something in common with the idea of the relation of two cost functions via a correction function as proposed by Colcombet [6].

Consider the tropical semiring $(\mathbb{N} \cup \{\infty\}, \min, +, \infty, 0)$ which is naturally ordered by the opposite of the usual order on $\mathbb{N} \cup \{\infty\}$. Now $s \leq s^2$ is not satisfied ($1 + 1 = 2$ is less than 1 with respect to the order of the semiring). Here we cannot find an analog of Theorem 11.2. Indeed, suppose M is a WTS over the tropical semiring with $weights(M) = \{\infty, 1\}$. Then $\sum weights(M) = \{\infty, 1\}$. If $weights(N) \subseteq \sum weights(M)$ and $weights(M/\pi) \subseteq \sum weights(M)$ is satisfied, then the only weight in a cascade product $N\omega(M/\pi)$ different from ∞ (the zero of the semiring) would be $1+1=2$. But 2 is less than 1 in the ordering of the semiring which contradicts the notion of a covering.

Example 11.1. Consider the WTS $M = (Q, \{a\}, \mu)$ with $Q = \{p_1, p_2, q_1, q_2\}$ over the semiring $F = ([0,1], \vee, \wedge, 0, 1)$ shown in Figure 11.1. Then $\pi = \{H_1, H_2\}$ with $H_1 = \{p_1, q_1\}$ and $H_2 = \{p_2, q_2\}$ is an admissible partition of M. The quotient WTS M/π is shown in Figure 11.2 in the upper left.

Now we put $\tau = \{K_1, K_2\}$ where $K_1 = \{p_1, p_2\}$ and $K_2 = \{q_1, q_2\}$. Then $\sim_\pi \cap \sim_\tau = \Delta_Q$ and $N = (\tau, \pi \times \{a\}, \mu')$, as constructed in the last proof, is shown in Figure 11.2 in the lower left. As usual, we do not depict transitions with weight 0 (here, like those caused by (H_2, a)). The WTS $N\omega(M/\pi)$ with $\omega = id_{\pi \times \{a\}}$ can be seen in Figure 11.2 on the right. Now $\eta : \tau \times \pi \to Q$ is given by

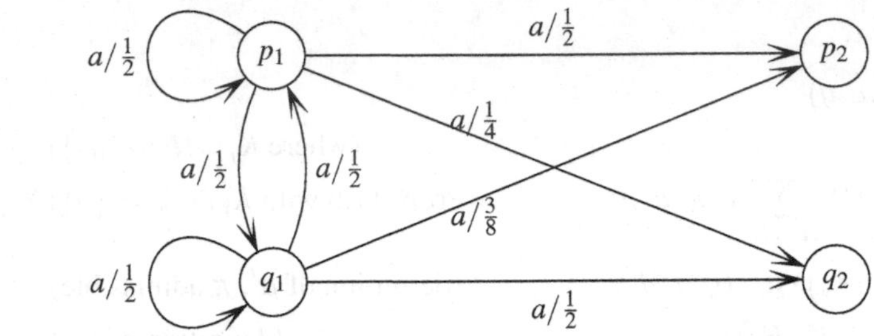

Fig. 11.1 A WTS M over the semiring $F = ([0,1], \vee, \wedge, 0, 1)$.

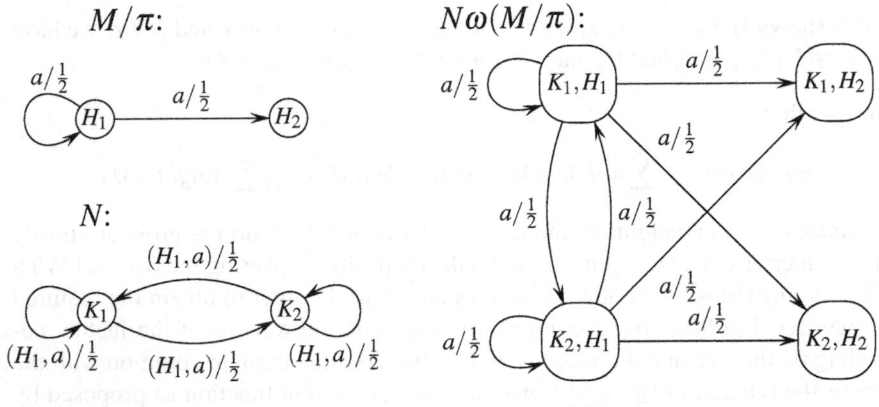

Fig. 11.2 A decomposition of the WTS M from Fig. 11.1 into a cascade product $N\omega(M/\pi)$.

$$\eta : (K_1,H_1) \mapsto p_1, \quad (K_2,H_1) \mapsto q_1, \quad (K_1,H_2) \mapsto p_2, \quad (K_2,H_2) \mapsto q_2,$$

whereas $\xi(a) = a$. Due to Proposition 11.2 and its proof, M is covered by $N\omega(M/\pi)$ via (η, ξ). Note that M is not strongly covered by $N\omega(M/\pi)$ because

$$\frac{1}{4} = \mu\big(\eta(K_1,H_1), a, q_2\big) < \mu_\omega\big((K_1,H_1), a, (K_2,H_2)\big) = \frac{1}{2}.$$

An admissible partition π of a WTS $M = (Q,A,\mu)$ is *maximal* if it is non-trivial and such that for all admissible partitions τ of M with $\sim_\pi \subseteq \sim_\tau$ we have $\sim_\tau = \sim_\pi$ or $\sim_\tau = Q^2$.

A WTS $M = (Q,A,\mu)$ is called *irreducible* if $|Q| > 1$ and there are no admissible partitions of Q other than the trivial one. As the next proposition shows, maximal admissible partitions and irreducible WTS are closely related.

Proposition 11.7. *Let $M = (Q,A,\pi)$ be a WTS and let π be an admissible partition of M. Then π is maximal if and only if M/π is irreducible.*

Proof. Let $\pi = \{P_1, \ldots, P_m\}$ be an admissible partition of M. Then the quotient WTS $M/\pi = (\pi, A, \mu_\pi)$ where μ_π is defined by (11.3).

We show: π is not maximal if and only if M/π is not irreducible.

First, let us assume that there is a non-trivial admissible partition ρ of M/π, i.e., $\rho = \{R_1, ..., R_n\}$ where every class R_j is a set of some classes P_i of π. Now we consider a partition $\pi' = \{P'_1, \ldots, P'_n\}$ of Q which is obtained by taking unions of classes of ρ, i.e.,

$$P'_k = \bigcup_{P_i \in R_k} P_i$$

for every $k \in \{1, \ldots, n\}$. Since ρ is non-trivial, $\sim_\pi \subsetneq \sim_{\pi'} \subsetneq Q^2$. Now we prove that π' is also admissible, hence, π is not maximal.

Let $p, q \in Q$ with $p \sim_{\pi'} q$. Let P_i and P_j be the classes in π such that $p \in P_i$ and $q \in P_j$. Due to $p \sim_{\pi'} q$, we have $P_i \sim_\rho P_j$. Since ρ is admissible, we have

$$\sum_{P \in R_k} \mu_\pi(P_i, a, P) = \sum_{P \in R_k} \mu_\pi(P_j, a, P)$$

for every element R_k of ρ.

Now we get for every element P'_k of π'

$$\begin{aligned}
\sum_{r \in P'_k} \mu(p, a, r) &= \sum_{P \in R_k} \sum_{r \in P} \mu(p, a, r) && (\text{since } P'_k = \bigcup_{P \in R_k} P) \\
&= \sum_{P \in R_k} \mu_\pi(P_i, a, P) && (\text{due to (11.3)}) \\
&= \sum_{P \in R_k} \mu_\pi(P_j, a, P) && (\text{see above}) \\
&= \sum_{P \in R_k} \sum_{r \in P} \mu(q, a, r) && (\text{due to (11.3)}) \\
&= \sum_{r \in P'_k} \mu(q, a, r) && (\text{since } P'_k = \bigcup_{P \in R_k} P)
\end{aligned}$$

which shows that π' is a non-trivial admissible partition of M.

To prove the converse, let us assume that there is an admissible partition $\pi' = \{P'_1, \ldots, P'_n\}$ of M with $\sim_\pi \subsetneq \sim_{\pi'} \subsetneq Q^2$. Now we can define a non-trivial partition $\rho = \{R_1, \ldots, R_n\}$ on M/π by

$$R_k = \{P \in \pi \mid P \subseteq P'_k\}$$

for every $k \in \{1, \ldots, n\}$. We prove that ρ is an admissible partition of M/π. Let $P_i \sim_\rho P_j$. Choose any $p \in P_i$ and $q \in P_j$. Then $p \sim_{\pi'} q$. We have for every class R_k of ρ and $a \in \Sigma$

$$\begin{aligned}
\sum_{P\in R_k} \mu_\pi(P_i,a,P) &= \sum_{P\in R_k}\sum_{r\in P} \mu(p,a,r) && \text{(due to (11.3))}\\
&= \sum_{r\in P'_k} \mu(p,a,r) && (\text{since } P'_k = \bigcup_{P\in R_k} P)\\
&= \sum_{r\in P'_k} \mu(q,a,r) && (\text{since } p \sim_{\pi'} q)\\
&= \sum_{P\in R_k}\sum_{r\in P} \mu(q,a,r) && (\text{since } P'_k = \bigcup_{P\in R_k} P)\\
&= \sum_{P\in R_k} \mu_\pi(P_j,a,P) && \text{(due to (11.3))}
\end{aligned}$$

which shows that ρ is admissible, hence, M/π is not irreducible.

Now we can show our main result which states that every WTS M is covered by an iterated cascade product of irreducible WTS of smaller size.

Theorem 11.3. *Let S be a naturally ordered semiring with $s \leq s^2$ for all $s \in S$. Then any WTS M over S with $|M| \geq 2$ can be covered by a cascade product of the form*

$$M \leq ((\ldots(N_1\,\omega_1\,N_2)\,\omega_2\,\ldots)\,\omega_{k-1}\,N_k)\,\omega_k\,M/\pi$$

where π is an admissible partition of M and $N_1, N_2, \ldots, N_k$, and M/π are irreducible WTS with $|N_1| < \ldots < |N_k| < |M|$. So, $k \leq |M| - 2$.

Proof. Let $M = (Q, A, \mu)$. If M is irreducible, then $M \leq M/\pi = M$ with $\sim_\pi = \Delta_Q$ and the statement is proven. If M is not irreducible, then there is a non-trivial admissible partition π of M. Since Q is finite, we can choose π maximal. By Proposition 11.7, M/π is irreducible and, by Theorem 11.2, there is a WTS N with $|N| < |M|$ such that $M \leq N\,\omega\,M/\pi$. Since π is non-trivial, we get $|N| < |M|$ (cf. the proof of Theorem 11.2). If N is irreducible, then we are done.

If N is not irreducible, then we repeat the process and find a maximal admissible partition π' of N such that $N \leq N'\,\omega'\,N/\pi'$ with $|N'| < |N| < |M|$. Due to Proposition 11.7, N/π' is irreducible. By Proposition 11.3(i), there is a cascade product $(N'\,\omega'\,N/\pi')\,\tilde{\omega}\,M/\pi$ with $N\,\omega\,M/\pi \leq (N'\,\omega'\,N/\pi')\,\tilde{\omega}\,M/\pi$. Applying transitivity of the covering relation, i.e., Proposition 11.2, we obtain $M \leq (N'\,\omega'\,N/\pi')\,\tilde{\omega}\,M/\pi$. Now we proceed inductively on the leftmost factor. The induction terminates after a finite number of steps (say k steps) since the size of the leftmost factor gets smaller with every step. Furthermore note that in every induction step all cascade products computed so far have to be updated in line with Proposition 11.3(i). Finally, we succeed in showing that $M \leq ((\ldots(N_1\,\omega_1\,N_2)\,\omega_2\,\ldots)\,\omega_{k-1}\,N_k)\,\omega_k\,M/\pi$ where M/π and N_i are irreducible WTS for all $i \in \{1,\ldots,k\}$ and $|N_1| < \ldots < |N_k| < |M|$.

Since we decrease by every inductive step the size of the new factor in the cascade product by at least one and each WTS of size 2 is irreducible, the number k of steps is bounded from above by $|Q| - 2$.

For the Boolean semiring $\mathbb{B}$, we get as a consequence that unweighted transition systems can be covered by a cascade product of smaller irreducible machines [14, Thm. 3.3.4].

Corollary 11.1 ([14, Thm. 3.3.4]). *Let $M = (Q, A, T)$ be an unweighted transition system with $|Q| \geq 2$. Then there is an admissible partition π of M such that*

$$M \leq ((\dots(N_1 \,\omega_1\, N_2)\,\omega_2 \dots)\,\omega_{k-1}\, N_k)\,\omega_k\, M/\pi$$

where $N_1, N_2, \dots, N_k$, and M/π are irreducible TS with $|N_1| < \dots < |N_k| < |M|$. So, $k \leq |M| - 2$.

Proof. Let $wts(M)$ be the WTS over $\mathbb{B}$ associated to the TS M, see Remark 11.1. By Theorem 11.3, there is an admissible partition π of $wts(M)$ such that

$$wts(M) \leq ((\dots(P_1 \,\omega_1\, P_2)\,\omega_2 \dots)\,\omega_{k-1}\, P_k)\,\omega_k\, wts(M)/\pi$$

where P_1, P_2, $\dots$, P_k, and $wts(M)/\pi$ are irreducible WTS with $|P_1| < \dots < |P_k| < |wts(M)|$. By Remark 11.6, $wts(M)/\pi = wts(M/\pi)$. Let $N_i = ts(P_i)$ for $i \in \{1, \dots, k\}$. If P_i is irreducible, so is N_i. The same holds true for M/π. By applying Remarks 11.1, 11.2, 11.4, and 11.6 inductively we get

$$wts(M) \leq wts\Big[((\dots(N_1 \,\omega_1\, N_2)\,\omega_2 \dots)\,\omega_{k-1}\, N_k)\,\omega_k\, M/\pi\Big].$$

Finally, we have by Remarks 11.1 and 11.2 that

$$M \leq ((\dots(N_1 \,\omega_1\, N_2)\,\omega_2 \dots)\,\omega_{k-1}\, N_k)\,\omega_k\, M/\pi$$

where N_1, N_2, $\dots$, N_k, and M/π are irreducible TS with $|N_1| < \dots < |N_k| < |M|$. Hence, $k \leq |M| - 2$.

Example 11.2. Consider the WTS M from Figure 11.1 over $F = ([0,1], \vee, \wedge, 0, 1)$. A decomposition of M due to Theorem 11.3 is $N\omega(M/\pi)$ given in Figure 11.2. Both components of the cascade product have size two, so they are irreducible.

11.7 What about Strong Coverings?

In [23, Thm. 6.14.21], Mordeson and Malik claim a decomposition result for WTS over the semiring $F = ([0,1], \vee, \wedge, 0, 1)$ analogous to Theorem 11.3, but with a strong and total covering.

The proof method is very similar to the one applied here. First, one has to find an admissible partition π of the original WTS $M = (Q, A, \mu)$. Then a WTS N is constructed in the same way as in the proof of Theorem 11.2 using another partition τ with $\sim_\pi \cap \sim_\tau = \Delta_Q$. However, they require an additional property, called μ-orthogonality [23, Def. 6.14.12], to guarantee a strong covering. Unfortunately, it is not clear how and if such a μ-orthogonal partition τ can be found, though, this is necessary to prove an analog for Theorem 11.3 (cf. [23, Thm. 6.14.15] and proof of [23, Thm. 6.14.21]).

We will show that for the WTS $M = (Q, \{a\}, \mu)$ from Figure 11.1 *no* non-trivial partition τ with $\sim_\tau \cap \sim_\pi = \Delta_Q$ and the associated WTS N can be found such that we get a strong covering of M by $N\omega(M/\pi)$. Recall that $\pi = \{H_1, H_2\}$ with $H_1 = \{p_1, q_1\}$ and $H_2 = \{p_2, q_2\}$ is an admissible partition of M. It is easy to see that it is the only maximal one. M/π is shown in the upper right of Figure 11.3.

There is no non-trivial partition τ of Q with $\sim_\pi \cap \sim_\tau = \Delta_Q$ such that for the WTS $N = (\tau, \pi \times \{a\}, \mu')$ induced by τ (see proof of Theorem 11.2) M is strongly covered by $N\omega(M/\pi)$. Indeed, suppose there was such a partition τ. By definition of μ' and since $\mu(p_1, a, q_2) = \frac{1}{4}$ and $\mu(q_1, a, p_2) = \frac{3}{8}$, there are both a transition with weight $\frac{1}{4}$ and one with weight $\frac{3}{8}$ in $N\omega(M/\pi)$. Due to the operations of the semiring and the fact that only the weight $\frac{1}{2}$ appears in M/π, there has to be a transition of weight $\frac{1}{4}$ and one of weight $\frac{3}{8}$ also in N. By definition of μ' and since $\mu(p_1, a, q_2) = \frac{1}{4}$ and $\mu(p_1, a, p_1) = \mu(p_1, a, q_1) = \mu(p_1, a, p_2) = \frac{1}{2}$, $\{q_2\}$ has to be an element of τ. Similarly, $\{p_2\}$ has to be an element of τ. Thus $\{p_1, q_1\}$ has to be an element of τ because otherwise τ would be trivial. But then $\sim_\pi \cap \sim_\tau \neq \Delta_Q$, a contradiction.

Thus, the way proposed in [23] for showing an analog of Theorem 11.3 for strong coverings is not successful.

As we have seen before, $M = (Q, \{a\}, \mu)$ from Figure 11.1 can be covered by a cascade product of two WTS of size two. But there is *no* strong covering of M by some $N\omega N'$ with $|N| = |N'| = 2$ as the following proposition shows.

Proposition 11.8. *Let $M = (Q, \{a\}, \mu)$ be the WTS from Figure 11.1 over the semiring $F = ([0,1], \vee, \wedge, 0, 1)$. Then there is no cascade product $N\omega N'$ with $|N| = |N'| = 2$ such that M is strongly covered by $N\omega N'$.*

Proof. Suppose there are WTS $N = (P, B, \mu_N)$, $N' = (P', B', \mu'_N)$ with $|P| = |P'| = 2$ and $\omega : P' \times B' \to B$, $\eta : P \times P' \to Q$, $\xi : \{a\} \to B'$ such that M is strongly covered by $N\omega N'$.

Since $\xi : \{a\} \to B'$, only one letter $\xi(a) = b' \in B'$ is of importance for our considerations. Note that $\eta : P \times P' \to Q$ has to be surjective, hence, η is total and bijective. Let $P = \{r_1, r_2\}$ and $P' = \{r'_1, r'_2\}$. Without loss of generality we put $\eta(r_1, r'_1) = p_1$. Moreover, we denote $\eta^{-1}(q_1) = (r_{q_1}, r'_{q_1})$ and similarly for p_2 and q_2. Now suppose that M is strongly covered by $N\omega N' = (P \times P', B', \mu_\omega)$ via (η, ξ). With $b = \omega(r'_1, b')$ we get

$$\frac{1}{2} = \mu(p_1, a, p_1) = \mu_\omega((r_1, r'_1), b', (r_1, r'_1)) = \underbrace{\mu_N(r_1, b, r_1)}_{\geq \frac{1}{2}} \wedge \underbrace{\mu'_N(r'_1, b', r'_1)}_{\geq \frac{1}{2}},$$

$$\frac{1}{2} = \mu(p_1, a, q_1) = \mu_\omega((r_1, r'_1), b', (r_{q_1}, r'_{q_1})) = \underbrace{\mu_N(r_1, b, r_{q_1})}_{\geq \frac{1}{2}} \wedge \underbrace{\mu'_N(r'_1, b', r'_{q_1})}_{\geq \frac{1}{2}},$$

$$\frac{1}{2} = \mu(p_1,a,p_2) = \mu_\omega((r_1,r'_1),b',(r_{p_2},r'_{p_2})) = \underbrace{\mu_N(r_1,b,r_{p_2})}_{\geq \frac{1}{2}} \wedge \underbrace{\mu'_N(r'_1,b',r'_{p_2})}_{\geq \frac{1}{2}},$$

$$\frac{1}{4} = \mu(p_1,a,q_2) = \mu_\omega((r_1,r'_1),b',(r_{q_2},r'_{q_2})) = \mu_N(r_1,b,r_{q_2}) \wedge \mu'_N(r'_1,b',r'_{q_2}).$$

Since η is bijective, $r_{q_2} \in P = \{r_1, r_{q_1}, r_{p_2}\}$ and, thus, $\mu_N(r_1,b,r_{q_2}) \geq \frac{1}{2}$. Similarly, $r'_{q_2} \in P' = \{r'_1, r'_{q_1}, r'_{p_2}\}$ and, thus, $\mu_N(r'_1,b,r'_{q_2}) \geq \frac{1}{2}$. We conclude that $\frac{1}{4} = \mu(p_1,a,q_2) \geq \frac{1}{2}$, a contradiction. Hence, M cannot be strongly covered by $N\omega N'$.

However, there is a strong covering of M by $N\omega(M/\pi)$ with π the admissible partition from above and $N = (P, \{b\}, \mu_N)$ a WTS of size three shown in the upper left of Figure 11.3. Below the cascade product $N\omega(M/\pi)$ with $\omega : \pi \times \{a\} \to \{b\}$ is depicted. Now $\xi = id_{\{a\}}$ and $\eta : P \times \pi \dashrightarrow Q$ is given by

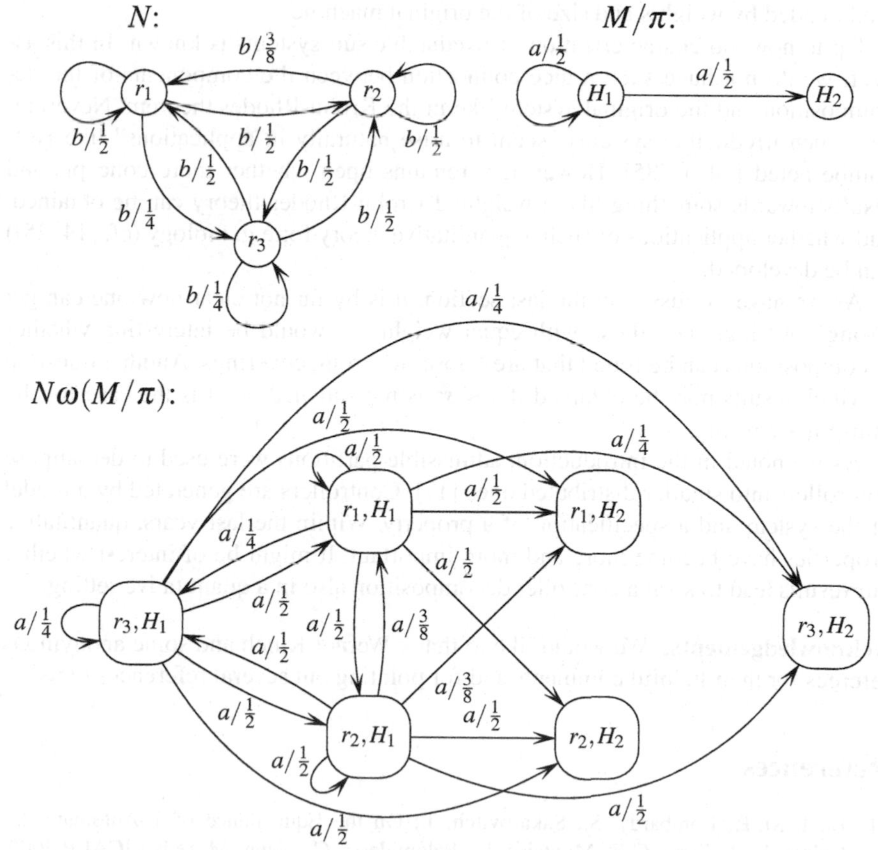

Fig. 11.3 A strong covering of the WTS M from Fig. 11.1 by a cascade product $N\omega(M/\pi)$.

$$\eta : (r_1,H_1),(r_3,H_1) \mapsto p_1,\ (r_2,H_1) \mapsto q_1,\ (r_1,H_2) \mapsto p_2,\ (r_3,H_2) \mapsto q_2.$$

Then (η,ξ) is a strong covering of M by $N\omega(M/\pi)$. Note that here η is partial and has to be partial. Thus the covering (η,ξ) is strong but not total.

However, as we have seen before, this strong covering does not result from a partition τ of Q. It is not clear how we can obtain the WTS N from M in a general way.

11.8 Conclusion

We have shown a decomposition result for weighted transition systems using cascade products and admissible partitions. Our approach was successful whenever the weights stem from naturally ordered semirings S with the additional requirement that $s \leq s^2$ for all $s \in S$. In our notion of covering, the transitions are not necessarily simulated with the same weight but the weights may become greater; however, they are bounded by weights and size of the original machine.

Up to now, no characterization of irreducible sub-systems is known. In this aspect, we do not have such a nice connection between the components of the decomposition and the original system like in the Krohn-Rhodes theorem. Nevertheless, such irreducible systems "seem to arise naturally in applications" like Holcombe noted [14, p. 85]. However, it remains open whether more concepts and results towards something like a weighted Krohn-Rhodes theory can be obtained, and whether applications of such a quantitative theory for e.g. biology (cf. [14, 25]) can be developed.

As we have discussed in the last section, it is by far not clear how one can get strong coverings, i.e., those with equal weights. It would be interesting whether decompositions can be found that are 'close' to strong coverings. Another question is which results may be obtained if $s \leq s^2$ is not satisfied, as it is the case for the min-plus-semiring.

As we noted in the introduction, admissible partitions were used to decompose controllers into smaller distributed ones [17]. Controllers are generated by a model of the system and a specification of a property. Within the last years, quantitative properties have become more and more important. It might be of interest whether our results lead to such a controller decomposition also in a quantitative setting.

Acknowledgements. We would like to thank Werner Kuich and some anonymous referees for their helpful comments and for pointing out several references to us.

References

1. Béal, M.-P., Lombardy, S., Sakarovitch, J.: On the Equivalence of ℤ-Automata. In: Caires, L., Italiano, G.F., Monteiro, L., Palamidessi, C., Yung, M. (eds.) ICALP 2005. LNCS, vol. 3580, pp. 397–409. Springer, Heidelberg (2005)

2. Béal, M.-P., Lombardy, S., Sakarovitch, J.: Conjugacy and Equivalence of Weighted Automata and Functional Transducers. In: Grigoriev, D., Harrison, J., Hirsch, E.A. (eds.) CSR 2006. LNCS, vol. 3967, pp. 58–69. Springer, Heidelberg (2006)
3. Bloom, S.L., Ésik, Z.: Iteration Theories: The Equational Logic of Iterative Processes. Springer, New York, Inc. (1993)
4. Buchholz, P.: Bisimulation relations for weighted automata. Theoretical Computer Science 393, 109–123 (2008)
5. Cho, S.J., Kim, J.-G., Lee, W.-S.: Decompositions of T-generalized transformation semigroups. Fuzzy Sets Syst. 122(3), 527–537 (2001)
6. Colcombet, T.: The Theory of Stabilisation Monoids and Regular Cost Functions. In: Albers, S., Marchetti-Spaccamela, A., Matias, Y., Nikoletseas, S., Thomas, W. (eds.) ICALP 2009. LNCS, vol. 5556, pp. 139–150. Springer, Heidelberg (2009)
7. Dömösi, P., Nehaniv, C.L.: Algebraic Theory of Automata Networks. In: SIAM Monographs on Discrete Mathematics and Applications, vol. 11, Society for Industrial and Applied Mathematics, Philadelphia (2004)
8. Droste, M., Kuich, W., Vogler, H. (eds.): Handbook of Weighted Automata. EATCS Monographs in Theoretical Computer Science. Springer (2009)
9. Droste, M., Meinecke, I., Šešelja, B., Tepavčević, A.: A Cascade Decomposition of Weighted Finite Transition Systems. In: Freund, R., Holzer, M., Mereghetti, C., Otto, F., Pelano, B. (eds.) Third Workshop on Non-Classical Models of Automata and Applications (NCMA 2011), pp. 137–152. Österreichische Computer Gesellschaft (2011). Short abstract in: Mauri, G., Leporati, A. (eds.) DLT 2011. LNCS, vol. 6795, pp. 472–473. Springer, Heidelberg (2011)
10. Eilenberg, S.: Automata, Languages, and Machines, vol. A. Academic Press (1974)
11. Eilenberg, S.: Automata, Languages, and Machines, vol. B. Academic Press (1976)
12. Ésik, Z.: A proof of the Krohn-Rhodes decomposition theorem. Theoretical Computer Science 234, 287–300 (2000)
13. Ginzburg, A.: Algebraic Theory of Automata. Academic Press (1968)
14. Holcombe, W.M.L.: Algebraic Automata Theory. Cambridge University Press (1982)
15. Kim, Y.-H., Kim, J.-G., Cho, S.-J.: Products of T-generalized state machines and T-generalized transformation semigroups. Fuzzy Sets Syst. 93, 87–97 (1998)
16. Kleene, S.: Representations of events in nerve nets and finite automata. In: Shannon, C., McCarthy, J. (eds.) Automata Studies, pp. 3–42. Princeton University Press (1956)
17. Krishnan, P.: Decomposing Controllers into Non-conflicting Distributed Controllers. In: Liu, Z., Araki, K. (eds.) ICTAC 2004. LNCS, vol. 3407, pp. 511–526. Springer, Heidelberg (2005)
18. Krohn, K., Rhodes, J.L.: Algebraic theory of machines, I. Prime decomposition theorem for finite semigroups and machines. Trans. Amer. Math. Soc. 116, 450–464 (1965)
19. Kuich, W., Salomaa, A.: Semirings, Automata, Languages. In: EATCS Monographs on Theoret. Comp. Sc., vol. 5. Springer (1986)
20. Maler, O.: A decomposition theorem for probabilistic transition systems. Theoretical Computer Science 145, 391–396 (1995)
21. Malik, D.S., Mordeson, J.N., Sen, M.K.: Admissible partitions of fuzzy finite state machines. International Journal of Uncertainty, Fuzziness and Knowledge-Based Systems 5(6), 723–732 (1997)
22. Malik, D.S., Mordeson, J.N., Sen, M.K.: Products of fuzzy finite state machines. Fuzzy Sets Syst. 92, 95–102 (1997)
23. Mordeson, J.N., Malik, D.S.: Fuzzy Automata and Languages – Theory and Applications. Computational Mathematics Series. Chapman & Hall/CRC (2002)

24. Reisz, R.D.: Decomposition theorems for probabilistic automata over infinite objects. Informatica, Lith. Acad. Sci. 10(4), 427–440 (1999)
25. Rhodes, J.L.: Applications of Automata Theory and Algebra: Via the Mathematical Theory of Complexity to Biology, Physics, Psychology, Philosophy, and Games. World Scientific Publishing Co., Inc. (2009)
26. Rhodes, J.L., Steinberg, B.: The q-Theory of Finite Semigroups. Springer (2008)
27. Schützenberger, M.: On the definition of a family of automata. Information and Control 4, 245–270 (1961)
28. Straubing, H.: Finite Automata, Formal Logic, and Circuit Complexity. Birkhäuser, Basel (1994)

References

1. Adhikari, M.R., Sen, M.K., Weinert, H.J.: On k-regular semirings. Bull. Cal. Math. Soc. 88, 141–144 (1996)
2. Aho, A.W., Ullman, J.D.: Introduction to Automata Theory, Languages and Computation. Addison Wesley, Reading (1976)
3. Ahsan, J.: Fully idempotent semirings. Proc. Japan Acad. Ser. A Math. Sci. 69, 185–188 (1993)
4. Ahsan, J.: Semirings characterized by their fuzzy ideals. J. Fuzzy Math. 6, 181–192 (1998)
5. Ahsan, J., Khan, M.F., Shabir, M.: Characterizations of monoids by the properties of their fuzzy subsystems. Fuzzy Sets and Systems 56, 199–208 (1993)
6. Ahsan, J., Khan, M.F., Shabir, M., Zaman, N.: Rings characterized by their fuzzy submodules. Inform. Sci. 74, 247–264 (1993)
7. Ahsan, J., Latif, R.M., Shabir, M.: Fuzzy quasi-ideals in Semigroups. J. Fuzzy Math. 9(2), 259–270
8. Ahsan, J., Latif, R.M., Shabir, M.: Representations of weakly regular semirings by sections in a presheaf. Communications in Algebra 21(8), 2819–2835 (1993)
9. Ahsan, J., Saifullah, K., Khan, M.F.: Fuzzy semirings. Fuzzy Sets and Syst. 60, 309–320 (1993)
10. Ahsan, J., Saifullah, K., Khan, M.F.: Semigroups characterized by their fuzzy ideals. Fuzzy Systems and Mathematics 9, 29–32 (1995)
11. Ahsan, J., Saifullah, K., Shabir, M.: Fuzzy prime ideals of a semiring and fuzzy prime subsemimodules of semimodules over a semiring. New Mathematics and Natural Computation 2(3), 219–236 (2006)
12. Ahsan, J., Shabir, M., Weinert, H.J.: Charaterizations of semirings by P-injective and projective semimodules. Comm. Algebra 26(7), 2199–2209 (1998)
13. Akram, M., Dudek, W.A.: Intuitionistic fuzzy left k-ideals of semirings. Soft Computing 12, 881–890 (2008)
14. Anthony, J.M., Sherwood, H.: Fuzzy groups redefined. J. Math. Analy. Appl. 64, 124–130 (1979)
15. Baik, S.I., Kim, H.S.: On Fuzzy k-Ideals in Semirings. Kangweon-Kyungki Math. 8(2), 147–154 (2000)
16. Beasley, L.B., Pullman, N.J.: Operators that preserve semiring matrix functions. Linear Algebra Appl. 99, 199–216 (1988)
17. Beasley, L.B., Pullman, N.J.: Linear operators strongly preserving idempotent matrices over semirings. Linear Algebra Appl. 160, 217–229 (1992)

18. Benson, D.B.: Bialgebras, Some foundations for distributed and concurrent computation. Fundamenta Informatica 12, 427–486 (1989)
19. Bhakat, S.K.: $(\in \vee q)$ –level subsets. Fuzzy Sets and Systems 103, 529–533 (1999)
20. Bhakat, S.K.: $(\in, \in \vee q)$ –fuzzy normal, quasinormal and maximal subgroups. Fuzzy Sets and Systems 112, 299–312 (2000)
21. Bhakat, S.K., Das, P.: On the definition of a fuzzy subgroup. Fuzzy Sets and Syst. 51, 235–241 (1992)
22. Bhakat, S.K., Das, P.: $(\in, \in \vee q)$ – fuzzy subgroups. Fuzzy Sets and Syst. 80, 359–368 (1996)
23. Bhakat, S.K., Das, P.: Fuzzy subrings and ideals redefined. Fuzzy Sets and Syst. 81, 383–393 (1996)
24. Bhattacharya, P., Mukharjee, N.P.: Fuzzy normal subgroups and fuzzy cosets. Inform. Sci. 34, 225–239 (1984)
25. Birkhoff, G.: Lattice Theory. Amer. Math. Soc. Colleq. Publications (1954)
26. Bourne, S.: The Jacobson radical of a semiring. Proc. Nat. Acad. Sci. (USA) 37, 163–173 (1951)
27. Bourne, S.: On the homomorphism theorems for semirings. Proc. Nat. Acad. Sci. (USA) 38, 118–119 (1952)
28. Bourne, S.: On multiplicative idempotents of a potant semiring, Proc. Nat. Proc. Nat. Acad. Sci. (USA) 42, 632–638 (1956)
29. Bourne, S.: On compact semirings. Proc. Japan Acad. 35, 332–334 (1959)
30. Bourne, S.: On the radical of a positive semiring. Proc. Nat. Acad. Sci. (USA) 45, 1519 (1959)
31. Bourne, S., Zassenhaus, H.: On a Wedderburn-Artin structure theory of a potent semiring. Proc. Nat. Acad. Sci. (USA) 43, 613–615 (1957)
32. Brown, B., Mc Coy, N.H.: Some theorems on groups with applications to ring theory. Trans. Amer. Math. Soc. 69, 302–311 (1950)
33. Camillo, V., Xiao, Y.: Weakly regular rings. Communications in Algebra 22(10), 4095–4112 (1994)
34. Carre, B.A.: Graph and Networks. Oxford University Press, Oxford (1979)
35. Conway, J.H.: Regular Algebra and Finite Machines. Chapman and Hall, London (1971)
36. Chum, Y.B., Kim, H.S., Kim, H.B.: A study on the structure of a semiring. J. Natural Sci. Res. Inst. (Yonsei Univ.) 11, 69–74 (1983)
37. Cuninghame-Green, R.A.: Minimax Algebra. Lect. Notes in Economics and Mathematical Systems, vol. 166. Springer (1979)
38. Das, P.S.: Fuzzy groups and level subgroups. J. Math. Analysis and App. 84, 264–269 (1981)
39. Davvaz, B.: $(\in, \in \vee q)$ –fuzzy subnearrings and ideals. Soft Computing 10, 3079–3093 (2006)
40. Dedekind, R.: Über die Theorie der ganzen algebraischen Zahlen, Supplement XI P.G.Lejeeume Dirichlet. Vorlesungenuber zahlentheorie 4, 4, Auf. Druck und Verlag Braunchweig (1894)
41. Dheena, P., Coumaressae, S.: Fuzzy 2-(0- or 1-) prime ideals in semirings. Bull. Korean Math. Soc. 43(3), 559–573 (2006)
42. Dixit, V.N., Kumar, R., Ajmal, N.: Fuzzy ideals and fuzzy prime ideals of a ring. Fuzzy Sets and Syst. 44, 127–138 (1991)
43. Droste, M., Kuich, W., Vogler, H.: Handbook of Weighted Automata. EATCS Monographs in Theoretical Computer Science. Springer (2009)

44. Dudek, W.A.: Intuitionistic fuzzy h-ideals of hemirings. WSEAS Trans. Math. 12, 1315–1331 (2006)
45. Dudek, W.A.: Special types of intuitionistic fuzzy left h -ideals of hemirings. Soft Computing 12, 359–364 (2008)
46. Dudek, W.A., Shabir, M., Anjum, R.: Characterizations of hemirings by their h-ideals. Comput. Math. Appl. 59, 3167–3179 (2010)
47. Dudek, W.A., Shabir, M., Ali, M.I.: (α,β)-Fuzzy ideals of Hemirings. Comput. Math. Appl. 58, 310–321 (2009)
48. Dutta, T.K., Biswas, B.K.: Fuzzy prime ideals of a semiring. Bull. Malays Math. Soc. 17, 9–16 (1994)
49. Dutta, T.K., Biswas, B.K.: Fuzzy k-ideals of semirings. Bull. Cal. Math. Soc. 87, 91–96 (1995)
50. Dutta, T.K., Kar, S.: On regular ternary semirings. Advances in Algebra. In: Proceedings of the ICM Satellite Conference in Algebra and Related Topics, pp. 343–355. World Scientific (2003)
51. Dutta, T.K., Kar, S.: On prime ideals and prime radical of Ternary Semirings. Bull. Cal. Math. Soc. 97, 445–454 (2005)
52. Dutta, T.K., Kar, S.: On semiprime ideals and irreducible ideals of ternary semirings. Bull. Cal. Math. Soc. 97, 467–476 (2005)
53. Dutta, T.K., Kar, S.: A note on regular ternary semirings. Kyungpook Math. J. 46, 357–365 (2006)
54. Eilenberg, S.: Automata, Languages and Machines. Academic press, New York (1974)
55. Feng, F., Zhao, X.Z., Jun, Y.B.: *-μ-semirings and *-λ-semirings. Theoret. Comput. Sci. 347, 423–431 (2005)
56. Feng, F., Jun, Y.B., Zhao, X.Z.: On *-λ-semirings. Inform. Sci. 177, 5012–5023 (2007)
57. Ghosh, S.: Matrices over semirings. Inform. Sci. 90, 221–230 (1996)
58. Ghosh, S.: Fuzzy k-ideals of semirings. Fuzzy Sets and Syst. 95, 103–108 (1998)
59. Glazek, K.: A Guide to Literature on Semirings and their Applications in Mathematics and Information Sciences with Complete Bibliography. Kluwer Acad. Publ, Dordrecht (2002)
60. Goguen, J.A.: L-fuzzy sets. J. Math. Anal. Appl. 18, 145–174 (1967)
61. Golan, J.S.: Making modules fuzzy. Fuzzy Sets and Systems 32, 91–94 (1989)
62. Golan, J.S.: The Theory of Semirings with applications in Mathematics and Theoretical Computer Science. Pitman Monographs and Surveys in Pure and Appl. Math, vol. 54. Longman, New York (1992)
63. Golan, J.S.: Semirings and their Applications. Kluwer, Dordrecht (1999)
64. Gondran, M., Minoux, M.: Graphs and Algorithmes. Edit Eyrolles, Paris (1979)
65. Hedayati, H.: Generalized fuzzy k-ideals of semirings with interval-valued membership functions. Bull. Malays Math. Sci. Soc. 32(3), 409–424 (2009)
66. Hedayati, H.: Equivalence Relations Induced byInterval Valued (S, T)-fuzzy h-ideals (k-ideals) of Semirings. World Applied Science Journal 9(1), 1–13 (2010)
67. Huang, H.K., Li, H.J., Yin, Y.Q.: The h-hemiregular fuzzy Duo hemirings. Int. J. Fuzzy Syst. 9(2), 105–109 (2007)
68. Hebisch, U., Weinert, H.J.: Semirings, Algebraic Theory and Applications in the Computer Science. World Scientific (1998)
69. Henriksen, M.: Ideals in semirings with commutative addition. Amer. Math. Soc. Notices 6, 321 (1958)
70. Henriksen, M.: The $a^{n(a)} = a$ theorems for semirings. Math. Japonica 5, 21–24 (1958)
71. Iizuka, K.: On Jacobson radical of a semiring. Tohoku Math. J. 11, 409–421 (1959)

72. Liu, W.J.: Fuzzy Invariant subgroups and fuzzy ideals. Fuzzy Sets and Systems 3, 133–139 (1982)
73. Liu, W.J.: Operations on fuzzy ideals. Fuzzy Sets Syst. 11, 31–41 (1983)
74. Jun, Y.B.: Generalizations $(\in, \in \vee q)$fuzzy subalgebras in BCK/BCI-algebras. Comput. Math. Appl. 58, 1383–1390 (2009)
75. Neggers, J., Jun, Y.B., Kim, H.S.: On *L*-Fuzzy ideals in semirings II. Czech. Math. J. 49, 127–133 (1999)
76. Jun, Y.B., Dudek, W.A., Shabir, M., Kang, M.S.: General types of (α, β) –fuzzy ideals of hemirings. Honam Math. J. 32(3), 413–439 (2010)
77. Jun, Y.B., Neggers, J., Kim, H.S.: Normal *L*-fuzzy ideals in semirings. Fuzzy Sets and Systems 82, 383–386 (1996)
78. Jun, Y.B., Neggers, J., Kim, H.S.: On *L*-fuzzy ideals in semirings I. Czech. Math. J. 48, 669–675 (1998)
79. Jun, Y.B., Xin, X.L., Roh, E.H.: A class of algebras related to *BCI*-algebras and semigroups 24, 309–321 (1998)
80. Jun, Y.B., Özürk, M.A., Song, S.Z.: On fuzzy *h*-ideals in hemirings. Inform. Sci. 162, 211–226 (2004)
81. Jun, Y.B., Kim, H.S., Özürk, M.A.: Fuzzy *k*-ideals in semirings. The J. of Fuzzy Maths. 13, 351–364 (2005)
82. Jun, Y.B., Song, S.Z.: Generalized fuzzy interior ideals in semigroups. Inform. Sci. 176(20), 3079–3093 (2006)
83. Karvellas, P.H.: Inversive semirings with commutative addition. J. Austral. Math. Soc. 18, 277–288 (1974)
84. Karvellas, P.H.: von Neumann Regularity in Semirings. Math. Nachr. 45, 73–79 (1970)
85. Kar, S.: On Quasi-ideals and Bi-ideals of Ternary Semirings. International Journal of Mathematics and Mathematical Sciences 18, 3015–3023 (2005)
86. Katsaras, A.K., Liu, D.B.: Fuzzy vector spaces and fuzzy topological spaces. J. Math. Anal. Appl. 58, 135–146 (1977)
87. Kaufman, A., Gupta, M.: Introduction to Fuzzy Arithmetic. von Nostrand-Reinhold, New York (1986)
88. Kim, C.B., Park, M.: *k*-fuzzy ideals in semirings. Fuzzy Sets and Syst. 81, 281–286 (1996)
89. Kim, C.B.: Isomorphism Theorems and fuzzy *k*-ideals of *k*-semirings. Fuzzy Sets and Syst. 112, 333–342 (2000)
90. Kim, C.B.: Quotient Semirings of a *k*-semiring by semiprimary *k*-fuzzy ideals. Journal of Fuzzy Logic and Intelligent Systems 14, 88–92 (2004)
91. Kolokoltov, V.N., Maslov, V.P.: Idempotent Analysis and its applications. Mathematics and its Applications, vol. 401. Kluwer (1997)
92. Kondo, M., Dudek, W.A.: On the Transfer Principle in fuzzy theory. Mathware Soft. Comput. 12, 41–55 (2005)
93. KaviKumar, Khamis, A., Jun, Y.B.: Fuzzy bi-ideals in ternary semirings. International Journal of Computational and Mathematical Sciences 3-4, 160–164 (2009)
94. Kumbhojkar, H.V.: Spectrum of prime fuzzy *L*-fuzzy *h*– ideals of a hemiring. Fuzzy Sets and Syst., doi:10.1016/j.fss. 2009.10.006
95. Kumbhojkar, H.V.: Spectrum of prime fuzzy ideals. Fuzzy Sets V1V1Z1Z1and Syst. 62, 101–109 (1994)
96. Kumbhojkar, H.V.: Some comments on spectrum of prime fuzzy ideals. Fuzzy Sets and Syst. 85, 109–114 (1997)
97. Kuroki, N.: Fuzzy bi-ideals in Semigroups. Comment Math. Univ. St. Paul 24, 21–26 (1975)

98. Kuroki, N.: On fuzzy ideals and fuzzy bi-ideals in Semigroups. Comment. Math. Univ. St. Paul 28, 17–22 (1979)
99. Kuroki, N.: On fuzzy ideals and fuzzy bi-ideals in semigroups. Fuzzy Sets and Syst. 5, 203–215 (1981)
100. Kuroki, N.: On fuzzy Semigroups. Inform. Sci. 53, 203–236 (1991)
101. La Torre, D.R.: On h-ideals and k-ideals in hemirings. Publ. Math. Debrecen 12, 219–226 (1965)
102. Lee, E.T., Zadeh, L.A.: Note on fuzzy Languages. Inform. Sci. 1, 421–434 (1969)
103. Lehmer, D.H.: A ternary analogue of Abelian groups. American Journal of Math. 59, 329–338 (1932)
104. Lopez-Permouth, S.R., Malik, D.S.: On categories of fuzzy Modules. Inform. Sci. 52, 211–220 (1990)
105. Ma, X., Zhan, J.: On fuzzy h-ideals of hemirings. J. Syst. Sci. Complex. 20, 470–478 (2007)
106. Ma, X., Zhan, J.: Generalized fuzzy h-bi-ideals and h -quasi-ideals of hemirings. Inform. Sci. 179, 1249–1268 (2009)
107. Ma, X., Zhan, J., Shum, K.P.: Generalized Fuzzy h-Ideals of Hemirings. Bull. Malays Math. Soc. 34(3), 561–574 (2011)
108. Mahmood, T.: Characterizations of Hemirings by $(\in_m, \in_m \vee q_n)$ −−fuzzy ideals (submitted)
109. Majumdar, S.: Theory of fuzzy modules. Bull. Cal. Math. Soc. 82, 395–399 (1990)
110. Malee, S., Chinram, R.: k-Fuzzy Ideals of Ternary Semirings. World Academy of Sciences, Engineering and Technology 67, 485–489 (2010)
111. Malik, A.S., Mordeson, J.N.: Fuzzy prime ideals of a ring. Fuzzy Sets Syst. 37, 93–98 (1990)
112. Mascle, J.P.: Torsion matrix semigroups and recognizable transductions. In: Kott, L. (ed.) Automata, Languages and Programing. LNCS, Springer, Berlin (1986)
113. Malik, D.S., Mordeson, J.N.: Algebraic Fuzzy Automata Theory. Arabian Journal for Science and Engineering, Section C. 25(2) (2000)
114. Mordeson, J.N., Malik, D.S.: Fuzzy Automata and Languages, Theory and Applications. Computational Mathematics Series. Chapman and Hall, Boca Raton (2002)
115. Mukherjee, T.K., Sen, M.K.: Prime fuzzy ideals in rings. Fuzzy Sets Systems 32, 337–341 (1989)
116. Mordeson, J.N., Malik, D.S.: Fuzzy commutative algebra. World Scientific, Singapore (1998)
117. Mukhopadhyay, T.P.: Characterization of regular semirings. Matematiqki Vesnik 48, 83–85 (1996)
118. Murali, V.: Fuzzy points of equivalent fuzzy subsets. Inform. Sci. 158, 277–288 (2004)
119. Nanda, S.: Fuzzy Modules over fuzzy rings. Bull. Cal. Math. Soc. 81, 197 (1989)
120. Negoita, C.V., Ralescu, D.A.: Applications of Fuzzy Sets to System Analysis, vol. 3. Birkhäuser, Basel (1975)
121. Neggers, J., Jun, Y.B., Kim, H.S.: On L-fuzzy ideals in semirings II. Czech. Math. J. 49, 127–133 (1999)
122. Pan, F.Z.: Fuzzy finitely generated modules. Fuzzy Sets Syst. 21, 105–113 (1987)
123. Pan, F.Z.: Fuzzy quotient modules. Fuzzy Sets Syst. 90, 85–90 (1988)
124. Paz, A.: Introduction to Probabilistic Automata. Academic Press, New York (1971)
125. Pu, P.M., Liu, Y.M.: Fuzzy topology I, neighborhood structure of a fuzzy point and Moore-Smith convergence. J. Math. Anal. Appl. 76, 571–599 (1980)
126. Ramamurthy, V.S.: Weakly regular rings. Canad. Math. Bull. 16, 317–321 (1973)
127. Rosenfeld, A.: Fuzzy groups. J. Math. Anal. Appl. 38, 512–517 (1971)

128. Sen, M.K., Adhikari, M.R.: On division hemirings. Bull. Calcutta Math. Soc. 83, 267–274 (1991)
129. Sen, M.K., Adhikari, M.R.: On k-ideals of semirings. Internat. J. Math. Math. Sci. 15(2), 347–350 (1992)
130. Sen, M.K., Mukhopadhyay, P.: Von Neumann regularity in semirings. Kyungpook Math. J. 35, 249–258 (1995)
131. Shabir, M.: Fully fuzzy prime semigroups. Int. J. Math. and Math. Sci., 163–168 (2005)
132. Shabir, M., Anjum, R.: Characterizations of hemirings by the properties of their k-ideals (submitted)
133. Shabir, M., Anjum, R.: Right k-weakly regular hemirings (submitted)
134. Shabir, M., Anjum, R.: On k-regular and k-intra-regular hemirings (submitted)
135. Shabir, M., Jun, Y.B., Nawaz, Y.: Characterizations of regular semigroups by (α,β) –fuzzy ideals. Comput. Math. Appl. 59, 161–175 (2010)
136. Shabir, M., Jun, Y.B., Nawaz, Y.: Semigroups characterized by $(\in,\in\vee q_k)$ –fuzzy ideals. Comput. Math. Appl. 60, 1473–1493 (2010)
137. Shabir, M., Mahmood, T.: Characterizations of hemirings by $(\overline{\in},\overline{\in}\vee\overline{q})$-fuzzy k-ideals (submitted)
138. Shabir, M., Mahmood, T.: Characterizations of hemirings by $(\in,\in\vee q)$-fuzzy ideals (submitted)
139. Shabir, M., Mahmood, T.: Characterizations of hemirings by $(\in,\in\vee q)$-fuzzy k-ideals (submitted)
140. Shabir, M., Mahmood, T.: Hemirings characterized by the properties of their fuzzy ideals with thresholds. Quasigroups and Related Systems 18, 111–128 (2010)
141. Shabir, M., Mahmood, T.: Spectrum of $(\in,\in\vee q)$-fuzzy prime h-ideals of a hemiring (submitted)
142. Shabir, M., Mahmood, T.: Characterizations of hemirings by interval valued fuzzy ideals. Quasigroups and Related Systems 19, 101–113 (2011)
143. Shabir, M., Mahmood, T.: Characterizations of hemirings by interval valued (α, β) –fuzzy ideals. East Asian Mathematical Journal 27(3), 349–372 (2011)
144. Shabir, M., Mahmood, T.: Characterizations of hemirings by $(\in,\in\vee q_k)$ –fuzzy ideals. Comput. Math. Appl. 61, 1059–1078 (2011)
145. Shabir, M., Mahmood, T.: Characterizations of hemirings by $(\overline{\in},\overline{\in}\vee\overline{q}_k)$ –fuzzy ideals (submitted)
146. Shabir, M., Mahmood, T.: On interval valued $(\overline{\in},\overline{\in}\vee\overline{q}_k)$ –fuzzy ideal (submitted)
147. Simmon, I.: Recognizable sets with multiplicities in the tropical semiring. In: Chytil, M.P., et al. (eds.) Mathematical Foundations for Computer Science 1988. LNCS. Springer, Berlin (1988)
148. Simmon, I.: The nondeterministic complexity of finite automaton. Notes, Herms, Paris, 384-400 (1990)
149. Starke, P.: Abstract Automata, 2nd edn. Academic Press, New York (1998); 1st edn. Birkhaeuser, Basel (1978)
150. Subramanian, H.: Von Neumann regularity in semirings. Math. Nachr. 45, 73–79 (1970)
151. Sun, G., Yin, Y., Li, Y.: Interval valued fuzzy h-ideals of Hemirings. Int. Math. Forum 5, 545–556 (2010)
152. Swamy, U.M., Swamy, K.L.N.: Fuzzy prime ideals of rings. J. Mat. Anal. Appl. 134, 94–103 (1988)
153. Vandiver, H.S.: Note on a simple type of algebra in which the cancellation law of addition does not hold. Bull. Amer. Math. Soc., 916–920 (1934)
154. Wechler, W.: The concept of fuzziness in automata and language theory. Akademic verlog, Berlin (1978)

155. Wu, L., Qiu, D.: Automata theory based on complete residuated lattice-valued logic. Reduction and minimization. Fuzzy Sets and Syst. 161(12), 1635–1656 (2010)
156. Williams, D.R.P.: *S*-Fuzzy Left *h*-Ideal of Hemirings. Int. J. Math. Sci. 1(2), 142–149 (2007)
157. Yin, Y.Q., Huang, X., Xu, D., Li, H.: The characterizationof *h* -semisimple hemirings. Int. J. of Fuzzy Syst. 11, 116–122 (2009)
158. Yin, Y., Li, H.: The characterizations of *h*-hemiregular hemirings and *h*-intra-hemiregular hemirings. Inform. Sci. 178, 3451–3464 (2008)
159. Yin, Y., Huang, X., Xu, D., Li, F.: The characterization of *h* -semisimple Hemirings. Int. J. Fuzzy Syst. 11, 116–122 (2009)
160. Zadeh, L.A.: Fuzzy Sets. Inform. Control 8, 338–353 (1965)
161. Zadeh, L.A.: The concept of linguistic variable and its applications to approximate reasoning-I. Inform. Control 18, 199–249 (1975)
162. Zeleznikov, J.: Orthodox semirings and rings. J. Austral. Math. Soc. 30, 50–54 (1980)
163. Zeleznikov, J.: Regular semirings. Semigroup Forum 23, 119–136 (1981)
164. Zhan, J.: On properties of fuzzy left *h*-ideals in hemirings with *t*-norms. Int. J. Mat. Math. Sci. 19, 3127–3144 (2005)
165. Zhan, J., Davvaz, B.: *L*-fuzzy *h*-ideals with operations in hemirings. Northeast Math. J. 23, 1–14 (2007)
166. Zhang, Y.: Prime *L*-fuzzy ideals and primary *L*-fuzzy ideals. Fuzzy Sets Syst. 27, 345–350 (1988)
167. Zhan, J., Dudek, W.A.: Fuzzy *h*-ideals of hemirings. Inform. Sci. 177, 876–886 (2007)
168. Zhan, L., Shum, K.P.: Intuitionistic (s,t)-fuzzy *h*-ideals in hemirings. East Asian Math. J. 22, 93–102 (2006)
169. Zhan, J., Tan, Z.: *T*-fuzzy *k*-ideals of semirings. Sci. Math. Japon. 58, 597–601 (2003)

Index